COGNITIVE ECOLOGY

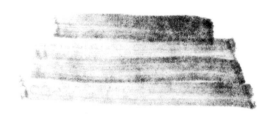

Cognitive Ecology

The Evolutionary Ecology of Information
Processing and Decision Making

Edited by
Reuven Dukas

THE UNIVERSITY OF CHICAGO PRESS
CHICAGO AND LONDON

36.00

Reuven Dukas is research assistant professor with the Nebraska Behavioral Biology Group, School of Biological Sciences, University of Nebraska.

The University of Chicago Press, Chicago 60637
The University of Chicago Press, Ltd., London
© 1998 by The University of Chicago
All rights reserved. Published 1998
Printed in the United States of America
07 06 05 04 03 02 01 00 99 98   1 2 3 4 5

ISBN: 0-226-16932-4 (cloth)
ISBN: 0-226-16933-2 (paper)

Library of Congress Cataloging-in-Publication Data

Cognitive ecology : the evolutionary ecology of information processing and decision making / edited by Reuven Dukas.
    p.   cm.
    Includes bibliographical references and index.
    ISBN 0-226-16932-4 (alk. paper). — ISBN 0-226-16933-2 (pbk. : alk. paper)
    1. Cognition in animals. 2. Evolution (Biology) 3. Ecology. I. Dukas, Reuven.
QL685.C5   1998
591.5'13—dc21   97-36597
                                                                                       CIP

CONTENTS

viii   Contents

ACKNOWLEDGMENTS

I wish to thank the numerous friends and colleagues who helped me during the long editing process. Most notably, M. Abrahams, P. Bednekoff, D. Blumstein, N. Clayton, L. Gass, T. Getty, D. Hugie, T. Laverty, D. Papaj, D. Sherry, W. Roberts, and D. Wilkie kindly reviewed a manuscript of a chapter. Three anonymous referees read and commented on the manuscript of the whole volume. S. Shettleworth co-organized with me the "Cognitive Ecology" symposium that inspired editing this volume, and the Animal Behavior Society provided financial support for that symposium. Finally, my editor C. Henry and additional staff of the University of Chicago Press provided friendly guidance.

CHAPTER ONE

Introduction

REUVEN DUKAS

Certain complex actions are of direct or indirect service under certain states of the mind . . . That some physical change is produced in the nerve-cells or nerves which are habitually used can hardly be doubted.

—Charles Darwin, 1872

I hold that if the doctrine of Organic Evolution is accepted, it carries with it, as a necessary correlate, the doctrine of Mental Evolution.

—G. Romanes, 1883

1.1 DEFINITION AND GOALS

The message of this volume is that animal cognition is a biological feature that has been molded by natural selection. Hence to understand cognition, one must study its ecological consequences and evolution. Cognition should be placed together with morphological, physiological, behavioral, and life-history characteristics as a topic to be thoroughly scrutinized by evolutionary biologists and ecologists.

To define cognitive ecology, it is useful to first define "cognition," "ecology," and the tightly linked term "evolution." Cognition may be characterized as the neuronal processes concerned with the acquisition and manipulation of information by animals. Cognitive processes most commonly determine behavior, but they also affect numerous decisions regarding changes in morphological, physiological, and life-history traits. Ecology is the study of interactions between organisms and their surrounding biotic and abiotic environments. Such interactions usually result in differential survival and reproduction of animals within a population due to individual variation in phenotypic traits. Typically, a sizable proportion of the phenotypic variation between individuals results from genetic variation. Thus, individual differences in survival and reproduction, or fitness, translate into changes in gene frequency and their associated phenotypic traits over generations. Such genetically based phenotypic change is evolution caused by natural selection.

As the product of cognition and ecology, cognitive ecology focuses on the effects of information processing and decision making on animal fitness. To that end, cognitive ecology employs formal evolutionary and ecological theory to (1) address ultimate questions about the optimal design of, constraints on, and function of cognitive traits; and (2) generate and test hypotheses on the mechanisms underlying acquisition and manipulation of information.

1.2 BACKGROUND

1.2.1 Evolutionary Biology and Psychology: A Brief History

In the latter half of the nineteenth century, biological discussions of evolution included explicit analyses of what we now term cognition. Similarly, many psychological studies had strong evolutionary foundations. Darwin already addressed cognitive issues to some extent in *The Origin of Species* (1859, chap. 8). He elaborated on many issues in subsequent writings, most notably *The Expression of Emotions by Man and Animals* (1872). Authors of later books on behavior, such as the evolutionary biologists Romanes (1882, 1883) and Morgan (1890, 1900), followed Darwin's lead by discussing the evolution of various cognitive traits. At the same time, leading psychologists such as James and Baldwin were strongly influenced by evolutionary thinking (Baldwin 1902; Richards 1987, chaps. 9 and 10). However, the psychology texts written by James and Baldwin did not emphasize evolution (James 1890; Baldwin 1893).

Early in the twentieth century, evolutionary biology and psychology diverged and underwent significant changes. At that time, the emerging field of Mendelian genetics stimulated thinking on the hereditary nature of human cognition and behavior. Public sensitivity to this topic was probably a major factor that steered away both evolutionary biologists and psychologists from cognitive issues (Richards 1987, chap. 11). In evolutionary biology, researchers simply focused on other questions. In psychology, behaviorism was established as the dominant paradigm (Watson 1924; Skinner 1938). Proponents of behaviorism maintained that animal behavior should be explained only in terms of empirically observed variables. They rejected both cognitive concepts dominating earlier psychological thinking and theories emphasizing genetic effects on behavior. The striking results of these social-scientific transitions were, first, that psychology established itself as a social science far removed from the life sciences, a position it still maintains today, and, second, that issues of information processing received little attention in either psychology or evolutionary biology (see Richards 1987 for a detailed review).

Since the mid-1950s, the cognitive approach has gradually become more influential in psychology (e.g. Broadbent 1958; Newell et al. 1958; Neisser 1967). This paradigm shift, however, has not altered the central biases of psychology. Contemporary cognitive psychologists still focus almost exclusively on human cognition and empirically derived laws and neglect evolutionary ecology. The tradition of behaviorism is still reflected, explicitly or implicitly, in research on animal psychology. Nevertheless, an increasing number of psychologists have recognized the effects of evolution and ecology on behavior (e.g. Bolles 1970; Seligman and Hager 1972; Kamil and Sargent 1981; Bolles and Beecher 1987; Kamil et al. 1987; Gallistel 1989; Shettleworth 1993), and this recognition has contributed to increased collaboration between psychologists and evolutionary ecologists.

In evolutionary biology, researchers continued to pay little attention to cognitive issues. Research in ethology addressed aspects of the evolution of behavioral traits (e.g. Tinbergen 1951; von Frisch 1967; Lorenz 1981); however, such research rarely focused on problems of acquisition and manipulation of information by animals. Until recently, such issues were in a unique psychological territory. Only with the emergence of behavioral ecology (Krebs and Davies 1978) has information processing gradually received increased attention from evolutionary ecologists. The construction of formal explanatory models of foraging behavior in the 1970s and 1980s crystallized thinking about the effect of behavior on fitness (Charnov 1976; Pyke et al. 1977; Stephens and Krebs 1986; Krebs and Davies 1991). These models enabled the realization that issues of learning and memory are crucial for understanding problems in behavioral ecology (e.g. Hughes 1979; McNair 1981; Kamil 1983; Pyke 1984; Stephens and Krebs 1986). Cognition, however, encompasses more than learning and memory. It comprises also perception of information, attention to certain information, forms of representation, and rules of manipulation of information. These issues, together with topics of learning and memory, have been subjected to intense research by evolutionary ecologists only in the past several years (e.g. Kacelnik et al. 1990; Real 1991, 1993; Dukas and Ellner 1993; Stephens 1993; Balda et al. 1996).

1.2.2 Other Disciplines

Issues related directly or indirectly to cognition are studied in several other fields, most notably neurobiology and computer science. The definition of cognition in section 1.1 referred explicitly to neuronal processes. This reference is significant because animal cognition is a product of the nervous system; natural selection has acted on networks of neurons during the evolution of cognitive systems. Hence, one must understand how the structure and activity of single neurons and networks of neurons enable and constrain cognitive processes. Neurobiology analyses proximate mechanisms of neuronal activity from the level of molecules to that of the whole brain and body. While research on ultimate mechanisms in cognitive ecology can benefit from all levels of neurobiological investigation, studies that consider the behavior of the whole animal may be the most relevant. Neuroethology is a subdiscipline devoted to the examination of neurobiological issues as they relate to the behavior of animals in their natural settings (e.g. Bullock 1984, 1993; Camhi 1984; Heiligenberg 1991). In addition, neurobiology texts, such as that of Dudai (1989), present a coherent blend of analyses from cellular to whole-animal activity.

Computer science has probably had significant effects on the development of modern cognitive thinking. The desire to develop machines with sophisticated information-processing capacities has stimulated computer scientists to consider how animals process information. Such research in computer science has also inspired psychologists and neurobiologists to take new quantitative approaches when analyzing cognition (e.g. Newell et al. 1958; Simon 1979; Fischler and Firschein 1987; Beer et al. 1993). Interactions among computer scientists, psychologists, and neurobiologists have led to the formation of the new field of cognitive science, which combines research on information processing in animals (mostly humans) with that in machines (e.g. Posner 1989; Stillings et al. 1995).

Other disciplines relevant to cognitive ecology are biopsychology (or physiological psychology) and psychophysics. Biopsychology concentrates on physiological mechanisms within the brain and interactions between the brain and body that underlie behavior (e.g. Graham 1990). Psychophysics is commonly defined as the quantitative study of perception. A central focus of psychophysics is the search for invariant laws of perception and the constraints these laws place on the underlying mechanisms of information processing (e.g. Baird and Noma 1978).

While all of these disciplines provide many insights into cognition, almost all fail to consider cognition as a characteristic that affects an animal's success in its natural environment. Such ecological evaluation of cognition, which is a central goal of cognitive ecologists, can add a unifying theoretical framework to the cognitive sciences.

1.3 WHY STUDY COGNITIVE ECOLOGY?

1.3.1 Cognition

The brain remains one the great enigmas of modern science. We now have a firm fundamental understanding of the physical world and of the basic components of life—such as molecular genetic material—and evolution. Obviously, physicists and biologists are striving to learn more, but elementary knowledge of our inanimate and animate surroundings is known. We still lack, however, a basic understanding of how the brain functions. We do know that cognition is produced by networks of neurons; this fact has been commonly accepted for more than a century. Scientists from various disciplines have accumulated a massive database on the brain, cognition, and behavior. Nevertheless, one cannot yet provide a fundamental explanation for the way an animal (even one as small as a bee) uses its brain to process and store large amounts of information so that it can react adaptively to its surroundings.

Because numerous competent researchers have not yet provided a satisfactory explanation for brain functioning, it must be highly complex. Nevertheless, animal brains are products of a long evolutionary process, and evolutionary ecologists are therefore in a unique position to provide key insights into brain functioning. The brain is not a device designed by an engineer for certain predetermined purposes; the brain evolved gradually from simpler structures that processed information. Genetically based variation in brains allowed natural selection to produce brains that were adapted to their environments. By understanding the ecological and evolutionary laws underlying that process of change, we can gain many insights into cognition. In short, cognition is a product of evolution; hence, it should be studied by evolutionary biologists and ecologists.

The importance of evolutionary thinking is illustrated by its influence on the study of senescence. Although the study of aging has long attracted researchers, Medawar's (1952) statement that aging is "an unsolved problem of biology" could not be rejected until recently, when research by evolutionary biologists provided a single ultimate theory for senescence (Williams 1957; Hamilton 1966; Kirkwood and Holliday 1979; Kirkwood and Rose 1991; Rose 1991; Stearns 1992). Many details of senescence require further study (e.g. Holliday 1995); however, the ultimate theory unifies findings of previous research and inspires new research. For instance, the evolutionary theory of senescence has shown that the processes of DNA repair, sexual reproduction, and senescence are tightly linked (Kirkwood et al. 1986; Bernstein and Bernstein 1991; Avise 1993; Lithgow and Kirkwood 1996). Understanding these relationships is neither obvious nor possible without an ultimate evolutionary-ecological theory.

What then can cognitive ecology contribute to the cognitive sciences? Currently, the various disciplines of cognitive science focus on proximate mechanisms at various levels, from molecular to whole-brain and whole-body activity. There is a strong bias toward the examination of human cognition and behavior and an attempt to explain behavior on the basis of proximate models of brain activity. Until recently, however, cognitive scientists have not paid sufficient attention to the fundamental fact that cognitive traits evolved under particular natural settings (examples of recent exceptions are P. Anderson 1989; J. Anderson 1990; Barkow et al. 1992). With consideration of the selection pressure on cognition, cognitive ecology can contribute intellectual coherence to the multidisciplinary study of cognition.

For instance, it is now well established that research on very simple animal models facilitates our understanding of complex species. This idea was illustrated by a 1991 cover of the journal *Science,* which showed a fruit fly providing support for a hang glider. This cover symbolized the striking fact that research on human genetics is heavily based on knowledge acquired from the genetic systems of less complex organisms such as fruit flies, yeast, and bacteria (Koshland 1991). At the present time, it is easier to accept the notion of genetic similarity between flies and humans than to consider their cognitive resemblances. However, resistance to the idea of cognitive resemblance is probably based on biased perception rather than on biological fact. Indeed, there appears to be general similarity between cognition in insects and that in vertebrates (e.g. O'Carrol 1993; Srinivasan et al. 1993). As Koshland (1991, 173) stated, "Anyone with a flyswatter knows that the flight information and landing computation of a fly would elicit the admiration of any flight controller. The learning and memory of a fly are fleeting and not likely to get a fly through first grade, but they exist and can be manipulated genetically and biochemically."

In spite of successful neurobiological and genetic research on learning in the fly (e.g. Dudai 1989, chap. 7; Ferrus and Canal 1994; Hall 1994; Tully 1996), results have not been tied with evolutionary and ecological thinking— although the natural history of learning in the fly has been investigated (e.g. Jaenike 1985; Papaj and Prokopy 1989; Prokopy et al. 1993). Cognitive ecology could integrate proximate knowledge from neurobiology, genetics, and cognitive psychology and provide an ultimate unified framework for studying the relatively simple, manageable model system of cognition in the fruit fly.

In addition, evolutionary ecology provides powerful tools that can be used to analyze information processing and behavioral decisions. Optimality models allow one to generate quantitative predictions about design and function. Such models are based on the fundamental assumption that structures and behaviors should be interpreted in terms of their contributions to fitness. Optimality models typically contain, implicitly or explicitly, three components. First, they contain some decision variable to be analyzed. Second, they contain assumptions about feasible choices that can be made and various factors that limit outcome. Finally, they include some currency assumptions that state how various choices are evaluated (Maynard Smith 1978; Stephens and Krebs 1986; Mangel and Clark 1988). Optimality models are usually not explicit about the genetics underlying the system in question. Quantitative genetics supplements such models by telling us how fast evolutionary changes may occur and whether predicted optima may be unattainable (Falconer 1989; Stearns 1992). These two classes of formal theory have been proved to be instrumental in various biological disciplines (Stephens and Krebs 1986; Krebs and Davies 1991, 1997; Rose 1991; Diamond 1992; Bulmer 1994); hence formal theory is likely to guide cognitive research as well.

1.3.2 Ecology, Evolution, Conservation, and Pest Management

The problem of evolution may be approached in two different ways. First, the sequence of evolutionary events as they have actually taken place in the past history of various organisms may be traced. Second, the mechanisms that bring about evolutionary changes may be studied. The first approach deals with historical problems, and the second with physiological. The importance of genetics for a critical evaluation of theories concerning the mechanics of evolution is fairly generally recognized.

—T. Dobzhansky, 1937

The most serious gap in the structure of Darwin's theory was his ignorance of how hereditary variations arise and persist. The development of genetics has been the single most important contribution to evolutionary thought since Darwin, for it has become the basis of the entire theoretical structure of the mechanisms of evolutionary change.

—D. J. Futuyma, 1979

Evolution can be inferred from patterns of phenotypic changes over generations. This is analogous to studying behavior by observing animals. Evolution and behavior, however, can also be studied at a fundamental level; this kind of study allows the establishment of an explanatory, predictive global theory. Indeed, detailed knowledge of genetic mechanisms have provided a solid foundation for modern evolutionary biology (e.g. Fisher 1930; Dobzhansky 1937; Futuyma 1979; Falconer 1989). I will argue that a thorough integrated study of cognitive mechanisms can enrich research in basic and applied evolutionary ecology.

Since the inception of behavioral ecology, one of the central goals of behavioral ecologists has been to demonstrate that explanatory models of individual foraging, mating, social, and other decisions give insights into population and community ecology (Krebs and Davies 1978, 1991, 1997). By understanding key individual behaviors, we can explain general patterns of interactions within and between species (e.g. Milinski and Parker 1991; Rosenzweig 1991; Clark 1993; Singer 1994; Werner 1994; Lima and Zollner 1996). Given that cognition underlies behavior, a fundamental theory of information processing can provide us with general principles of behavior of individuals, populations, and communities.

Specifically, with experimental manipulation, one can infer the decision rules that underlie behavior (e.g. chaps. 5, 8–10). An alternative or supplementary approach would enable the deciphering of the neuronal computations that underlie a decision rule with direct measurement of neuronal activity (e.g. Hammer 1993; Hatsopoulos et al. 1995). Currently, information on actual neuronal computation provides us only with an intuition of what neurons can perform. Ultimately, however, the fundamental understanding of neuronal computation could guide behavioral research by providing a detailed account of actual mechanisms of and constraints on manipulation of information.

For instance, models of interactions between parasitoids and their hosts typically assume fixed behavior by the parasitoid. However, long-term research on the population regulation of the red scale insect, *Aonidiella aurantii,* by its wasp parasitoid, *Aphytis melinus* (reviewed in Murdoch 1994), suggests that each wasp processes information about several internal and external variables in order to determine her oviposition decisions. The wasp's behavior is far from fixed; it varies in space and time on the basis of her perceptions of the environment and the choices she makes given these perceptions. Hence, the wasp's capacity to perceive various stimuli, integrate information about environmental variables, and translate her representation of the environment into adaptive decisions are all relevant for explaining her dynamic pattern of interactions with her host (see Mangel 1989; Mangel and Roitberg 1992; Collier et al. 1994; Murdoch 1994).

Similar arguments can be made for the crucial function of cognition in evolution. Mayr (1982, 612) stated, "Many if not most acquisitions of new structures in the course of evolution can be ascribed to selection forces exerted by newly acquired behaviors . . . Behavior, thus, plays an important role as the pacemaker of evolutionary change. Most adaptive radiations were apparently caused by behavioral shifts" (see also Wcislo 1989; West-Eberhard 1989). Many newly acquired behaviors are guided by cognition. Thus, by understanding the cognitive mechanisms that lead to new behaviors, we can explain patterns of evolution and adaptive divergence.

Cognitive ecology can also contribute to the applied disciplines of conservation biology and pest management. Conservation biologists focus on predicting and mitigating the effects of habitat degradation on natural populations. Currently, many conservation policies implicitly assume that humans and the targeted species perceive the environment identically. However, various perceptual constraints and cognitive biases expressed by that targeted species may result in the failure of a well-meaning preservation scheme. For example, designers of corridors connecting habitat fragments must consider what the animals will likely perceive as a feasible route. An even more elementary issue is the maximal distance at which a certain species can perceive critical environmental features that affect survival and reproduction (Lima and Zollner 1996; see also Shepherdson 1994; Curio 1996; Clemmons and Bucholz 1997).

Most modern research on the management of insects and other animals that damage crops focuses on the use of ecologically and behaviorally relevant methods that replace or reduce pesticide application. This approach is known as integrated pest management (e.g. Reuveni 1995). To successfully employ behaviorally based techniques—such as the use of attractants and deterrents or the release of a pest's natural enemy—one must acquire a thorough understanding of the cognitive aspects that shape the behavior of the pest and its predator/parasitoid. For instance, parasitoid wasps learn various parameters that are associated with their hosts, and their learning must be considered when the trapping or commercial release of wasps is designed (Prokopy and Lewis 1993).

Evolutionary biologists, ecologists, and behavioral ecologists must appreciate findings of cognitive research even if these findings do not seem currently relevant. Scientists cannot predict many outcomes and future utilities of knowledge. For example, laser was not invented for use in the compact disc player or delicate surgery; 50 years ago, designers of computers did not dream of their current uses; and the relativity theory was not thought of for its application in constructive or destructive nuclear energy. In biology, molecular tools, such as DNA fingerprinting, have revolutionized our understanding of many phenomena. Nobody could have predicted the effect of molecular biology on ordinary behavioral and evolutionary-ecological research. In birds and other animals, behavioral ecologists have used DNA fingerprinting to measure the frequency of extrapair copulation (e.g. Dixon et al. 1994), which has far reaching evolutionary consequences (Westneat and Sherman 1993; Webster et al. 1995). Molecular methods have changed earlier views on the phylogenetic relationship among animals (e.g. Sibley and Ahlquist 1990), and such methods have empowered researchers to examine basic evolutionary issues such as adaptive divergence (e.g. Bradshaw et al. 1995; Mitchell-Olds 1995) and phenotypic plasticity (Schmitt et al. 1995; Pigliucci 1996). Thus, given the central function of cognition in animal life, one must recognize that cognitive ecology has the *potential* to profoundly affect future biological research.

In summary, the cognitive system has been shaped by natural selection; hence, formal evolutionary and ecological investigation is a powerful way to decipher cognitive design. Ecological interaction is a product of individual behavior; that behavior is determined by capacities and biases of and constraints on information processing. By considering cognition, we can explain general patterns of behavior that affect ecological interactions and their evolution. The essential components of cognition will now be reviewed briefly.

1.4 THE COGNITIVE SYSTEM

To produce a really good biological theory one must try to see through the clutter produced by evolution to the basic mechanisms lying beneath them, realizing that they are likely to be overlaid by other, secondary mechanisms. What seems to physicists to be a hopelessly complicated process may have been what nature found simplest, because nature could only build on what was already there.

—Francis Crick, 1988

In almost any other important biological field . . . it is possible to present the main theories historically and to show a steady progression from a large number of speculative ideas to one or two highly probable, main hypotheses.

—A. Comfort, 1979

The study of cognition is not at a mature state: basic mechanisms and main hypotheses are not well identified. Hence, the following reflects the current partial understanding of cognitive mechanisms. Because there is only partial understanding, only the most fundamental processes confirmed in neurobiological studies will be emphasized. The reception of stimuli and behavioral decision making will be presented first and last, respectively; however, this order is somewhat arbitrary. Many stages of information processing occur in parallel, and decisions may often precede the reception of certain stimuli; for example, an animal may decide to attend to some stimuli and ignore others.

The first step of information processing is the collection of stimuli such as light, sound, or smell with specialized receptors, which translate the incoming information into the internal signals of the nervous system. The amount of information encountered by an animal is very large. For example, Douglas and Martin (1996) estimated that the potential visual information load for a fly is about 0.5 megabytes per second. Therefore, an important initial function of the cognitive system is to reduce the amount of information that must be processed given a set of stimuli and its relevance to fitness (see Barlow 1972; van Hateren 1992). In general, there is a negative correlation between the amount of information processed simultaneously and its quality, or resolution. Because of severe constraints on the amount of information that can be processed simultaneously, the brain attends only to a very small proportion of the stimuli encountered at any given time (see detailed review in Dukas, this vol. chap. 3).

All species examined so far seem to possess short-term information storage, or working memory. The neurons that process a certain pattern of incoming information are able to retain a representation of that pattern at least for some short period. Representation means a neural code of environmental parameters (e.g. Gardner 1993; Deadwyler and Hampson 1995; Frester and Spruston 1995). Most information from working memory probably vanishes, but some relevant data is consolidated into long-term memory. The addition of new information into long-term memory is referred to as learning; learning is defined as the acquisition of neural codes of environmental parameters, which could potentially guide behavior (see Dudai 1989). Learning and memory are discussed throughout this volume.

Obviously, the brain is not a video monitor that represents images of the environment; it is a highly complex parallel-computing device. The translation of information received by the sensory system into neural codes involves various computational processes of individual neurons and networks of hundreds or thousands of neurons. Most of these computational processes are not understood (e.g. Gardner 1993; Deadwyler and Hampson 1995; Frester and Spruston 1995; Logothetis and Sheinberg 1996). Nonetheless, understanding these computations is essential for the ultimate understanding of cognition and behavioral decisions, such as the ones discussed in chapters 6, 8, 9, and 10. Neuronal computations are involved essentially in anything an animal does. Examples are the visual identification and classification of various objects, evaluation of the quality and quantity of different resources, and orientation in space and time.

A clear example of neuronal computation was reported recently by Hatsopoulos et al. (1995). They examined a wide-field, movement-sensitive visual neuron in the locust brain. The computation carried out by this neuron allows the locust to detect objects or predators in its path and initiate an escape response. Hatsopoulos et al. showed that the neuron's responses can be described by multiplying the velocity of an approaching image edge ($d\theta/dt$, or the rate of change in the angle between the edge of the object and the focus of expansion) by an exponential function of the size of the object's image on the retina ($e^{-\alpha\theta}$, where α is a positive constant). To visualize this computation, consider an object approaching at a constant velocity. When the object is far (small θ), $d\theta/dt$ increases faster than $e^{-\alpha\theta}$ decreases, so the response of the neuron increases with time. As the object gets closer, however, the neuronal response peaks and then decreases because of the exponential factor. The product of neuronal multiplication peaks before the object reaches its maximum size during approach; therefore, the response of the neuron enables anticipation of collision. The importance of this study and similar ones is that they report how neurons *actually* compute critical information (see Churchland et al. 1990; Koch 1990). Moreover, studies like that of Hatsopoulos et al. show what kind of information processing indeed occurs inside a simple insect brain. Complex information processing is carried out by insects, and thus, they may be used as models for the study of cognition and its ecological and evolutionary bases.

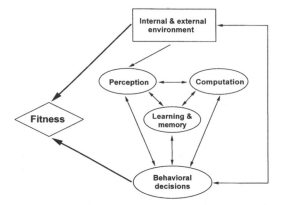

Figure 1.1. The central cognitive processes and their relation to fitness.

The numerous computations conducted in the brain are translated into motor instructions that determine behavior (fig. 1.1). Ultimately, it is the behavior itself that influences an animal's success in social interactions, foraging, escaping predation, mating, and reproduction. However, neither the cognitive processes that underlie behavior nor the effect of cognition on fitness should be ignored by researchers in a fully developed science of animal cognition and behavior.

1.5 PREVIEW

All of the following ten chapters integrate ultimate evolutionary and ecological thinking with knowledge of proximate mechanisms of information processing and associated behavioral decisions. However, as emphasized in section 1.4, many of the mechanisms that underlie cognition and the selective regimes that act on cognitive functions are not yet well understood. Hence, this volume may provide a foundation for further research on the evolutionary ecology of cognition. To this end, all chapters provide essential background information about their respective topics and suggest promising lines of future integrated research.

The volume presents various stages of information processing, from perception of stimuli to central behavioral decisions. An ecological analysis of cognition, however, must unite cognitive stages with each other and with behavioral decisions and their fitness consequences. Thus, many of the chapters refer to various aspects of cognition and behavior. The first step in information processing is the gathering of sensory cues from one's surroundings. Many biological cues are animal traits that are outcomes of coevolution between the animal with the trait (the signaler) and the animal perceiving the trait (the receiver). In chapter 2, Enquist and Arak analyze the signal form and focus on receiver biases and the probable effects these biases have on the evolution of signals. Enquist and Arak introduce neural-network models, which they use to model mechanisms of signal recognition and the outcomes of coevolution between senders and receivers. They also introduce the notion that the fundamental machinery for information processing has inherent constraints and biases. In chapter 3, I focus on some of these limitations and the effects they may have on factors such as diet choice, antipredator behavior, and temporal patterns of activity. The combined message of chapters 2 and 3 is that even though the brain has remarkable computational capacity, its biases and boundaries have had great effects on the evolution of biological traits that have been used for signaling and on the central behavioral characteristics that determine fitness.

Some types of information processing only use genetically encoded behavioral programs. However, most (and perhaps all) animals are also able to modify some of their behavior on the basis of recorded experience. Such behavioral plasticity is being studied intensely at all levels of analyses, from molecular biology to population ecology. Chapters 4 and 5 deal with a few of the many aspects of plasticity. In chapter 4, I relate the literature of phenotypic plasticity in evolutionary ecology to that of animal learning. I then address general questions about the evolution of learning and the interaction between various life-history traits and learning. Song learning in birds has been extensively studied in the laboratory and the field. In chapter 5, Beecher, Campbell, and Nordby use some of the established knowledge of song learning to integrate proximate cognitive mechanisms of song acquisition and repertoire with field data on the function and adaptive significance of song.

Many animal activities require knowledge of spatial location. For example, hymenopterans such as ants and bees forage at distances too far for use of direct visual or olfactory cues about nest location. Hence, they must use some navigation technique. In chapter 6, Dyer discusses various navigational abilities of animals in terms of the various integrated processes that have evolved to support spatial navigation and the evolutionary factors that may have shaped these processes. In chapter 7, Sherry also deals with spatial orientation by focusing on mechanisms of spatial memory and their underlying neurobiology. Environmental variation in time and space is such an integral part of daily life that we tend to forget its importance. Many animals, however, show high sensitivity to environmental variation, which may have a profound effect on fitness. In chapter 8, Bateson and Kecelnik review extensive empirical evidence for animal sensitivity to variance and discuss functional and mechanistic explanations for these data.

Overall, individual fitness is the outcome of a few central behaviors: feeding, escaping predation, and reproduction. Rigorous research on foraging ecology has been conducted for 30 years. In chapter 9, Ydenberg builds on the accumulated knowledge by discussing how foragers use information to make foraging decisions and how risk of predation affects such decisions. Some animals forage alone, but many show some social interaction. This interaction may involve a choice of partner(s). In chapter 10, Dugatkin and Sih argue that studies of partner choice in various domains—such as foraging, mating, and antipredatory defense—should be integrated into a single framework. A unified analysis can identify common and distinct cognitive mechanisms that are used for partner choice in various domains. Finally, in chapter 11, I assess what type of research will most effectively achieve the goals of cognitive ecology, which are (1) providing a unifying theory for the study of cognition and (2) evaluating how cognition affects animal behavior and ecology at the levels of individuals, populations, and communities.

ACKNOWLEDGMENTS

I thank P. Bednekoff, C. Clark, and J. McKinnon for comments on the manuscript and V. Stein, R. Nahatlatch, R. Clayoquot, and B. Tower for helping me formulate some of the ideas presented here.

LITERATURE CITED

Anderson, J. R. 1990. *The Adaptive Character of Thought.* Hillsdale, N.J.: Erlbaum.
Anderson, P. A. V., ed. 1989. *Evolution of the First Nervous Systems.* New York: Plenum.
Avise, J. C. 1993. The evolutionary biology of aging, sexual reproduction, and DNA repair. *Evolution* 47:1293–1301.
Baird, J. C., and E. Noma. 1978. *Fundamentals of Scaling and Psychophysics.* New York: Wiley.
Balda, R. P., A. C. Kamil, and P. A. Bednekoff. 1996. Predicting cognitive capacities from natural histories: Examples from four species of corvids. *Current Ornithology* 16:33–66.
Baldwin, J. M. 1893. *Elements of Psychology.* New York: Holt.
———. 1902. *Development and Evolution.* London: Macmillan.
Barkow, J. H., L. Cosmides, and J. Tooby, eds. 1992. *The Adapted Mind: Evolutionary Psychology and the Generation of Culture.* Oxford: Oxford University Press.
Barlow, H. B. 1972. Single units and sensation: A neuron doctrine for perceptual psychology? *Perception* 1:371–394.
Beer, R. D., R. E. Ritzmann, and T. McKenna, eds. 1993. *Biological Neural Networks in Invertebrate Neurobiology and Robotics.* San Diego: Academic.
Bernstein, C., and H. Bernstein. 1991. *Aging, Sex, and DNA Repair.* San Diego: Academic.
Bolles, R. C. 1970. Species-specific defense reactions and avoidance learning. *Psychological Review* 77:32–48.

Bolles, R. C., and R. C. Beecher, eds. 1987. *Evolution and Learning.* Hillsdale, N.J.: Erlbaum.

Bradshaw, H. D., S. M. Wilbert, K. G. Otto, and D. W. Schemske. 1995. Genetic mapping of floral traits associated with reproductive isolation in monkeyflowers (*Mimulus*). *Nature* 376:762–765.

Broadbent, D. E. 1958. *Perception and Communication.* Oxford: Pergamon.

Bullock, T. H. 1984. Comparative neuroscience holds promise for quiet revolutions. *Science* 225:473–478.

———. 1993. *How Do Brains Work?* Boston: Birkhauser.

Bulmer, M. 1994. *Theoretical Evolutionary Ecology.* Sunderland, Mass.: Sinauer.

Camhi, J. M. 1984. *Neuroethology: Nerve Cells and the Natural Behavior of Animals.* Sunderland, Mass.: Sinauer.

Charnov, E. L. 1976. Optimal foraging: The marginal value theorem. *Theoretical Population Biology* 9:129–136.

Churchland, P. S., C. Koch, and T. J. Sejnowski. 1990. What is computational neuroscience? In E. L. Schwartz, ed., *Computational Neuroscience,* 46–55. Cambridge: MIT Press.

Clark, C. W. 1993. Dynamic models of behavior: An extension of life history theory. *Trends in Ecology and Evolution* 8:205–209.

Clemens, J. R., and R. Bucholz, eds. 1997. *Behavioral Approaches to Conservation in the Wild.* New York: Cambridge University Press.

Collier, T. R., W. W. Murdoch, and R. M. Nisbet. 1994. Egg load and the decision to host-feed in the parasitoid *Aphytis melinus. Journal of Animal Ecology* 63:299–306.

Comfort, A. 1979. *The Biology of Senescence.* 3d ed. London: Livingstone.

Crick, F. 1988. *What Mad Pursuit.* New York: Basic Books.

Curio, E. 1996. Conservation needs ethology. *Trends in Ecology and Evolution* 11:260–263.

Darwin, C. 1872. *The Expression of the Emotions in Man and Animals.* London: Murray.

———. 1859. *The Origin of Species by Means of Natural Selection.* London: Murray.

Deadwyler, S. A., and R. E. Hampson. 1995. Ensemble activity and behavior: What's the code? *Nature* 270:1316–1318.

Diamond, J. M. 1992. The red flag of optimality. *Nature* 355:204–206.

Dixon, A., D. Ross, S. L. C. O'Malley, and T. Burke. 1994. Paternal investment inversely related to degree of extra pair paternity in the reed bunting. *Nature* 371:698–700.

Dobzhansky, T. 1937. *Genetics and the Origin of Species.* New York: Columbia University Press.

Douglas, R. J., and K. A. C. Martin. 1996. The information superflyway. *Nature* 379:584–585.

Dudai, Y. 1989. *The Neurobiology of Memory: Concepts, Findings, Trends.* Oxford: Oxford University Press.

Dukas, R., and S. Ellner. 1993. Information processing and prey detection. *Ecology* 74:1337–1346.

Falconer, D. S. 1989. *Introduction to Quantitative Genetics.* 3d ed. New York: Wiley.

16   Reuven Dukas

Ferrus, A., and I. Canal. 1994. The behaving brain of a fly. *Trends in Neuroscience* 17:479–485.

Fischler, M. A., and O. Firschein. 1987. *Intelligence: The Eye, the Brain, and the Computer.* Reading, Mass.: Addison-Wesley.

Fisher, R. A. 1930. *The Genetical Theory of Natural Selection.* Oxford: Oxford University Press.

Frester, D., and N. Spurston. 1995. Cracking the neuronal code. *Science* 270:756–757.

Frisch, K. von. 1967. *The Dance Language and Orientation of Bees.* Cambridge, Mass.: Harvard University Press.

Futuyma, D. J. 1979. *Evolutionary Biology.* Sunderland, Mass.: Sinauer.

Gallistel, C. R. 1989. Animal cognition: The representation of space, time and number. *Annual Review of Psychology* 40:155–189.

Gardner, A., ed. 1993. *The Neurobiology of Neural Networks.* Cambridge: MIT Press.

Graham, R. B. 1990. *Physiological Psychology.* Belmont, Calif.: Wadsworth.

Hall, J. C. 1994. The mating of a fly. *Science* 264:1702–1714.

Hamilton, W. D. 1966. The moulding of senescence by natural selection. *Journal of Theoretical Biology* 12:12–45.

Hammer, M. 1993. An identified neuron mediates the unconditioned stimulus in associative olfactory learning in honeybees. *Nature* 366:59–63.

Hateren, J. H. van. 1992. Real and optimal neural images in early vision. *Nature* 360:68–70.

Hatsopoulos, N., F. Gabbiani, and G. Laurent. 1995. Elementary computation of object approach by a wide-field visual neuron. *Science* 270:1000–1003.

Heiligenberg, W. 1991. The neural basis of behavior: A neuroethological view. *Annual Review of Neuroscience* 14:247–267.

Holliday, R. 1995. *Understanding Ageing.* Cambridge: Cambridge University Press.

Hughes, R. N. 1979. Optimal diets under the energy maximization premise: The effects of recognition time and learning. *American Naturalist* 113:209–221.

Jaenike, J. 1985. Genetic and environmental determinants of food preference in *Drosophila tripunctata. Evolution* 39:362–369.

James, W. [1890] 1950. *The Principles of Psychology.* New York: Dover.

Kacelnik, A., D. Brunner, and J. Gibbon. 1990. Timing mechanisms in optimal foraging: Some applications of scalar expectancy theory. In R. N. Hughes, ed., *Behavioral Mechanisms of Food Selection,* 61–82. Berlin: Springer-Verlag.

Kamil, A. C. 1983. Optimal foraging theory and the psychology of learning. *American Zoologist* 23:291–302.

Kamil, A. C., and T. D. Sargent, eds. 1981. *Foraging Behavior.* New York: Garland.

Kamil, A. C., J. Krebs, and H. R. Pulliam, eds. 1987. *Foraging Behavior.* New York: Plenum.

Kirkwood, T. B. L., and R. Holliday. 1979. The evolution of ageing and longevity. *Proceedings of the Royal Society of London,* ser. B 205:531–546.

Kirkwood, T. B. L., and M. R. Rose. 1991. Evolution of senescence: Late survival sacrificed for reproduction. *Philosophical Transactions of the Royal Society of London,* ser. B 332:15–24.

Kirkwood, T. B. L., R. F. Rosenberger, and D. J. Galas, eds. 1986. *Accuracy in Molecular Processes.* London: Chapman and Hall.

Koch, C. 1990. Biophysics of computation: Toward the mechanisms underlying, information processing in single neurons. In E. L. Schwartz, ed., *Computational Neuroscience,* 97–113. Cambridge: MIT Press.

Koshland, D. E. 1991. Flying into the future. *Science* 254:173.

Krebs, J. R., and N. B. Davies, eds. 1978. *Behavioural Ecology.* Oxford: Blackwell Scientific.

———. 1991. *Behavioural Ecology.* 3d ed. London: Blackwell Scientific.

———. 1997. *Behavioural Ecology.* 4th ed. Oxford: Blackwell Scientific.

Lima, S. L., and P. A. Zollner. 1996. Towards a behavioral ecology of ecological landscapes. *Trends in Ecology and Evolution* 11:131–135.

Lithgow, G. J., and B. L. Kirkwood. 1996. Mechanisms and evolution of aging. *Science* 273:80.

Logothetis, N. K., and D. L. Sheinberg. 1996. Visual object recognition. *Annual Review of Neuroscience* 19:577–621.

Lorenz, K. Z. 1981. *The Foundations of Ethology.* New York: Springer-Verlag.

Mangel, M. 1989. Evolution of host selection in parasitoids: Does the state of the parasitoid matter? *American Naturalist* 133:688–705.

Mangel, M., and C. W. Clark. 1988. *Dynamic Modeling in Behavioral Ecology.* Princeton, N.J.: Princeton University Press.

Mangel, M., and B. D. Roitberg. 1992. Behavioral stabilization of host-parasite population dynamics. *Theoretical Population Biology* 42:308–320.

Maynard Smith, J. 1978. Optimization theory in evolution. *Annual Review of Ecology and Systematics* 9:31–56.

Mayr, E. 1982. *The Growth of Biological Thought: Diversity, Evolution, and Inheritance.* Cambridge, Mass.: Harvard University Press.

McNair, J. L. 1981. A stochastic foraging model with predator training efects. II. Optimal diets. *Theoretical Population Biology* 19:147–162.

Medawar, P. B. 1952. *An Unsolved Problem in Biology.* London: H. K. Lewis.

Milinski, M., and G. A. Parker. 1991. Competition for resources. In J. R. Krebs and N. B. Davies, eds., *Behavioral Ecology,* 137–168. London: Blackwell Scientific.

Mitchell-Olds, T. 1995. The molecular basis of quantitative genetic variation in natural populations. *Trends in Ecology and Evolution* 10:324–328.

Morgan, L. C. 1890. *Animal Life and Intelligence.* Boston: Ginn.

———. 1900. *Animal Behaviour.* 2d ed. London: Arnold.

Murdoch, W. W. 1994. Population regulation in theory and practice. *Ecology* 75:271–287.

Neisser, U. 1967. *Cognitive Psychology.* New York: Appleton-Century-Crofts.

Newell, A., J. C. Shaw, and H. A. Simon. 1958. Elements of a theory of human problem solving. *Psychological Review* 65:151–166.

O'Carrol, D. 1993. Feature-detecting neurons in dragonflies. *Nature* 362:541–543.

Papaj, D. R., and R. J. Prokopy. 1989. Ecological and evolutionary aspects of learning in phytophagous insects. *Annual Review of Entomology* 34:315–350.

Pigliucci, M. 1996. How organisms respond to environmental changes: From phenotypes to molecules (and vice versa). *Trends in Ecology and Evolution* 11:168–173.

Posner, M. I., ed. 1989. *Foundations of Cognitive Science.* Cambridge: MIT Press.

Prokopy, R. J., and W. J. Lewis. 1993. Application of learning to pest management.

18   Reuven Dukas

In D. R. Papaj and A. C. Lewis, eds., *Insect Learning: Ecological and Evolutionary Perspectives,* 308–342. New York: Chapman and Hall.

Prokopy, R. J., S. S. Cooley, and D. R. Papaj. 1993. How well can relative specialist *Rhagoletis* flies learn to discriminate fruit for oviposition. *Journal of Insect Behavior* 6:167–176.

Pyke, G. H. 1984. Optimal foraging theory: A critical review. *Annual Review of Ecology and Systematics* 15:523–575.

Pyke, G. H., R. H. Pulliam, and E. L. Charnov. 1977. Optimal foraging: A selective review of theory and test. *Quarterly Review of Biology* 52:137–154.

Real, L. A. 1991. Animal choice behavior and the evolution of cognitive architecture. *Science* 253:980–986.

———. 1993. Toward a cognitive ecology. *Trends in Ecology and Evolution* 8:413–417.

Reuveni, R., ed. 1995. *Novel Approaches to Integrated Pest Management.* Boca Raton, Fla.: CRC.

Richards, R. J. 1987. *Darwin and the Emergence of Evolutionary Theories of Mind and Behavior.* Chicago: University of Chicago Press.

Romanes, G. 1882. *Animal Intelligence.* London: Kegan Paul, Trench.

———. [1883] 1900. *Mental Evolution in Animals.* London: Kegan Paul, Trench.

Rose, M. R. 1991. *Evolutionary Biology of Aging.* New York: Oxford University Press.

Rosenzweig, M. L. 1991. Habitat selection and population interactions: The search for mechanism. *American Naturalist* 137(suppl.):5–28.

Schmitt, J., A. C. McCormac, and H. Smyth. 1995. A test of the adaptive plasticity hypothesis using transgenic and mutant plants disabled in phytochrome-mediated elongation responses to neighbours. *American Naturalist* 146:937–953.

Seligman, M. E. P., and J. L. Hager, eds. 1972. *Biological Boundaries of Learning.* New York: Appleton.

Shepherdson, D. 1994. The role of environmental enrichment in the captive breeding and reintroduction of endangered species. In P. J. S. Olney, G. M. Mace, and A. T. C. Feistner, eds., *Creative Conservation: Interactive Management of Wild and Captive Animals,* 167–177. London: Chapman and Hall.

Shettleworth, S. J. 1993. Where is the comparison in comparative cognition? Alternative research programs. *Psychological Sciences* 4:179–184.

Sibley, C. G., and J. E. Ahlquist. 1990. *Phylogeny and Classification of Birds: A Study in Molecular Evolution.* New Haven, Conn.: Yale University Press.

Simon, H. A. 1979. *Models of Thought.* New Haven, Conn.: Yale University Press.

Singer, M. C. 1994. Behavioral constraints on the evolutionary expansion of insect diet: A case history from checkerspot butterflies. In L. A. Real, ed., *Behavioral Mechanisms in Evolutionary Ecology,* 279–296. Chicago: University of Chicago Press.

Skinner, B. F. 1938. *The Behavior of Organisms.* New York: Appleton-Century-Crofts.

Srinivasan, M. V., S. W. Zhang, and B. Rolfe. 1993. Is pattern vision in insects mediated by "cortical" processing? *Nature* 362:539–540.

Stearns, S. 1992. *The Evolution of Life Histories.* Oxford: Oxford University Press.

Stephens, D. W. 1993. Learning and behavioral ecology: Incomplete information and environmental predictability. In D. R. Papaj and A. C. Lewis, eds., *Insect Learning,* 195–218. New York: Chapman and Hall.

Stephens, D. W., and J. Krebs. 1986. *Foraging Theory.* Princeton, N.J.: Princeton University Press.

Stillings, N. A., S. E. Weisler, C. H. Chase, M. H. Feinstein, J. L. Garfield, and E. L. Rissland. 1995. *Cognitive Science: An Introduction.* 2d ed. Cambridge: MIT Press.

Tinbergen, N. 1951. *The Study of Instinct.* Oxford: Oxford University Press.

Tully, T. 1996. Discovery of genes involved with learning and memory: An experimental synthesis of Hirschian and Benzerian perspectives. *Proceedings of the National Academy of Sciences of the United States of America* 93:13460–13467.

Watson, J. B. 1924. *Behaviorism.* New York: People's Institute.

Wcislo, W. T. 1989. Behavioral environments and evolutionary change. *Annual Review of Ecology and Systematics* 20:137–169.

Webster, M. S., S. Pruett-Jones, D. F. Westneat, and S. J. Arnold. 1995. Measuring the effects of pairing success, extra pair copulations and mate quality on the opportunity for sexual selection. *Evolution* 49:1147–1157.

Werner, E. E. 1994. Individual behavior and higher-order species interactions. In L. A. Real, ed., *Behavioral Mechanisms in Evolutionary Ecology,* 297–324. Chicago: University of Chicago Press.

West-Eberhard, M. J. 1989. Phenotypic plasticity and the origins of diversity. *Annual Review of Ecology and Systematics* 20:249–278.

Westneat, D. F., and P. W. Sherman. 1993. Parentage and the evolution of parental behavior. *Behavioral Ecology* 4:66–77.

Williams, G. C. 1957. Pleiotropy, natural selection, and the evolution of senescence. *Evolution* 11:398–411.

CHAPTER TWO

Neural Representation and the Evolution of Signal Form

MAGNUS ENQUIST AND ANTHONY ARAK

2.1 INTRODUCTION

The study of animal communication has been a major area of interest in biology for more than a century. Signals used by animals for communication are sometimes the most striking aspects of their morphology or behavior. The remarkable color patterns of butterflies, coral reef fish, and birds; the facial expressions of monkeys and people; the fantastically varied sounds of insects, frogs, and birds; the distinctive smell of a skunk; and the sweet scents of flowers—all capture the imagination and demand explanation. Despite many years of intensive study, understanding how a particular signal form has evolved—how it has come to look, sound, or smell the way it does—remains a major challenge to biologists. One need only consider the diversity of displays in any particular taxonomic group, for example birds or insects, to understand that none of our existing theories satisfactorily explains the incredible variation in color patterns, sounds, and behavior that constitute signals in different species.

Our understanding of biological communication systems has advanced greatly with the application of techniques of evolutionary game theory. This application has proved useful for clarifying important strategic questions about animal communication. Strategic questions concern how the interests of different players in an interaction determine what actions each should take in order to maximize fitness (see Dugatkin and Sih, this vol. chap. 10). Is it advantageous for individuals to give away information to other players, particularly if players' interests conflict? It is now clear that it is advantageous for an individual to provide information to others (Zahavi 1975, 1977; Enquist 1985; Grafen 1990) despite earlier assertions that an individual should not (Maynard Smith 1974, 1982; Dawkins and Krebs 1978; Caryl 1979; Wiley 1983). Models in evolutionary game theory generate predictions about how signal cost should vary according to an individual's state and about whether discrete or continuous signals will be favored in certain circumstances (Grafen and Johnstone 1993). Game theory, however, makes no specific predictions about the details of signal design, nor does it offer any clue as to why signals are so diverse in form (fig. 2.1).

Similar limitations apply to genetic models. In recent years, genetic models have been extensively developed for understanding the evolution of elaborate sexual displays in animals (reviewed by Andersson 1994). Although several plausible mechanisms have been discovered that permit the evolution of costly displays in males and corresponding preferences for such traits in females, these models cannot be generalized to account for the evolution of costly signals in other contexts. For example, some traits of flowers seem as conspicuous and extreme as the most remarkable secondary sexual traits of animals. Yet, because flowers and the pollinators they attract belong to different species, there can be no genetic coupling between signal traits and preferences, which seems to be essential in models of sexual selection (Andersson 1994). A much simpler mechanism between flowers and pollinators is probably at work.

A very different approach to the study of the evolution of signal form is to ask how signals are designed for reaching their target destination (e.g. Wiley 1983). Because the form of every animal signal is a kind of energy—chemical, electromagnetic, mechanical, or electrical—the signal is constrained by physical laws that govern its production and transmission through natural environments. There is now abundant evidence that physical laws exert strong effects on signal form (Gerhardt 1983). In many cases, very specific predictions can be made about the signal form that will be favored in particular environments with use of the principles of signal-detection theory (Green and Swets 1966). Studies relating the structure of bird song to the type of habitat in which the song is used are good examples of the use of these principles (Morton 1975; Hunter and Krebs 1979; Wiley and Richards 1983). Additional factors affecting signal transmission—such as the position and orientation of the signaler relative to that of the receiver—and sources of noise in the environment also strongly influence the design of signals (Gerhardt 1983).

In signal design, "detectability" can provide at best only a partial explanation of the enormous diversity of biological signals. Detectability fails to account for the fact that many displays are much more complex and conspicuous than they need be to ensure detection by the receiver. For example, when competing for food in winter, the great tits *Parus major* use highly exaggerated postures to threaten one another (Blurton Jones 1968). These displays are used over distances that are negligible when compared to those at which the birds normally react to a predator; this kind of display suggests that constraints on signal transmission are unlikely causes of exaggeration (Hurd et al. 1995). In same bird species, highly complex songs, which often incorporate mimetic elements, are difficult to explain purely as adaptations that ensure detection (see Beecher et al., this vol. chap. 5). Observations like these suggest that some conspicuous displays are intended to persuade the receiver to react in

Figure 2.1. Structures used by birds for courtship display. With strategic models, it is predicted that signals conveying information about male quality can evolve, but the problem of why there is so much diversity in these signal forms remains. (*a*) Display and facial engorgement of the wattled pheasant, *Lophura bulweri*. (*b*) Highly modified feathers of the standard-wing nightjar, *Macrodipteryx longipennis*, which are dropped shortly after courtship. (*c*) Frontal display of Temminck's tragopan *Tragopan temmincki* with the blue and red lappet, or bib, fully expanded. (*d*) Balloon display, which includes a large inflated and pendulous throat sac, of the Australian bustard, *Ardeotis australis*. (*e*) Bower built by the male MacGregor's bowerbird, *Amblyornis macgregoriae*.

a certain way and not to overcome background noise in the environment (Dawkins and Krebs 1978).

Thus, to understand how signals can be designed for maximum effect, we have to look beyond the medium through which the signal is traveling and into the receiver's nervous system (Guilford and Dawkins 1991). What factors influence an individual's selective response to stimuli? Historically, this question has received much attention from workers in several disciplines–including ethology, psychology, sensory physiology, and artificial intelligence. Until recently, however, the idea that cognitive mechanisms serve a crucial function in signal evolution has been nearly ignored by evolutionary biologists. The reason for this neglect may stem from a widespread opinion that details of proximate mechanism are irrelevant when questions about the adaptive significance of behavior are addressed. This opinion, however, is clearly wrong when the evolution of signal form is examined. The receiver's brain is as much a part of the signal's environment as is the external medium through which the signal must first travel to reach the receiver's sensory organs.

How then may we study the influence of the receiver on the evolution of signal form? Several approaches seem possible. First, one can carry out comparative studies on variation in signals and receiver responses in a group of closely related species. The data from these comparative studies, when supplemented with information about the phylogeny of the group, may enable one to reconstruct part of the historical sequence of the evolutionary change in signal form. Basolo's (1990) and Ryan's (1990) studies on the evolution of male courtship signals in certain species of fish and frogs are good examples of this approach. In these studies, it was suggested that female preference for particular male-signaling traits may sometimes arise in a lineage prior to the appearance of the male trait. The male trait subsequently appears and is favored because it exploits a preexisting sensory bias in females (Ryan 1990, 1994). This approach can sometimes reveal the origin of signals, but it enables no general prediction about the form that signals should take. Many signaling traits almost certainly evolved more gradually from incidental movements and structures; this process was called ritualization by early ethologists (Huxley 1966). In this chapter, these gradual changes in signal form that result from the coevolutionary interaction between signaler and receiver will be discussed and particular attention will be paid to the function of receiver mechanisms.

In part, we will use a theme frequently adopted by psychologists in their search for common cognitive mechanisms that underlie an individual's response to stimuli (Pearce 1987, 1994; Eysenck and Keane 1995). The ability of an animal to recognize a signal and respond appropriately depends critically on the storing of some representation of the signal in memory. Thus, the nature of such internal representations and the manner in which stimuli are compared with these representations to produce a response are crucial issues. We will present theoretical arguments and empirical evidence to support the view that recognition mechanisms are not perfect in all respects; they contain slight biases. These biases may act as potent forces on the evolution of signal form.

What can be predicted about the evolutionary consequences of such biases? We need a way of modeling gradual changes in receiver responsiveness to answer this question. Simple models of memory formation on the basis of the development of generalization gradients can be used (Leimar et al. 1986). A more general approach uses models known as artificial neural networks (Rumelhart et al. 1985; Hinton 1992; Churchland and Sejnowski 1992; Haykin 1994). Such models take their inspiration directly from knowledge of the structure and function of biological brains; these models have been used successfully to study aspects of biological cognition at a variety of levels. Building on a recent tradition within psychology (e.g. McClelland and Rumelhart 1985; Rumelhart et al. 1985), we use artificial neural networks as a general model of stimulus-response contingencies to examine how such relationships relate to representation in memory. Thus, our aim is not to model any particular cognitive mechanism in detail (e.g. vision or hearing in primates); we seek to understand certain general behavioral aspects of pattern recognition that are independent of species group and sensory modality.

Artificial neural networks are reasonably accessible tools for simulating changes in signals and receiver mechanisms over evolutionary time (Arak and Enquist 1993, 1995; Enquist and Arak 1993, 1994; Hurd et al. 1995; Johnstone 1994). We use these models to show how some common features of signals—such as their conspicuousness, distinctiveness of form, contrasting elements, and various symmetries—may emerge as outcomes of simple evolutionary processes. We also show how selection pressures that emerge from the nervous system can be combined with strategic and transmission factors to provide a more complete understanding of biological communication. Finally, we return to a controversial question first broached by Darwin (1871): Do other animals possess an aesthetic sense that is akin to that in humans?

2.2 COMMON QUALITIES OF SIGNAL FORM

Perhaps the most striking aspect of animal communication is the enormous variety of forms that signals take. Among closely related species, signals are usually more dramatically variable in form than other traits. For example, plumage color in birds often varies to a much greater extent within genera than other morphological features such as beak, wing, or body shape; this fact is most useful for bird-watchers. In some taxa, there is also considerable intraspecific variation in signal form. This variation may be expressed between different geographical areas or within a single population.

What generalizations can be made about signal form? Because the available methods for producing signals are different, signal form depends to some extent on the taxonomic group under consideration. We will focus on general characteristics of signal form that are more or less independent of the systematic position of the species (e.g. Cott 1957; Eibl-Eibesfeldt 1975; Brown 1975; Smith 1977) (fig. 2.2).

A signal is typically made up of a number of distinct elements. Frequently, some of these elements are exaggerated (e.g. Brown 1975; Andersson 1994). Examples of these exaggerated elements include elongated tails and plumes in birds, enlarged petals in flowers, saturated colors, loud calls or songs, expressive movements, and large quantities of pheromones. Another common quality of signals is a degree of simplicity. The types and number of components and their complexity are often limited. When behavior patterns are included as signal elements, these behavior patterns usually consist of a few postures or motions performed at a typical intensity (Morris 1957, Barlow 1977). By tracing the evolution of signals with use of the comparative method, one often sees that signal components have become either more pronounced or have disappeared during the process of ritualization (e.g. Eibl-Eibesfeldt 1975). One example of the disappearance of a signal component is in the courtship behavior of peacocks, which probably originated from feeding behavior; in courtship, the behavior of pecking the ground has disappeared (Schenkel 1956). One could say generally that signals have become more "digital" during their evolution.

Signal components often relate to one another in a nonrandom way. For instance, visual patches may appear conspicuous because they contrast with neighboring patches, which are complementarily colored or outlined with a contrasting border. Components are commonly repeated, such as in the case of the distinctive black and yellow stripes of wasps. In visual displays, movements may be repetitive, and acoustic signals often consist of repetitions of elements in several different time scales. Finally, in many visual signals, components are symmetric or arranged symmetrically. Recently, much attention has been paid to such symmetries (e.g. Møller 1990; Enquist and Arak 1994; Watson and Thornhill 1994).

There is some evidence that the function and context of signals have an influence on signal form, but more comparative work is needed. Huxley (1938) and Cott (1957) stated that colors designed to warn are often more bold and conspicuous than colors designed to invite sex. It is often difficult, however, to infer the function of a signal just by considerating its form. The same signal may be used in more than one context, such as in fighting and courtship (e.g. Beer 1975). The widely diverse body color of coral reef fish exemplifies the problem of relating function and form. These signals have a number of functions depending on species, sex, and age (e.g. Eibl-Eibesfeld 1975; Andersson

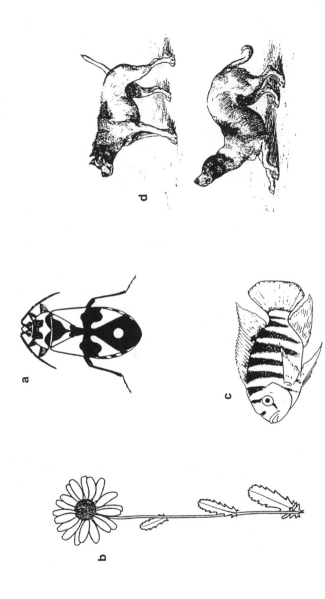

Figure 2.2. Qualities of signal form. (a) Contrasting colors in the seed bug, *Lygaeus equestris*. (b) Radial symmetry in the oxeye daisy, *Chrysanthemum leucanthemum*. (c) Spatial repetition in body color in the cichlid fish, *Tilapia mariae*. (d) Displays of threat and submissiveness in the dog. (Examples of exaggeration are shown in fig. 2.1.)

1994); however, except in certain cases of deceit, the particular appearance of these signals seems to provide few clues as to their function.

So far, we have considered only the form of the signal in isolation. A full understanding of signal form requires that we make reference to the usual background of the signal or signal element and other stimuli reaching the receiver. It is possible for signals to be conspicuous in isolation, but signals are conspicuous mostly because they are different from the background (Cott 1957). In some species, the whole animal appears conspicuous because of reverse countershading or uniform body color that contrasts with the background and thus enhances body shape. Movement is another obvious feature of conspicuousness in visual signaling. When an object moves within a stationary visual field, it stands out immediately from the background. Signals with movement often occur suddenly, and the speed and direction of movement is controlled so that it will contrast with random movements of surrounding objects.

Another general feature of signals is their distinctiveness (e.g. Brown 1975; Smith 1977). Signals that elicit different responses are often, but not always, very different in appearance. In the most extreme situation, noted by Darwin (1872), signals designed to provoke opposite reactions are opposite in form. Darwin, who referred to this as the principle of antithesis, pointed to the extreme difference in posture between a hostile dog and the same animal "in a humble and affectionate frame of mind." Antithesis is also seen in gulls, the beak is shown off in aggressive displays, but it is hidden, by turning away the head, in appeasing displays (Tinbergen 1959). Some visual signals that do not involve movement also demonstrate antithesis. For example, the fish *Neetroplus nematopus* has two morphs; the social, schooling morph is gray with a black bar and the solitary, breeding, territorial morph is gray with a white bar (Baylis 1974, cited in Hailman 1977). A signal may differ from other signals and the background in several simultaneous ways, giving a strong impression of redundancy. Signals often are more distinct from one another than is necessary so that accurate discrimination is ensured. For example, the plumage patterns of male ducks characteristically consist of many different patches of color covering different areas of the body. This complexity provides redundant visual cues for species recognition.

We considered signals that are conspicuous in various ways; however, there are many exceptions. In primates, subtle facial expressions provide effective and precise communication (e.g. Zeller 1980), and similar subtlety is observed in the communication systems of many cooperatively breeding birds and mammals. With such observations, it is suggested that conspicuousness is not always necessary to ensure detection. Signals designed for long distances will, of course, be conspicuous when close to the sender; however, many conspicuous signals are used only at very close range, and in some cases, close-range signals are more complex than those used over greater distances. In birds, for example, visual displays designed to attract attention from a distance consist mainly of simple patterns and saturated colors covering most of the body; signals used at close range tend to be more complex in design, more elaborate in color, and often involve smaller features that are appropriate for near vision (Huxley 1938; Cott 1957; but see Wiley 1983). The same is true for many acoustic signals; the long-range advertisement calls used by certain frogs and grasshoppers tend to be much simpler than those used at close range for territorial defense or courtship (Wells 1988; Ragge 1965). These signal differences suggest that the receiver perceives different aspects of a signal at different distances. For instance, males of the green tree frog, *Hyla cinerea,* produce a complex call with two major frequency bands and a distinctive temporal structure. At very low sound levels that correspond to long distance, females are attracted with just the low-frequency peak or even a continuous low-frequency noise band, which lacks a temporal structure (Gerhardt 1976; Ehret and Grerhardt 1980). As a female gets closer to a male and the sound level increases, however, she selectively responds to calls with both low- and high-frequency bands; subtle temporal differences also come into play (Gerhardt 1978). Clearly, in the case of the green tree frog signal design is influenced by the properties of the receiver's nervous system as well as environmental constraints that affect transmission.

2.3 THE PROBLEM OF RECOGNITION

Traditionally, most investigators of animal communication assumed that a close match existed between the form of a communication signal and the receiver mechanisms designed to detect and respond to the signal. Lorenz (1935) developed the idea that particular sounds, scents, or colors have evolved in a special way to elicit particular responses from a receiving animal. Lorenz called these signals social releasers and implied that they have evolved in parallel with special devices in the receiver, innate releasing mechanisms (IRMs), which allow recognition of such signals (see Baerends 1982 for a full discussion). An IRM was defined by Tinbergen (1951) as a "special neurosensory mechanism that releases the reaction and is responsible for its selective susceptibility to a very special combination of sign stimuli." Thus, IRMs and releasers were considered to be mutually adapted to one another (e.g. Eibl-Eibesfeldt 1975). One animal gives a stereotypic signal, its releaser, to which another animal is especially responsive by way of its equivalent IRM. Experimental psychologists have used related but independent concepts when speaking of analyzers, or filters, which are tuned to specific aspects of a stimulus (Sutherland 1969; Baerends and Kruijt 1973).

Sensory physiologists sometimes compare the evolution of a biological communication sensor with the problem confronting an engineer who is designing a specialized communication receiver (e.g. Hopkins 1983). This analogy implies that natural selection builds sensory mechanisms in animals according to strict engineering principles. In particular, the specificity of the mechanism is expected to be precisely adjusted; the receiver reacts to a signal when the signal is present and does not react when the signal is absent. Perfect recognition is also implicit in most recent models of the evolution of signals (Arak and Enquist 1993).

To what extent will animal sensory systems evolve to behave in such an idealized fashion? In principle, perfect recognition necessitates that an error-free sensory system be designed and maintained over evolutionary time.

Certainly, there are some cases that seem to demonstrate the precise selectivity that is expected on the basis of engineering principles. For example, the olfactory sensors on the antennae of some male moths are exquisitely designed to enable detection of pheromones emitted by conspecific females; these sensors are highly sensitive and selective. In the silkworm moth, *Bombyx mori,* males respond to the female odor of the pheromone bombykol at distances of 1 km or more (Kaissling 1971); however, various synthetic stereoisomers, when used in place of bombykol, are 100 to 1,000 times less effective (Schneider et al. 1967). Comparable specificity is also seen in specialized electroreceptors (called "Knollenorgans") of certain African electric fishes; these receptors function as precisely tuned filters that are matched to the species-specific signal (Hopkins 1977).

These examples, however, appear to be exceptions. The sole function of the specialized receptors of moths and electric fish is to detect the presence of a *single* desired stimulus; these receptors are not used for other, more general sensory tasks. Sensory systems typically have multiple functions and do not just receive and process communication signals. The visual function of the brain, for example, has been likened to a service industry, which enables navigation, balance, and object recognition and provides guidance in social interactions (Harris and Humphreys 1994). The design of such a system inevitably involves a large number of compromises. A balance must be struck between the ability to generalize and the need for specificity, speed and accuracy, the size of memory and the cost of space, and production and maintenance of nervous tissue. A complex organ could not be expected to perform flawlessly each specific task it confronts (see Dukas, this vol. chap. 3).

Consider the problem of recognizing visual stimuli. To humans, recognition of complex visual stimuli seems to be simple and effortless; however, this apparent easy recognition belies the fact that the processes of transforming and interpreting visual information are extraordinarily complex. In fact, some of these complexities became clear only when workers in artificial intelligence attempted to program computers to perceive their environment. Even when the environment was artificially simplified (e.g. consisting only of white solids) and the task was apparently simple (e.g. deciding how many objects there were), computers required very complicated programs to succeed. There are no computers that can match more than a small fraction of the visual-perception skills possessed by many animals (Boyle and Thomas 1988).

One reason visual recognition presents such a problem for theorists is that any object can give rise to a vast number of images on the retina depending on the distance, angle, and orientation of the object from the observer; the prevailing light conditions; and details of the background. In fact, it is unlikely that an organism will experience exactly the same pattern of activation across its retina more than once during its lifetime, and there is serious doubt that recognition can be achieved simply by matching a pattern of stimulation against a preexisting neural template. Even if we assume that the incoming information is reduced in some way (as most theorists do), ultimately we are still faced with the problem of how a filter can be designed that will always generalize in the correct way.

Not uncommonly, because of imperfect information about the environment, the receiver may commit errors that result in suboptimal behavior. For example, if an animal learns that the color red is generally associated with poisonous food, on at least some occasions, the animal may reject perfectly edible food. Likewise, a dangerous object may be explored if it happens to resemble an object that is usually good to eat. Of course, the signaler may turn the receiver's uncertainty to its advantage (Dawkins and Krebs 1978). In Batesian mimicry, palatable prey escape attack by closely resembling unpalatable species. In the case of the water mite *Neumania papillator,* more interesting mimicry is seen (Proctor 1992). Males attract females by mimicking the wave-vibration patterns of copeopod prey, and females approach the signaling males exactly as if they were prey. Often a female will grab a male before she discovers that he is not prey, and it is only then that the male begins to court her. Ryan (1994) reviewed other examples of courtship signals that exploit responses in receivers, which probably evolved in other contexts.

Another kind of uncertainty arises when an animal responds to novel stimuli (Arak and Enquist 1993). During evolution, responses are tuned only toward stimuli that occur regularly. Although many novel stimuli may be ignored, we can expect some novel stimuli to be highly effective in eliciting responses from the receiver. Imagine a simplified situation in which an animal must discriminate between two stimuli that differ in value along a single dimension, for instance color. One stimulus represents a food item that the animal can eat (e.g. a yellow banana) and the other stimulus represents an object that is inedible (e.g. a green banana). We assume that the sensory input represented by the two stimuli is filtered through the animal's nervous system. If the system detects a ripe banana, it sends a signal to the brain's motor control center

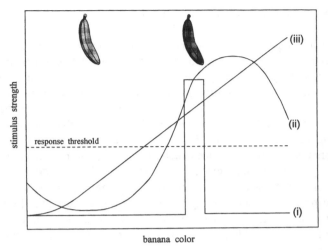

Figure 2.3. Tuning of a hypothetical recognition mechanism that enables distinction between two stimuli that vary along one dimension. Attraction to bananas is assumed to depend on color. Inedible bananas are green (*left*), and ripe edible bananas are yellow (*right*). The response of picking a banana is assumed to occur when stimulus strength exceeds a certain threshold in a dedicated recognition neuron or template. Curve i represents perfect recognition; curves ii and iii represent alternative mechanisms that lead to the same behavioral consequences. The mechanisms of curves ii and iii, however, also result in a strong response to bananas of other colors, like orange or red, should they arise in nature.

that causes the animal to grab the banana; otherwise, the system does not respond. How should the sensory filter be tuned so that an appropriate behavioral response is elicited?

In figure 2.3, the curves i–iii represent a series of alternative tuning curves. Curve i is the idealized case in which perfect recognition occurs. The animal always grabs bananas that are of a color within the range of wavelengths that correspond to yellow in the visual spectrum; the animal ignores bananas of all other colors. However, there is no particular evolutionary reason why the filtering mechanism represented by curve i should be favored. The mechanisms represented by curves ii and iii, for example, function equally well. In these cases, there are no evolutionary forces that favor any particular recognition strategy; the recognition mechanism during evolution will be subject to random drift.

This instability of the recognition mechanism means that banana color, a signal which attracts frugivores, is also not stable over evolutionary time. If the recognition mechanism represented by curve ii or iii is used by the animal, then orange or red bananas may invade the population of bananas should they arise by mutation. In effect, there can be a mismatch between the current form of the signal and the stimulus most preferred by the receiver. The receiver's preference may be described as "hidden" because the preference is not expressed until a red banana arises (Basolo 1990; Ryan 1990; Arak and Enquist 1993).

A hidden preference can be described in more general terms as an evolutionary game that is played between the signaler and receiver. In such a game, the signaler tries to elicit a certain response from the receiver, and the receiver is faced with the problem of recognizing and reacting to the signal. It is assumed that a number of appearances a are available to the signaler and a number of recognition strategies r are available to the receiver. An evolutionarily stable strategy, or ESS (Maynard Smith 1982), for such a game would consist of a stable equilibrium pair of strategies (a^*, r^*), where a^* is the best reply to r^* and vice versa (this ensures equilibrium). For the equilibrium pair of strategies to be stable, we also require that if the population deviates slightly from the equilibrium, selection returns it to (a^*, r^*).

If the number of sets of signaling strategies and recognition strategies are restricted, such a game may have an ESS; however, for a given level of cost, signals take on many possible appearances, and recognition of any one of such appearances may be achieved by many different mechanisms, $r_1, r_2 \ldots r_n$. This means that a pair of strategies (a^*, r^*) cannot be maintained at a stable equilibrium because there are no forces that will favor r^* from among all those r that correctly recognize a^*. Consequently, r^* may drift, and it follows that a^* will no longer be the best reply to the receiver's new r. Thus, a signal's appearance may change continuously as it tries to match changes occurring in the recognition mechanism.

How close to the optimal signal, a^*, any particular appearance, a, will be, depends on, among other things, the rate at which the recognition strategy, r, changes and the rate at which new appearances of the signal arise in the population. The proximity of a to a^* also depends critically on the size of the strategy space for signal appearances. In reality, signals may differ along many dimensions simultaneously, and the number of possible appearances that could arise by mutation is almost infinite. If a signaler is to find the best possible signal to use against a given recognition strategy, all possible signals must be tested. Clearly, this is not feasible—except perhaps for the most simple signals, which are constrained to a few discrete states. In situations where complex signals are used, we should expect mismatches between receiver preferences and signal form to be common. When such mismatches occur, strong selection may be exerted on signals that cause signal form to evolve rapidly in the direction of the bias.

In conclusion, there are no theoretical reasons to expect selection to produce ideal matches between sensory input and behavioral responses or favor any particular recognition strategy over others when strategies solve a given discrimination problem equally well. Thus, game theory does not offer a solution to the question of what particular stimulus-response relationship may evolve in nature. To understand this matter further, knowledge about the functioning of real recognition systems is required.

2.4 EVIDENCE FOR RECEIVER BIAS

2.4.1 Receptor Bias

Most evidence for biases in peripheral sensory receptors is in studies of the auditory sensitivity of frogs to acoustic signals used in mate attraction (reviewed by Ryan and Keddy-Hector 1992). In several species, there is a slight mismatch between the tuning of the peripheral auditory system and the mean dominant frequency of the male's call. For example, in the tungara frog, *Physalaemus pustulosus,* the most sensitive frequency of the basilar papilla (an inner ear organ) is 2,130 Hz, but the mean dominant frequency of the "chuck" component of the male's call is 2,550 Hz (fig. 2.4). This mismatch is responsible for a female's preference for chucks with lower frequencies than the population mean. The peak sensitivity of the basilar papilla in the closely related species *Physaelaemus coloradorum* is statistically identical to that of *P. pustulosus,* yet the calls of the former species lack the chuck component altogether. When females of *P. coloradorum* were given a choice between the normal conspecific male call, which is a whine, and this same call with chucks of *P. pustulosus* added, the females preferred calls with chucks. Moreover, phylogenetic evidence suggests that the preference for chucks preceded the evolution of chucks in the *P. pustulosus* species group. Thus, it is likely the chuck evolved to exploit a preexisting bias in the tuning of the female's auditory system. Similar mismatches between auditory tuning and call frequency have been documented in the gray tree frog, *Hyla versicolor* (Gerhardt and Doherty 1988), and in the cricket frog, *Acris crepitans* (Ryan et al. 1992). Females of these species also prefer calls with a lower frequency than the mean frequency of the population.

2.4.2 Supernormal Stimuli

Although filtering mechanisms in the peripheral sensory organs undoubtedly affect the evolution of signal form, these filtering mechanisms usually represent only the first step in the processing of sensory stimuli. A great deal of evidence suggests that biases also occur in the central nervous system, where mechanisms affect the recognition of complex stimuli. Some early evidence was obtained by ethologists in studies with use of dummies to investigate the properties of stimuli that are important for eliciting responses from animals. Not uncommonly, researchers produced dummies that provoked stronger re-

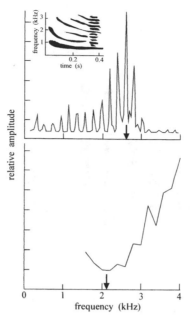

Figure 2.4. *Top:* Fourier transform of the chuck component of the call of the tungara frog, *Physalaemus pustulosus,* shows the distribution of energy across frequencies. Arrow indicates the frequency that contains maximum energy. *Inset:* Sonagram of a whine and chuck call shows the frequency distribution over time. *Bottom:* Frequency sensitivity of the basilar papilla from multiunit recordings of the torus semicircularis. Arrow indicates the frequency at which the basilar papilla is most sensitive; the basilar papilla appears to be most sensitive to frequencies that are lower than the peak frequency in the chuck call. This sensitivity provides the basis for the female preference for chucks of a lower frequency than the mean frequency in nature (from Ryan et al. 1990).

sponses than those provoked with the natural stimuli the dummies were meant to represent. This phenomenon was first discovered by Koehler and Zagarus (1937) in a study of egg recognition by birds. Ringed plovers, *Charadrius hiaticula,* were found to prefer white eggs with black spots over normal eggs, which are light brown with darker brown spots. Other examples of such apparently supernormal stimuli were soon discovered: chicks of the herring gull, *Larus argentatus,* pecked more readily at a long, thin, red rod with three white bars at the tip than a realistic copy of the parent's bill and head (Tinbergen and Perdeck 1950); female oyster catchers, *Haematopus ostralegus,* invariably preferred giant dummy eggs over their own (Tinbergen 1951); and male silver-washed fritillary butterflies, *Argynnis paphia,* showed stronger attraction to female dummies with faster wing-beat frequencies than real females (Magnus 1958). More recent examples include female widowbirds, *Euplectes progne,*

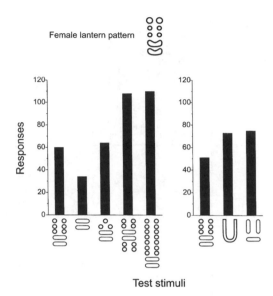

Test stimuli

Figure 2.5. Responses to supernormal stimuli in the glow worm, *Phausis splendidula*. Males advertise their presence to females with a luminescent pattern on their bodies; receptive females signal back with a different pattern. Graphs show the frequency of male responses to various artificial light patterns in two sets of experiments. The first pattern in each graph mimics the natural stimulus. Some artificial patterns are much more effective in eliciting responses from males than the pattern of the real female or its artificial equivalent (from Schaller and Schwalb 1960).

that preferred males with supernormal long tails (Andersson 1982), female grackles, *Quiscalus quiscula,* that responded more frequently to song repertoires of supernormal complexity (Searcy 1992), and female zebra finches, *Taeniopygia guttata,* that preferred males adorned with red leg bands and fancy white hats (Burley, unpublished, cited in Trivers 1985) (fig. 2.5).

In many cases, the most effective stimuli seem to be, in a sense, caricatures of reality: These stimuli exaggerate key features of the natural stimulus. Many studies have shown that a stimulus that is bigger, longer, louder, or brighter is more likely to elicit a response or be preferred over a stimulus with a mean size or intensity (Ryan and Keddy-Hector 1992). Thus, it is sometimes argued that stimuli of greater quantity are more effective because they cause greater neural stimulation; however, a simple relationship between the strength of a stimulus and its effectiveness cannot always be assumed. For instance, an incubating herring gull responds most strongly to model eggs that are green, but a gull that cannibalizes other birds' eggs prefers red eggs (Baerends and Kruijt 1973). In this case, then, supernormal stimuli do something more than simply cause greater general neural stimulation. Because the form of the stimulus that is maximally effective is specific in each behavioral context, we must assume in the case of the gull that biases originate within central mechanisms that are responsible for categorizing stimuli differentially. In other words, the gull's selective responsiveness to color indicates there are biases that operate in different directions for different responses; in some situations, a bias results in preferential responsiveness to green (the color to which the bird's eye is maximally sensitive), and in other situations a bias results in a preference for red. The bias for green occurs when gulls are retrieving eggs and attending the nest, and the bias for red occurs when gulls are removing broken egg shells from the nest or robbing eggs from other gulls or when juveniles are begging for food (see Baerends and Kruijt 1973).

2.4.3 Intensity Generalization and Peak Shift

In studies of the mechanisms that underlie learning, experimental psychologists made some discoveries that parallel the findings of ethologists, particularly in relation to supernormal stimuli. Typically, when an animal learns to respond to a single stimulus, such as a tone of a particular frequency or a light of a particular wavelength, the animal then tends to generalize its responses to nearby stimuli. The pattern of generalization is usually a bell-shaped curve with a maximum peak that corresponds to the training stimulus. An exception occurs, however, when the stimulus varies in intensity. Under these circumstances, the animal's response typically increases as the intensity of the stimulus increases, even when the stimulus intensity is above that of the training stimulus (Mackintosh 1974) (fig. 2.6c).

Another important result emerges from experiments in which an animal is trained to discriminate between two stimuli that vary in one dimension. During training, the animal's behavioral response to a certain stimulus (the positive stimulus) is rewarded, and its response to another stimulus (the negative stimulus) is not rewarded or is associated with negative consequences. As in the simple case when an animal learns to respond to a single stimulus, the animal tends to generalize from negative or positive stimuli to nearby stimuli in the stimulus space. The stimulus, however, that evokes the strongest response from the animal after learning does not coincide with the positive stimulus; this stimulus is located beyond the positive stimulus in a direction opposite to that of the negative stimulus (fig. 2.6b). This effect, known as peak shift (Hanson 1959; Mackintosh 1974; Rilling 1977), appears to be a very general property of discrimination learning, and peak shift has been interpreted as an outcome of the interaction between excitatory and inhibitory gradients that surround individual positive and negative stimuli (Spence 1937) (fig. 2.7).

These results have important implications for the evolution of signal form (Staddon 1975; Leimar et al. 1986; Ten Cate and Bateson 1988; Enquist and

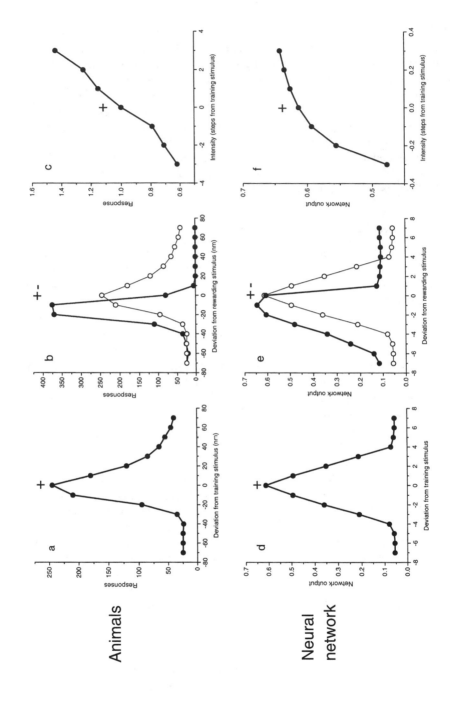

Animals

Neural network

a   Deviation from training stimulus (nm)

b   Deviation from rewarding stimulus (nm)

c   Intensity (steps from training stimulus)

d   Deviation from training stimulus

e   Deviation from rewarding stimulus

f   Intensity (steps from training stimulus)

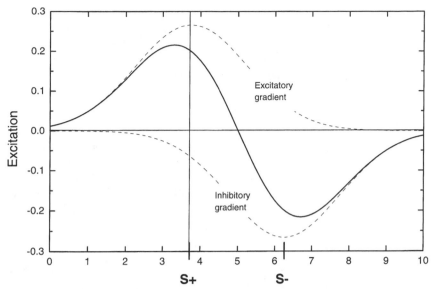

S+   S-

Figure 2.7. Spence's (1937) model of intradimensional discrimination learning. Excitatory and inhibitory stimulus-generalization gradients are assumed to establish independently around the positive (S+) and negative (S−) stimuli. The level of excitation on the y-axis corresponds to the tendency to respond to stimuli that vary along a single dimension (e.g. frequency of sound) shown on the x-axis. The peak-shift effect (overgeneralization) is predicted from the net gradient (*solid line*), which is calculated by subtracting the inhibitory gradient from the excitatory gradient. Note how the maximum and minimum of the net gradient shifts in relation to the maximum and minimum of the excitatory and inhibitory gradients. In this case, one predicts the animal responds maximally after training to a tone that is lower in frequency than the positive stimulus.

Figure 2.6. Comparison of generalization in animals and artificial neural networks. (*a*) Generalization curve of pigeons trained to peck a key in response to light of 550 nm wavelength (positive stimulus, S+). (*b*) Pigeons are trained to discriminate between two wavelengths of 550 nm (positive stimulus, S+) and 555 nm (negative stimulus, S−). After training, the peak response of pigeons is produced with colors of shorter wavelength than that of the original positive stimulus (solid circles). Curve with open circles is the same as shown in graph *a* for comparison. (*c*) Intensity generalization in dogs trained to react to a fixed tone of a given amplitude and then tested with the same tone at different amplitudes. (*d–f*) Comparable results obtained with neural networks. In (*d*) and (*e*), the stimulus is a cross in which the vertical bar is moved horizontally to different positions along the horizontal bar. In (*f*), the network was "trained" to react to a gray square on a black background. Testing was then performed with squares of different values on a gray scale, from white to black, on a black background (intensity 0 = black, intensity 1 = white). Animal examples are from Hanson's (1959) study of pigeons and Razran's (1949) study of dogs. Network examples are unpublished results from Enquist and Arak.

Arak 1993; Weary et al. 1993). As in the case of responses to supernormal stimuli, biases that result from discrimination learning provide a mechanism that may lead to the exaggeration of signals or, more generally, cause signals to become increasingly different from those other stimuli to which the animal should not respond in the same way (see section 2.6.3).

2.4.4 Biases in the Rate of Learning and What Can Be Learned

Certain variations of a stimulus that give rise to faster learning than other variations may affect the evolution of signal form (see also Dukas, this vol. chap. 4). What biases occur in learning processes? First, not all associations can be learned; there are innate predispositions in learning (Shettleworth 1972, 1994; Roper 1983). For instance, it is very difficult to condition animals to associate noise or light with poisonous food, but taste is easily associated with poisonous food (Garcia and Koelling 1966). How do modifications of a stimulus influence learning? Laboratory studies of classical conditioning have shown that certain stimuli lead to more rapid or stronger learning (reviewed by Mackintosh 1974). Typically, stimuli become more effective as they become more conspicuous against the background (Mackintosh 1974; Schuler and Roper 1993) Also, within limits, the speed or strength of learning increases with stimulus intensity. An increasing degree of novelty in the stimulus also seems to enhance learning (Shettleworth 1972; Roper 1993). Finally, previous experience can interfere with later learning. If an animal has been conditioned to react to a stimulus, this conditioning tends to prevent the learning of new components that are subsequently added to the stimulus. This prevention of learning is known as blocking (Mackintosh 1974). Dukas (this vol. section 3.5.2.1) discusses related issues of interference. In conclusion, there are clear similarities between what is easily discriminated and what is easily learned. Generally, animals take a long time to learn fine discriminations, but responses to more intense, conspicuous patterns that differ sharply from other stimuli are quickly acquired.

2.4.5 Compound Stimuli

Many signals used by animals are highly complex and composed of a number of distinct components. The different components of a stimulus may differ in their effectiveness to elicit a response from an animal. For instance, if an animal is trained to respond to a compound stimulus, conditioning occurs mainly to the more intense or conspicuous component. Conditioning to other components is less than the conditioning that results if each component is used alone in a learning experiment. The element to which conditioning mainly occurs is said to overshadow the other components (Mackintosh 1974). Often, the lack of one component or a reduction in its intensity can be compensated for by making another component more intense. Modifying the components of a stimulus can also produce supernormal effects. If an animal is trained to produce a particular response to several stimuli that each contain different components, the animal then may respond more strongly to a new stimulus that combines components from several different training stimuli (e.g. Weiss 1972).

2.5 NEURAL REPRESENTATION AND RECOGNITION OF SIGNALS

The reaction to any stimulus depends on the process of comparing a new pattern of sensory input to previously acquired information that is stored in the nervous system. Numerous theories have been proposed to explain how this comparison may occur, and the problems of pattern recognition are the subject of considerable debate and controversy (e.g. Pearce 1994; Shanks 1994; Eysenck and Keane 1995). Because this subject is central to understanding how signals may be designed best to elicit responses from a receiver, we will outline briefly some of the main ideas.

2.5.1 Templates, Prototypes, Features, and Statistical Machines

One way that recognition could work is to assume that a master copy or template exists in the nervous system for each pattern of input that is to be recognized by the animal. Recognition is then achieved by comparing each sensory input with these templates. Thus, every stimulus configuration is stored in long-term memory with its category label or associated outcome. This recognition is termed instance theory or exemplar theory and is linked to the notion of multiple-trace memories (Pearce 1994; Shanks 1994). The problem with these theories becomes readily apparent when one considers the huge number of memory stores that would be required to recognize each of the numerous objects that can be identified separately by the animal (perhaps many millions for human beings). In addition, it is not clear how information could be retrieved from such memories in a reasonable time. Finally, there is the problem of how simple matching of stimuli against a stored template allows for the great natural variation in stimulus conditions (e.g. the same stimulus seen in different orientations or on different backgrounds).

Given these problems, one may abandon template matching altogether and assume instead that the degree of similarity between input and memory determines response strength. This is essentially the position adopted by proponents of prototype theories of sensory representation. Prototype theories offer a more economical way of stimuli representation with storage of only the typical or average instance of each stimulus category. Recognition is then based on comparing received input with the prototype. This process could, of course, explain generalization. Experiments, however, have shown that details of particular instances are often remembered, which leads some authors to suggest that memory of detail makes prototype theory unlikely. On the other hand, there is also experimental evidence suggesting that information is stored in a more refined way than the remembering of all details of every experience (Shanks 1994).

Feature theory is sometimes proposed as an alternative to the template or prototype theories of sensory representation. According to the model in feature theory, sensory input is not analyzed in raw form; however, there is an intervening level of processing that operates on sensory input, and features such as straight lines and curves are identified. This set of features is then compared with information stored in memory. Feature theory explains how visual stimuli that vary greatly in size, position, orientation, and other minor details may, nevertheless, be identified as instances of the same pattern. Feature theory, however, does not explain why pattern recognition does not depend solely on listing the features of a stimulus. For example, the letter "A" consists of two oblique lines and a dash, but these features can be presented in a way such that they are not perceived as an A—for example, \/-. Thus, features and the relationships between features appear to be important for recognition (Eysenck and Keane 1995).

Finally, a different idea of representation and processing of sensory input is derived from the fact that humans and other animals are sensitive to contingencies in their environment and adjust their behavior accordingly. This observation has spawned the idea that the central nervous system operates as a statistical machine (Shanks 1994). In such a system, memories would contain conditional probabilities, which are used in statistical decision making and are constantly updated as a result of experience.

2.5.2 Neural Networks

The theories of recognition we discussed are mostly far removed from the biological realities of brain functioning. Within the last decade, however, theorists have made efforts to develop models that take account of our knowledge of the underlying structure and functioning of the brain. One such approach has been labeled "connectionism" and is based on the idea of parallel distributed processing (PDP). According to connectionist theory, information about stimuli is stored in a distributed fashion across a vast network of nerve cells, and the configuration of connection strengths between the cells in the network determines how the mechanism reacts to sensory information. Connectionist models have proved to be powerful explanatory tools when applied to a wide variety of phenomena such as concept learning, the development of motor skills, language acquisition, and studies of amnesia and brain damage (McCleeland et al. 1986; Churchland and Sejnowski 1992; Churchland 1995).

We will emphasize connectionist ideas here for two reasons. First, we judge these ideas to offer more plausible models of memory and pattern recognition than the other theories mentioned. The connectionist approach also shares many predictions at the behavioral level with some other theories of recognition discussed. For example, networks of neurons appear to store specific and general information about stimuli; networks of neurons enable analysis of objects in terms of their features and facilitate statistical decision making. Thus, connectionism may provide a unifying theory of sensory representation. Our second reason for favoring the connectionist approach is that it offers the only model of recognition and memory that can be easily applied with a high degree of practicality to problems of biological communication. In particular, it is possible to use connectionist models in evolutionary simulations.

In practice, connectionists use models called artificial neural networks. Such networks consist of a number of elementary neuronlike units, or nodes, which are connected together so that a single unit has many links with other units. Units affect other units by exciting or inhibiting them. Each unit that receives stimulation usually takes the weighted sum of all input and produces a single output to another unit if the weighted sum exceeds some threshold value. Networks can have different structures, working principles, and aims. One popular architecture is the multilayered network that consists of a single layer of input cells, one or more intermediate layers of "hidden units," and a layer of output units (fig. 2.8).

These networks can model recognition without recourse to the kinds of explicit rules found in conventional computers, which rely on serial processing. These networks "learn" the association between different inputs and outputs with modification of the weights of the links between units in the network (these weights can be thought of as the network's equivalent of synaptic strengths). A variety of techniques may be used to "train" the network (Haykin 1994). Once the network has learned to produce a particular response after the presentation of a particular stimulus, the network can exhibit behavior that appears to have been produced with a rule of the form "IF such-and-such is the case, THEN do so-and-so." However, no such rule is followed explicitly in the model. The specific input-output relations of the network simply *emerge* as a consequence of the particular configuration of weights that occur between the separate links in the network.

What are the advantages of the connectionist approach? We will discuss some features that have particular relevance to the study of biological communication (for general discussions see McClelland et al. 1986; Rumelhart et al. 1986; Bechtel and Abrahamsen 1991; Shanks 1994; Churchland and Sejnowski 1992; Churchland 1995).

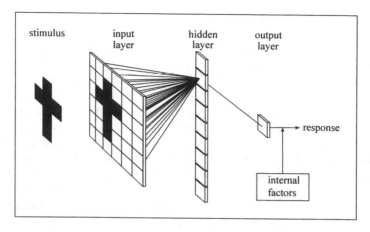

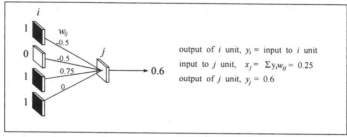

Figure 2.8. *Top:* A simple feed-forward artificial neural network that consists of three layers. Each cell in one layer is connected to all cells in the next layer, and each connection has a weight that modifies the stimulus strength passing between the cells. In some examples, it is assumed that the output is modified further by internal factors (e.g. motivational variables), which are independent of the stimulus. For purposes of clarity, only those connections that arrive at and emanate from a single cell in the hidden layer are shown. In reality, each cell in the hidden layer is connected to all 36 cells of the input layer and to the single output cell. *Bottom:* Diagram shows how the outputs from a number of cells in one layer *i* are combined to determine the overall activation of a cell *j* in the next layer. The output from a cell is a sigmoid function of the total input to the cell. One example of a suitable sigmoid transfer function is given. A reaction occurs if the activity of the cell in the output layer exceeds a threshold of 0.5. If an internal factor is used, the sum of the internal factor and the activity of the output cell should exceed 0.5. The internal factor used varied according to a normal distribution with $\mu = 0$ and a suitable σ. By adding more cells to the output layer, more than one response can be modeled.

2.5.2.1 Neural Plausibility

One of the major attractions of connectionist networks is that they appear to resemble the network of neurons and synapses found in animal nervous systems. This similarity may be superficial, but there are indications it may not be. In one study (Lekhy and Sejnowski 1988), a multilayered network learned to recognize the curvatures of geometrical shapes, and hidden units within the network responded to edge and bar patterns when they were presented in particular orientations. The property of these hidden units corresponds closely to those of the simple and complex cells identified by Hubel and Wiesel (1962) within the visual system of vertebrates. Other potentially important similarities between connectionist networks and the detailed physiological functioning of the brain's visual system are discussed by Bruce and Green (1990).

Of course, artificial neural networks do not capture all features of neural architecture and processing in real nervous systems. For example, neural networks do not approximate the particular pattern of connectivity between neurons found in the brain or simulate differences between various neurotransmitters. A common functional problem in networks is retroactive interference that is highly unrealistic; old information can be almost totally overwritten with new information (see Dukas, this vol. section 3.5.2.1). It is also difficult to see how neural networks can account for all aspects of symbolic reasoning. These kinds of limitations have sometimes produced strong criticism of the connectionist approach in general (Fodor and Pylyshyn 1988; Pinker and Prince 1988) and the application of neural networks to evolutionary problems in particular (Dawkins and Guilford 1995). Dawkins and Guilford argued that the very simplicity of the models used to simulate recognition processes makes them inadequate. One can argue, however, that to substitute a more complex, but poorly understood model of the world in place of the poorly understood world itself is likely to generate greater inadequacy in the model. In the end, the only way to understand how a complex structure such as the brain actually functions is to abstract pieces from it or study simplified models in which responses are more transparent. The crucial test of any such simplified model then is the extent to which it captures essential features of the behavior of real animals. If the model does, then the fact that the model is neurally inspired should be considered advantageous.

2.5.2.2 Powerful Discriminatory Ability

The coding of sensory stimuli as a vector of activation across a wide range of units allows neural networks to deal naturally with stimuli that vary along several dimensions simultaneously; this is the case for most animal signals. This method of coding by networks allows them to discriminate far more subtleties of any sensory situation than can be typically expressed with a discrete set of symbols such as words. Think, for instance, of our ability to recognize almost instantly a familiar face when viewed from almost any angle. Although faces can be described with their constituent parts—the size of the nose, the shape of the lips, etc.—our capacity for verbal description falls a long way short of our capacity for sensory analysis. Churchland (1995) neatly summarized, "The bank teller's determined but inevitably vague description of the face of the bank robber will likely fail to distinguish that face from a hundred thousand others, and yet the teller might be able to recognize and discriminate the robber's face exactly, when she finally lays eyes on him."

According to the connectionist scenario, each familiar face is coded as a particular pattern of activation in a special area of the visual cortex—perhaps five or six synaptic steps downstream from the retina. Each pattern of activation is a kind of signature that is specific for each face. The pattern's elements correspond to various abstract dimensions of observed faces. As a gross over-simplification, one may imagine that faces are coded according to just three dimensions of variation: eye separation, mouth fullness, and nose width (Churchland 1995). Figure 2.9a shows just such a face-coding space generated in these three dimensions. In this coding space, very similar faces are coded very close together, and vastly different faces are coded far apart. Of course, this is highly unrealistic; our coding space for faces probably has many more than three dimensions. For the purpose of argument, suppose that human brains encode faces in ten dimensions and are sensitive to only five increments of discrimination along each dimension. Then we should be able to discriminate 5^{10}, or roughly 10 million, different faces. Thus, the advantages of combinational coding of each face as a pattern of activation over many neurons becomes clear. We no longer need to worry that we may run out of neuronal resources to recognize our colleagues, friends, relatives, acquaintances, and enemies!

It is easy to see how the same principles might be applied to the problem of discriminating animals' communication signals. Figure 2.9b shows the coding space for plumage patterns of ducks in three arbitrary dimensions of variation. The distinct plumage pattern of each species is coded as a unique point in this vector space and corresponds to a unique pattern of activation in the neural network. From a duck's perspective, only a restricted number of all possible patterns of activation may be relevant to its social life—the patterns corresponding to male, female, similar heterospecifics with whom it must avoid interbreeding, and perhaps a few competitors. Even with the crude system of coding stimuli along three dimensions, impressive representational power results.

2.5.2.3 Fast Retrieval

Neural networks may perform many thousands or millions of computations simultaneously instead of in laborious sequence. This enables animals to react very quickly to a large range of problems and to recognize most objects at a glance. Obviously, neural networks provide an important benefit in situations that require rapid reactions to a signal.

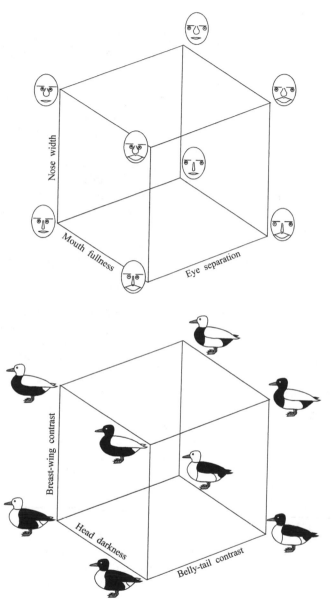

Figure 2.9. (a) A simple face-coding space with three dimensions of variation (redrawn from Churchland 1995). (b) A comparable coding space for the plumage patterns of ducks.

2.5.2.4 General versus Specific Information

People and other animals typically behave as if they have both general and specific information stored in memory, and memory theorists must account for the relationship between these two kinds of stored information. For example, we may have stored information about the specific breeds of dogs we have encountered, but we also have more general and abstract information about what a dog is. Many animals show individual recognition of conspecifics such as kin or mates and discriminate between their own and other species. The connectionist approach provides a reasonably convincing account of how general information can emerge from specific information (see McClelland et al. 1986 and Rumelhart et al. 1986 for details). In essence, presentation of numerous specific stimuli (e.g. different dogs or different conspecifics) leads, by way of an averaging process, to a set of attributes (e.g. size, color, or patterning) that corresponds in some sense to a typical dog or a typical conspecific. In other words, networks can simultaneously represent categories and specific examples within each category (McClelland and Rumelhart 1985).

2.5.2.5 Content Addressability and Default Assignment

In distributed representations, a partial representation of an entity is sufficient to reinstate the whole entity. Thus, if an animal gives a signal and only part of the signal is seen by the receiver, the signal may still be sufficient to elicit an appropriate response. The internal representation of the signal is said to be content addressable. When only part of the normal stimulus is applied, the information that is missing is filled in by default on the basis of information about previous occurrences of the signal that are stored in memory. This feature of memory may be especially important for species that communicate in "noisy" environments because this feature enables receivers to respond appropriately to signals when only part of the signal is perceived. In some circumstances, however, signalers can exploit receivers' tendencies to classify signals by default assignment, and this exploitation allows deception to evolve (Dawkins and Krebs 1978). For example, if an animal has learned that tooth baring usually precedes a vicious attack by a stronger individual, this learning could favor the appearance of tooth baring in individuals that have no intention to attack, if tooth baring makes it more likely that the opponent will flee. Because the receiver lacks information about its opponent's true intentions and fighting ability, the receiver will tend to categorize the likely outcome by default assignment.

2.5.2.6 Automatic Generalization

In a manner that is related to the content addressability of distributed representations, similar patterns of input can produce similar responses. Thus, a neural network that learned to recognize eight familiar emotional states (astonishment, delight, pleasure, relaxation, sleepiness, boredom, misery, and anger) from photographs of 20 human subjects subsequently identified the emotional expressions of novel subjects with a surprising level of accuracy. This identification peaked at 80% for the four positive emotions; however, poor levels of accuracy were achieved with the negative emotions. The sole exception was anger, which was correctly identified 85% of the time. When the same set of photographs was shown to humans, they also scored lowest on the negative emotions, but anger again was a notable exception. Overall, humans showed a much better ability to generalize than the network, perhaps because of the crudeness of the network compared with the complexity of the human visual system; the network's retina contained just 4,096 cells and its hidden layer just 80 cells (Cottrell et al. 1991). The ability to recognize facial expressions and body language and respond appropriately is obviously crucial to species other than humans. This ability has been a recurring theme in ethological studies of communication since Darwin's (1872) study in "Expression of Emotions in Man and Animals."

2.5.3 Bias in Neural Networks

Of particular relevance to the evolution of signals is the question of whether networks possess the same kinds of response biases that are exhibited by animals when they are confronted with novel stimuli. If so, how close is the resemblance between the behavior of the model and that of the real animal?

The phenomena of peak shift discussed in section 2.4.3 can be demonstrated with a simple network that was trained to react to a pattern of a cross but not to several other random stimuli. In the network, a reaction was defined as an activation level that exceeded an arbitrary threshold in the output cell, and no reaction was any activation in the output cell below this threshold. The network was trained by an evolutionary procedure; the network's connection weights were mutated at random and a series of new networks were generated that were then subjected to artificial selection. The criterion for selection was that the network making the highest number of correct reactions survived to the next round of mutation; all other networks were terminated. After producing a network that performed this simple task with negligible error, the network's responses to novel variants of the training stimulus were tested. In this instance, the novel variants were patterns resembling a cross in which the vertical bar was displaced horizontally to different distances and in different directions along the horizontal bar. The resulting generalization curve was bell-shaped, and the maximum point coincided with the original positive stimulus (fig. 2.6d).

Subsequently, the network was trained to discriminate between two similar stimuli. The positive stimulus was a symmetric cross and the negative stimulus was a cross showing some degree of asymmetry. After training, the network was tested with novel variants of the stimuli as before. The generalization curve showed a maximum point that was displaced at some distance from that of the original positive stimulus (in a direction away from the negative stimulus). Thus, the network exhibited the property of peak shift (see fig. 2.6e). In further experiments, peak-shift effect became stronger as the positive and negative stimuli became more similar and thus more difficult to discriminate.

Intensity generalization was also demonstrated with a network that was first trained to react to a simple stimulus (in this case, a gray square on a black background) and subsequently tested on its responses to other shades of gray not presented during the training phase. In this case, the probability of response increased with increasing intensity along the gray scale (i.e. whiter shades) and there was no reduction of response even at the most extreme gray values (fig. 2.6f).

The behavior of these networks almost exactly recapitulated the results found in discrimination trials with animals under similar circumstances (fig. 2.6a–c). Yet, the network used to generate these results consisted of only 46 interconnected cells; the brains of pigeons and dogs contain many millions of cells! Moreover, in animals and artificial neural networks, the direction of the bias can be predicted on the basis of prior knowledge about the discrimination task.

Finally, consider an example of discrimination involving stimuli that vary along several dimensions (e.g. patterns designed to crudely resemble flowers). Several networks were trained to discriminate between a pattern of a flower with one elongated petal (positive stimulus) and a pattern of a flower with petals of equal length (negative stimulus) (fig. 2.10). After training, the networks were presented with novel patterns. Different networks generalized their responses to novel stimuli in different ways. For example, Net 1 showed the strongest reaction to novel patterns (d, h, i, j) that resembled flowers with petals even longer than those of the positive stimulus and did not react to patterns of flowers with short petals (b) or petals missing (a, e). Net 2, however, showed no reaction to some of the patterns most preferred by Net 1 (g, h) but instead reacted more strongly to others (k, l, m). Finally, it should be noted that some strange-looking patterns, which did not at all resemble flowers to us (e.g. n) resulted in strong responses.

When discriminating between multidimensional stimuli, the less certain outcome of generalization probably reflects each network's idiosyncratic coding strategy in its internal representation of positive and negative stimuli. The positive and negative training stimuli differed along two principal dimensions: the illumination of the central pixel and the illumination of the pixel two units below the center. In retrospect, it appears that the two different networks were

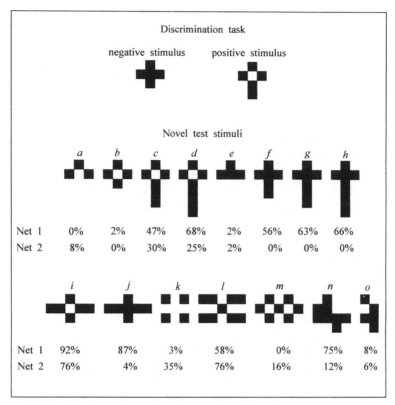

Figure 2.10. Results from two networks trained to discriminate between two multidimensional patterns that crudely resemble flowers. After training the networks to solve the discrimination task with negligible error, their reactions to novel stimuli were tested. Each novel stimulus was presented in numerous positions and orientations on the artificial retina. The percentages of presentations that resulted in a supernormal response (i.e. a response that is stronger than the mean response to the positive stimulus) from the network are shown (adapted from Arak and Enquist 1993).

attaching different weights to each of these variable dimensions when comparing novel stimuli with the positive stimulus. Net 1 appeared to generalize most strongly to patterns with longer petals, and Net 2 generalized most strongly to novel patterns that showed a strong contrast between the central and outer elements of the pattern.

These results confirm that there is more than one solution to any discrimination problem. Moreover, different networks (representing different solutions) appear to generalize in very different ways when confronted with novel stimuli. A somewhat speculative human analogy is the disagreement that commonly occurs between people judging the resemblance of a newborn baby to its parents. While some observers feel that the baby most resembles the father, others sometimes see a much stronger resemblance to the mother. This difference is not surprising if humans, like the neural networks discussed, assign different weights to different cues when making decisions about similarity.

In conclusion, the biases in responses that emerged in these simple experiments can be interpreted as the outcome of interference between stored representations (e.g. memories) of different stimuli that a network recognizes. When a network is trained to respond differentially to stimuli that are similar, the memories that become established are not independent of one another. This occurs because information about each stimulus is not stored at a unique location, as in a conventional serial computer; information is distributed as a pattern of connection weights within the common network. Thus, each connection weight may serve a function in the representation of several stimuli. When a network is trained to respond in a certain way to a novel stimulus, its connection weights are adjusted, and this adjustment may affect how the network subsequently responds to other stimuli that it has seen.

2.5.4 Innate versus Learned Representations

In reality, an individual's behavior is a result of a complex interaction between genetic information and the unique experience of the individual with its environment. Certain behavior patterns show great flexibility and variation among individuals; other behavior patterns seem to vary little among individuals over a wide range of natural circumstances, and these behavior patterns appear almost fully formed when an animal first finds itself in the appropriate situation. Traditionally, ethologists have tended to refer to the former patterns as learned behavior and the latter as innate behavior. The convention within ethology had been to attribute the reactions to "sign stimuli" to simple template matching and the reactions to learned stimuli to a separate mechanism (Lorenz 1935, 1978; Baerends 1982).

Is there any justification for this view? It appears not (Schneirla 1966; Lehrman 1970). More recent research has shown that many reactions to stimuli traditionally considered innate often involve an imprecise recognition mechanism organized with use of genetic information that is subsequently replaced or refined by learning (Hailman 1967; Hogan 1994). Many biologists have abandoned altogether the distinction between learned and innate behavior. Instead, it is more reasonable to view behavior as a set of traits that vary along a continuum, from those unaffected to those greatly modified by learning (Dukas, this vol. chap. 5).

The most parsimonious view is that responses to stimuli that depend on genetic or learned information are produced with the same neuronal mechanism. In the simulations that follow, we assume (1) the parameters of the network that influence reaction to stimuli (i.e. the particular configuration of connection weights) are inherited; and (2) changes occur only because of mutations. We do not attempt to deny, however, that changes in responsiveness occur also by learning. We simply are most interested in the potential, longer-term, evolutionary consequences of response biases to signal form.

Likewise, we make no distinction between biases that have a genetic origin and biases that develop as a consequence of learning. Indeed, Hogan et al. (1975) suggested that responses to supernormal stimuli (traditionally assumed to be innate) and peak-shift effects in learning may have a common mechanism.

2.6 COEVOLUTION OF SIGNALS AND RECEIVER MECHANISMS

In the remainder of this chapter, we will investigate the effects of receiver bias on the coevolution of signals and receiver mechanisms. We will use artificial neural networks to model the biological recognition/decision mechanism, and we will simulate several situations in which this mechanism coevolves with one or more signals. In simulations, senders and receivers meet in interactions. The sender gives the signal, the receiver responds, and then a payoff is handed out to each player. The receiver must discriminate several categories of stimuli, including signals. The sender attempts to elicit a desired response from the receiver. Several techniques can be used to simulate coevolution. A computationally efficient method is to use an iterative procedure with one sender and one receiver. In each step of the iteration, mutants are produced and the best mutant or the original player go to the next step (for details see Arak and Enquist 1994). It is also possible to carry out simulations with an entire population of senders and receivers and include factors such as biological variation, drift, and recombination. Most of the following results are based on iterations, but qualitatively similar results are produced with population simulations.

We start by exploring the consequences that receiver biases may have on signal form. We assume that signals are without cost and consider situations that are strategically simple (the effects of signal cost and strategic factors are discussed in sections 2.7 and 2.8). We identify two basic evolutionary forces that emerge from the recognition mechanism when senders and receivers interact (fig. 2.11). One force, which we refer to as repulsion, tends to make signals different from other stimuli; repulsion is relevant in situations in which receivers need to discriminate between different signals or between a signal and other stimuli in the background. The other force, which we refer to as attraction, causes signal form to converge on certain patterns; attraction is relevant in situations in which animals should give the same response to several different signals.

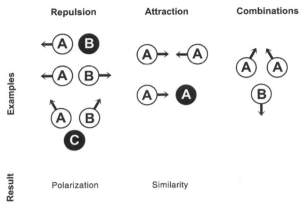

Figure 2.11. Forces of selection that act on signals caused by the recognition mechanism. Two basic forces are identified: a repulsion force tends to make a signal different from other stimuli and an attractive force results in signal similarity. Figure exemplifies situations that commonly arise in nature. A full understanding of signal form requires that one take into account the nature of the recognition problem, strategic factors affecting the interests of players, and constraints on signal form. *White circles*-signals, *black circles* = nonevolving stimuli. Letters in each circle indicate the behavioral response desired by the sender.

2.6.1 Repulsion

Imagine a simple signal in the form of a square that can take on any shade of gray. First, let the square be medium gray. The problem for the receiver is to detect whether the signal is present or not on a white background. It is easy to train an artificial network to do this, even if the signal is presented in different orientations or sizes (a simple filter can also accomplish this). After training, the network invariably produces a stronger output when the signal is darker and a weaker output when the signal is lighter than the original medium gray square (i.e. intensity generalization occurs). Repulsion of the signal from the white background can be demonstrated by letting both the network and the signal evolve. The end result is a completely black square that is maximally different from the white background (fig. 2.12a). In this case, the speed of change of the signal mainly depends on the rate of signal mutations.

A slightly more elaborate example shows that repulsion is a general phenomenon and not just the effect of increasing stimulus intensity. We imagine two stimuli representing two species of flowers with petals of different length; species X has four petals of equal length, but species Y has one petal slightly shorter than the others (fig. 2.12b). The network, representing the recognition system of an insect, must correctly recognize flowers of species X to obtain a reward (nectar). Species Y does not produce the reward. To make the situation

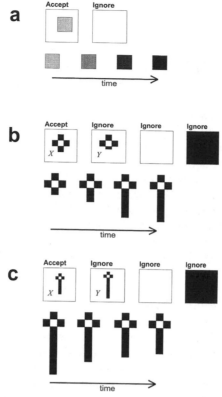

Figure 2.12. Results of simulations of coevolution between signals and receivers. Repulsion of signal away from stimuli that the receiver is under selection to ignore is demonstrated. The receiver mechanism was modeled with an artificial neural network of the same type as described in figure 2.8. During coevolution, the network and the correct stimuli were allowed to change, and the stimuli to be ignored were held constant. In all cases, there was complete common interest between receivers and senders. In each generation, mutant signals and networks were produced and improvements were retained (iterations). The networks were not pretrained and initial weights were randomly assigned. In *a*, the network consisted of 10 × 10 input cells and 10 cells in the hidden layer. The discrimination problem was to identify a gray square when present on a white background. The network was mutated by adding an increment to each of five randomly selected weights. Increments were randomly drawn from a normal distribution with $\mu = 0$ and $\sigma = 0.2$. The signal was mutated by adding an increment drawn from a normal distribution with $\mu = 0$ and $\sigma = 0.02$. In *b*, the network was asked to discriminate the rewarding flower species *X* with a longer petal from another species *Y* with a shorter petal and irrelevant stimuli. In *c*, the problem posed in *b* was reversed. The rewarding flower *X* had a shorter petal than the nonrewarding flower. The network was mutated by adding an increment to 10 randomly selected weights. Increments were randomly drawn from a normal distribution with $\mu = 0$ and $\sigma = 0.2$. The petal length was mutated by adding an increment drawn from a normal distribution with $\mu = 0$ and $\sigma = 0.5$. In all three cases, a clear repulsion effect is seen. In *a*, coevolved patterns are shown after 0, 5, 20, and 100 iterations. In *b* and *c*, patterns are shown corresponding to 0, 1,000, 3,000, and 10,000 iterations. Little or no change occurred in patterns after the final generation shown.

slightly more complex, it is assumed that the network must not only distinguish between species X and Y, but must differentiate X with a nonstimulated and a fully stimulated retina. Thus, the network is trained to react in the presence of stimulus X, but not to any other stimuli, including Y (the reaction is assumed to be one that simulates an approach by the insect). The stimuli are pasted onto the modeled retina of a neural network in numerous positions and rotations to simulate a more lifelike situation in which stimuli may appear in different parts of the insect's visual field.

After training the network to perform the discrimination with negligible error, the network's reactions to novel stimuli are then investigated. Most novel stimuli are not effective in eliciting an approach reaction from the network, but some stimuli are supernormal, including patterns resembling species X but with slightly lengthened lower petals. Flowers with lower petals that are even shorter than those in species Y produce a very weak response, which suggests that generalization occurs in both directions along the dimension of lower petal length. These results indicate that the recognition mechanism may potentially exert selection pressure on the signal; in this case, there is a bias that favors flowers with longer petals.

To investigate potential evolutionary consequences of this bias, the signal and the receiver mechanisms are allowed to adapt to one another in a coevolutionary process. An initially naive network is presented with a variety of mutant flowers of species X, each with a lower petal of a different length. The mutant pattern that elicits the strongest reaction from the network is then selected. In the next step, the network is allowed to mutate by random change of some of its connection weights, and a variety of such mutated networks are tested for their ability to discriminate between the new signal (selected in the previous step) and images of species Y and other irrelevant stimuli. The best network is then kept, and additional mutated versions of the signal are presented to the network. This procedure is reiterated 10,000 times. The result is a gradual increase in lower petal length from 2.0 to 9.0 units in species X (see fig. 2.12b). To create a control, we reverse the problem. Flowers with slightly shorter petals are assumed to be most rewarding to the network (fig. 2.12c). As before, the signal and the network are allowed to coevolve. This coevolution results in a gradual decrease in petal length from 13.0 to 6.5 units in the rewarding flower (see fig. 2.12c).

The simulation technique described has been used to investigate how signals may evolve when they vary in more than one dimension (e.g. Enquist and Arak 1994; Hurd et al. 1995). When the receiver had to discriminate a multidimensional signal made up of random shades of gray from a white background, the discriminating receiver drove the coevolving signal toward black, the antithesis of the white background (fig. 2.13a) (Hurd et al. 1995). This is the same result that is obtained with a signal of only one shade of gray (fig. 2.12a).

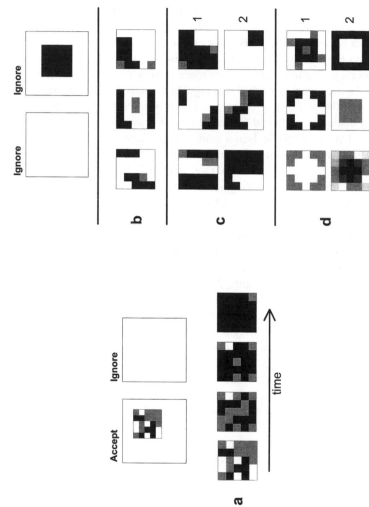

Figure 2.13. (a) A signal evolves to become increasingly dissimilar from a pure black stimulus and a white background. (b) When the receiver discriminates against both a pure black stimulus and a white background, signals evolve contrasting patterns of black and white. When two signals evolve to elicit different responses (1 and 2) from the receiver, they evolve into opposite forms. In c, patterns are pasted onto the model retina. In c, patterns are pasted onto the model retina by translation only (vertical and horizontal movements of the pattern are pasted onto the retina). In d, signals are also presented in different orientations. Results of three different simulation runs are shown in each case (from Hurd et al. 1995).

In another case, the signal was prevented from taking on the antithetical form by introducing a second irrelevant stimulus, a pure black pattern. The receiver was then selected to discriminate the coevolving signal from both the white background and the pure black pattern. In this case, the signal did not evolve into a uniform medium gray, as may be expected, but into a highly contrasting pattern composed of patches of black and white (fig. 2.13*b*). Such a result could provide an explanation for why the plumage of many species of birds consists of several distinct patches of contrasting colors. The need to be distinguished from not just one but several similar species may impose a selection pressure on signalers that favors more complex plumage patterns.

Animals often use a variety of different signals to elicit different responses from a receiver. Therefore, how a more complex signaling system that involves several distinct signals and several different responses may evolve is of interest. Hurd et al. (1995) devised a neural network capable of more than one type of response and added a second coevolving signal to which the receiver must respond with a different behavior. Although the two signals were initially identical (uniform gray), they rapidly evolved into opposite forms, despite being processed by the same network (see fig. 2.13*c* and *d*). These results were explained in terms of the evolution of signals within a signal space, each axis of which represented a particular dimension of the signal. The dimensions may have represented the features of the signals that were extracted and coded by the network as a distinct pattern of activation in the cells of the intermediate layer. Figure 2.14 shows examples of such signal spaces with the locations of signals that are antithetical to pure black and pure white signals. In the case of one dimension (when the signal is represented by one square), a coevolving gray signal is most distinct from the black and white stimuli. In the case of two dimensions (two squares), the pattern most different from pure black or white is a combination of these extremes: one black and one white square. When the signal consists of three dimensions, the most dissimilar patterns contain one white, one gray, and one black square. Thus, as the number of dimensions increases, the resulting signals increase in complexity. The general principle, however, remains: Evolving signals that elicit different responses become strongly polarized in form.

Figure 2.15 illustrates how such a process can be conceptualized in the case of plumage in birds, assuming there are just three dimensions of variation. The anticipated change in the plumage in males of species B from that in species A emerges if there is selection for species B females to avoid mating with species A males, but not vice versa. This situation may occur, for example, if two similar species, separated for a long time, subsequently came into close contact. The change in the plumage of species B males would then be a case of character displacement in the area of sympatry (Futuyma 1986; Andersson 1994).

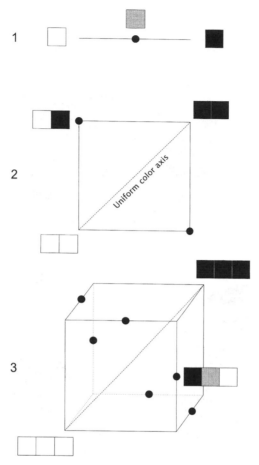

Figure 2.14. Signal spaces in one, two, and three dimensions. In all three spaces, the location of patterns that are maximally different (Euclidean distance) from a pure black and a pure white pattern are indicated with black circles. In the case of one dimension, the pattern is gray. In the case of two dimensions (two squares), the most dissimilar patterns contain one white and one black square. In the case of three dimensions, the most dissimilar patterns contains one white, one gray, and one black square. Thus, the complexity of the patterns increases as the number of dimensions increases (from Hurd et al. 1995).

Empirical support for these predictions comes from many studies that show signals used to elicit different responses exhibit very different forms. See, for example, Darwin's dog (see fig. 2.2c). Another example comes from Blurton Jones's (1968) study of great tit displays. The displays can be depicted in a signal space, in which separate axes represent body position, head position, and wing extension. The threat displays actually observed in nature lay

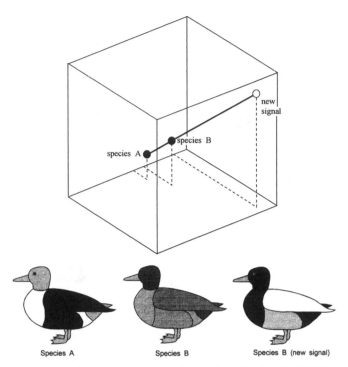

Figure 2.15. The effect observed in the coevolutionary simulation (see fig. 2.12) is generalized to a situation in which species identity is coded by a complex, multidimensional signal.
In situations in which species identity is likely to be confused, mate-recognition signals
should diverge to become more dissimilar from confusing stimuli. In the illustration, it is
assumed that species B females initially experience difficulty in discriminating male conspecifics from species A males, but not vice versa. The effect shown corresponds to character displacement.

at the extreme ends of these axes, in the corners of the cube of the signal
space.

2.6.2 Attraction

In section 2.6.1, we illustrated how biases in recognition mechanisms may
give rise to evolutionary forces of repulsion that cause signals to diverge from
other stimuli and from each other. The basis for this evolutionary force was
that the receiver was required to respond differently to different stimuli. Another aspect of recognition is that many different stimuli may require the same
response. For example, similar feeding behavior may be required for dealing
with different types of food, or escape behavior may be required in a variety
of different threatening situations. Moreover, a particular stimulus may have

some natural variability, for instance in its shape or color. Finally, and perhaps most importantly for animals with vision, exactly the same object may generate different images on the animal's retina, depending on the object's orientation and distance and the conditions of illumination, etc. These variable factors result in the evolution of signals toward some optimum appearance that allows the receiver to identify the signals correctly in a variety of circumstances.

Biological displays are often strikingly symmetric. Moreover, there is growing evidence in numerous species—from earwigs to humans—that receivers are sensitive to symmetries of various kinds, including symmetries in flowers, sexual signals, novel objects, and art (Rensch 1957, 1958; Berlyne 1971; Møller 1992, 1993; Radesäter and Halldórsdóttir 1993; Swaddle and Cuthill 1994; Perrett et al. 1994; Lehrer et al. 1995; Møller 1995). Because of instabilities in development, a normally symmetric signaling trait may be subject to random deviations from symmetry that result in many different variants (i.e. fluctuating asymmetries) (Van Valen 1962). Moreover, as explained, even a single variant will be encountered in many different positions and orientations, giving rise to a range of different retinal images (Enquist and Arak 1994). The need to recognize all such signal variants and discriminate them from other irrelevant stimuli may result in preferences for symmetry in the receiver and thus the appearance of symmetric signals (Enquist and Arak 1994; Johnstone 1994).

Consider a simple situation in which a receiver must recognize a suite of stimuli varying in degree of symmetry along a single dimension (Enquist and Johnstone, 1997). It is assumed that all stimulus variants take the basic form of a cross, but differences occur in the distance of the vertical bar from the center of the horizontal bar. (fig. 2.16a). With distances to the left of center defined as negative values, perfect symmetry as 0, and distances to the right as positive values, a continuum of variation is defined, from strong left asymmetry to symmetry to strong right asymmetry.

We considered the simplest possible situation, in which networks are trained to respond to only two images that form a left/right mirror-symmetric pair. These two images may be thought of as variants of a single display arising from instabilities in development (i.e. fluctuating asymmetries) or as views of a single asymmetric object from different sides. The degree of asymmetry of the two images was varied, and the effects of this variation on the response of networks after training was recorded.

Figure 2.16b shows the generalization gradient of trained networks (i.e. their probability of response to stimuli that varied from left asymmetry to symmetry to right asymmetry) for different pairs of training stimuli. When the two training stimuli were both perfectly symmetric (and hence identical), training gave rise to a bell-shaped gradient with a peak at perfect symmetry. A similar gradient was obtained when the two stimuli were slightly asymmetric,

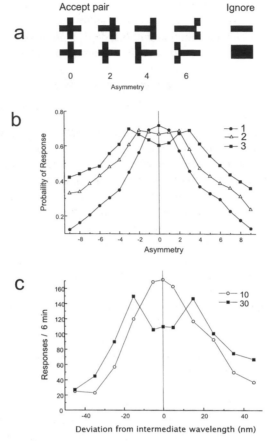

Figure 2.16. (*a*) Networks were trained to respond to each of two patterns that formed a mirror-symmetric pair (positive stimuli); networks were trained not to respond to two other patterns (negative stimuli). Eight networks were used and each network was trained with a different pair of positive stimuli that varied in degree of asymmetry from 0 to 7. In the pair examples given, the upper pair members show right asymmetry (positive values) and the lower members show left asymmetry (negative values). (*b*) After training, the networks' responses to novel patterns with different degrees of asymmetry were measured. Generalization curves are shown for networks trained with pairs 1, 2 and 3. When trained to recognize a pair of patterns that deviated only slightly from symmetry (curve 1), networks tended to generalize most strongly to patterns that displayed perfect symmetry. As the degree of asymmetry in the training patterns increased, the generalization gradient became flatter (curves 2 and 3). When the mirror-symmetric pair deviated strongly from a perfectly symmetric pattern, the network seemed to learn to recognize each pattern of the pair as a unique stimulus. (*c*) Pigeons trained to respond to two different colors that deviate by 10 and 30 nm showed similar behavior. When the difference between the two colors was small, subjects generalized most strongly to a color that was intermediate; however, when the color difference was large, subjects generalized as if they had formed unique representations of each color (pigeon data from Kalish and Guttman 1957).

although the peak gradually flattened as the degree of asymmetry was increased. Finally, when the training stimuli were highly asymmetric, a gradient with two symmetrically situated peaks was obtained. Even in this case, however, the two peaks were closer to perfect symmetry than the training stimuli (i.e. the networks responded more strongly to patterns that were less asymmetric than the training patterns). For the artificial neural networks, whether a single- or double-peaked gradient was obtained depended not only on the degree of asymmetry of the training patterns but on the details of the training procedure. If training was prolonged, the network appeared to learn to recognize and represent each stimulus of the mirror-image pair as a unique pattern; however, when training was more cursory, the network did not appear to categorize each member of the pair uniquely. The same effect was found in pigeons trained in a similar way (fig. 2.16c).

The results illustrated in figure 2.16 are similar to those found by Johnstone (1994) who used patterns that resembled birds' tails. Both studies showed that a network trained to recognize variants of a signal with small, random deviations from perfect symmetry may subsequently generalize in such a way that the network reacts more strongly to a perfect, symmetric signal than to any of the training exemplars. Similar results were also obtained in a simulation in which a network was trained to recognize more complex stimuli, irrespective of their rotation (Enquist and Arak 1994). In this case, the networks subsequently gave the highest responses to novel patterns with strong rotational symmetries. Moreover, when the stimuli were allowed to coevolve with the network, marked rotational symmetries appeared in the patterns. These simulations did not always produce symmetries that are visually interesting to humans. A common outcome was a homogeneous pattern (e.g. a uniformly colored square); however, mathematically such patterns were highly symmetric (Stewart and Golubitsky 1992).

The results of these studies on symmetry reflect a more general finding that emerges from studies carried out with artificial neural networks and humans. After training on a particular recognition task, networks and humans often responded most vigorously not to any of the particular training exemplars but to patterns that, in some sense, represented combinations of features of the training exemplars (fig. 2.17). Preferences for symmetry therefore represent a specific example of the process of categorization, which occurs naturally in networks when they are trained to recognize different stimulus variants as belonging to the same stimulus category.

2.6.3 Conclusions

The simulation studies described suggest that neural mechanisms involved in the perception and categorization of stimuli may act as a potent force that drives the evolution of signaling traits. Some common features of signals—

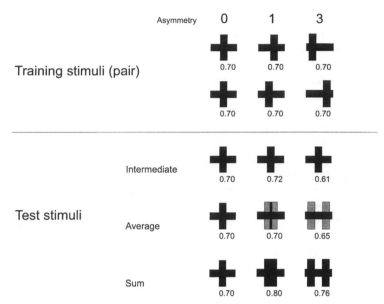

Figure 2.17. After networks were trained to recognize both members of a pair of mirror-asymmetric patterns, networks' responses to novel stimuli were tested. In each test, the novel stimuli were the intermediate, average, and sum of the training pair. The output of the network is below each pattern (the threshold for recognition was 0.5). Networks responded most vigorously to novel, symmetric stimuli that incorporated the summed elements of the training stimuli (data from Enquist and Johnstone, in press).

such as conspicuousness, distinctiveness, contrast between signal elements, and various kinds of symmetries—emerge naturally from models that are based on neural networks. In brief, these results can be explained as the consequence of interference between different stimuli to which the network is exposed during selection. This interference generates small biases in responses, which the signal then exploits. Repulsion and attraction may operate simultaneously as exemplified in figure 2.13d.

Dramatic changes in signals do not take place in a single step; signal changes appear as the outcome of cumulative small changes that occur as the signal coevolves with the recognition mechanism. Typically, such coevolutionary forces do not lead to stability because senders and receivers successively try to adapt to each other and biases in the recognition mechanism are never completely exhausted. The exaggeration of signal elements commonly observed in nature may result from selection of signals that are different from competing stimuli that the receiver should not to react to. These competing stimuli may be other biological signals or other inanimate stimuli that constitute noise. The results of our simulations are therefore consistent with some of the predictions of signal-detection theory (Green and Swets 1966; Wiley 1983); however, there are also important differences. Arguments for detectability suggest that signals should only become conspicuous enough to be reliably detected. In our simulations, signals continued to evolve toward greater conspicuousness; this evolution was driven by biases in the system, even when the networks could discriminate these signals almost perfectly from competing stimuli.

In the simulations described, we deliberately made minimal assumptions about the kinds of selection forces that act on senders and receivers. The extent to which signals become exaggerated in nature will almost certainly depend on additional factors not included in these simple models. Most importantly, signal costs and the diverging interests of players are expected to have a dramatic impact on the evolution of signal form. We will consider these strategic factors in the next section.

2.7 CONFLICTS OF INTEREST AND THE EVOLUTION OF SIGNALS

The classic explanation for the evolution of distinctive, conspicuous signals in animals is that natural selection favors the use of unambiguous signals that are more efficient releasers of receiver behavior (Huxley 1966; Cullen 1966; Eibl-Eibesfeldt 1975). The process of ritualization, by which signals evolve, was viewed as a kind of mutual adaptation between sender and receiver: A signal changes, becoming more efficient in eliciting responses, while the receiver becomes more sensitive to the signal. Dawkins and Krebs (1978; Krebs and Dawkins 1984) were among the first to point out a flaw in the concept of ritualization. If there is complete common interest between sender and receiver, signals would be expected to evolve not toward greater conspicuousness but greater subtlety. This evolution is expected because receivers always benefit by responding, favoring heightened sensitivity to the signal; the heightened sensitivity of receivers, in turn, allows senders to use inconspicuous, cost-minimizing displays.

How then can we explain the common occurrence of conspicuous and apparently costly signals? One possibility is that senders are forced to use costly signals to overcome environmental transmission constraints. Indeed, many signals used in long-range communication appear to be costly, and their characteristics can be interpreted in terms of signal-detection theory (Wiley 1983). More puzzling, however, is that many signals used over very short distances are also highly conspicuous—although detectability in these circumstances is reliable. This observation is difficult to reconcile with the classic ethological view of communication as a mutually beneficial interaction between sender and receiver.

Conflicts of interest arise between individuals during most kinds of biological communication. It is not always in the receiver's best interest to respond to a signal. Additional factors, such as the receiver's internal state, influence its fitness. For example, an animal's response to threat signals when competing for food is likely to depend on motivational factors, such as hunger, recent fight success, and so on. Even in the absence of conflict between sender and receiver, there may be several senders competing for the same reaction from the receiver. Excluding the possibility of direct physical combat, senders can compete with use of signals. This kind of competition appears to be common in nature; males often display in groups to attract females, and flowers compete with their neighbors to attract insect pollinators.

To model how conflicts of interest affect the evolution of signals, we consider the evolution of signals used for mate attraction in detail. Courtship signals are diverse and often spectacular; they may combine several components such as sound, color, and behavior. Many components, but not all, are likely to be costly. We imagine a situation in which males compete for the attention of females for matings on a lek. In order to mate, a female has to approach one of the signaling males. We assume that for whatever male is chosen there are no fitness consequences for the female. The female, however, must discriminate males of her own species from other stimuli, including other females, immature males, members of other species, and inanimate objects. Males are assumed to be equal in all respects except for tail length and body color. It is assumed that increasing tail length is associated with a survival cost and color has no cost.

Games in a population of males and females were simulated with use of neural networks that modeled females' recognition-decision mechanisms. Each female made her choice among one or a group of randomly selected males and a number of irrelevant stimuli. Figure 2.18 shows the results of coevolution between the female's recognition mechanism and male signal form. In one simulation, males and females met in pairwise interactions, and there was no conflict between them: Females were selected to mate with a conspecific male whenever the female detected a male (fig. 2.18*b:* no conflict). Under these circumstances, the cost-free attribute, color, always evolved and the costly attribute, tail length, remained unchanged or actually decreased in size. A closer look at the dynamics of the process revealed that tails first increased in size; however, as females became better at discriminating between conspecific males and other stimuli, tail size began to decrease again. When the cost associated with tail size was removed, tails evolved to an extreme length. Common interest between sender and receiver did not promote the evolution of costly signals; only cost-free signals evolved. In a second set of simulations, competition between males was introduced (see fig. 2.18*b:* sender-sender conflict) and females chose the most stimulating male. In this

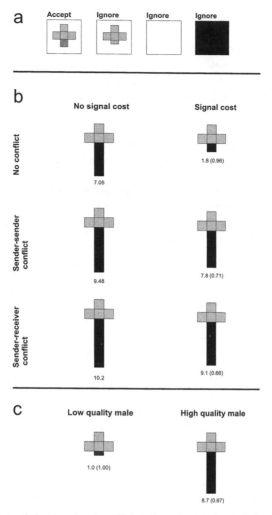

Figure 2.18. Effects of signal cost and conflict on the outcome of coevolution between senders and receivers. The signal represented the male's tail feathers that varied in color and length. Signal cost did or did not increase as feather length increased, but cost was always unaffected by tail color. (*a*) The female receiver had to correctly identify her own species in the presence of another species (with a very short, light gray tail) and some irrelevant stimuli. (*b*) The cost-free trait, color, always evolved rapidly in the direction opposite from that of the stimulus the female was under selection to ignore. If a survival cost was associated with tail length, tails did not increase in size as long as there were no conflicts between players (row 1). Survival as a function of tail size t was calculated as z^{t-1}, with $z = 0.95$ (optimal tail length for survival equals one). When males competed for females (row 2) or females showed resistance to male courtship (row 3) long tails evolved, despite their cost. (*c*) If the cost of a given tail length was higher for low-quality ($z = 0.8$) than high-quality ($z = 0.95$) males, low-quality males evolved shorter tails than high-quality males. Results in *b* and *c* come from population simulations with use of 10,000 individuals of both sexes. Male fitness was calculated by multiplying survival with expected success from interactions with females. In each generation, 10 individuals were mutated.

case, both cost-free and costly attributes evolved. Next, we introduced conflict between males and females (fig. 2.18c: sender-receiver conflict). In nature, conflict arises because it is not always advantageous for a female to mate. At any given moment, the female may not yet be ready to mate, she may have mated already, or she may be occupied with some other activity such as feeding. Thus, there are internal and external factors, separate from the male's signal, that influence the female's response. For the purposes of simplification, these additional factors were modeled as a single, normally distributed variable, which varied independently of the signal (fig. 2.8). The receiver was assumed to react to a signal when the response cell's activation level—caused by the stimulus plus the factor m—exceeded a certain threshold value. In this case, it was assumed that there was only one male signaler displaying to a female at any one moment. The evolution of the costly signal, tail length, was similar to that in the case of sender-sender conflict. Apparently, when the female was less motivated to mate, the male compensated by producing a more costly, but more stimulating signal. In reality, this compensation may occur during the individual male's lifetime—as the male encounters females that differ in motivation—or as the outcome of an evolutionary process in response to typical conditions. In the latter case, we may predict a correlation across species between the development of the signal (i.e. its cost) and the resistance of females to courting males. In other words, the use of costly signals by males should be associated with coyness in females. Finally, we used the simulation to explore how variation in the quality of individual signalers influences the exaggeration of signals. When a given increase in tail length was more costly for certain individuals, only the high-quality individuals evolved a costly signal (see Fig. 2.18c). Contrary to "handicap" models of signaling (reviewed by Andersson 1994), we did not assume that receivers are sensitive to differences in the underlying qualities of senders. The result emerged simply because individuals of low intrinsic quality could not afford to pay the costs of exaggerating their signals to the same extent that higher-quality individuals could. In conclusion, the results support earlier published findings (Arak and Enquist 1995) and show the circumstances under which both costly and cost-free signal components can evolve. Even in the absence of quality differences between senders, costly signals may evolve if conflicts of interest exist between players and receivers are under selection to recognize the signal from similar, but inappropriate stimuli.

2.8 Implications for Strategic Models of Signaling

By comparison with the mechanistic models presented, strategic models of communication offer few valuable insights into the question of why signals take on the forms they do. To be fair, this is not a question strategic models are designed to answer; they enable the identification of conditions that allow particular signaling and response strategies to coexist at evolutionary equilibrium. Strategic models are less useful when questions about the evolution of signal form are considered because, as we have shown, it is possible that there is no evolutionarily stable strategy that precisely defines the physical structure of any signal. Instead, it seems likely that signaling and response strategies are subjected to a process of almost continual change: As the signal evolves to exploit biases in the response mechanism, the receiver reacts by changing its response properties; in so doing, the receiver generates further biases that can be exploited by the signaler. A similar idea is suggested by the arms race analogy of communication proposed by Dawkins and Krebs (1978). Implicit in the notion of the signaler's manipulation or persuasion of the receiver is the assumption that there is a lack of equilibrium between the signal form and the receiver's response strategy.

A currently popular idea, derived directly from the strategic modeling approach, is that signals that convey information to the receiver about some aspect of the signaler's underlying quality, such as general viability, can evolve. This idea has been particularly influential in recent discussions of sexual-selection theory (Andersson 1994); one may be easily persuaded that all cases of exaggerated courtship signaling are concerned solely with advertising quality! Another focus of strategic modeling is the question of why animals should so blatantly advertise their unprofitability to predators. Animals that display striking warning colors expose themselves to the risk of attack by naive predators; this risk, coupled with the observation that these animals are frequently killed by displaying their colors, has led to the suggestion that aposematic coloration is stabilized through kin selection (Guilford 1990; Schuler and Roper 1992). An important omission in these strategic models is the idea that the perceptual system itself can act as a source of selection on signals, in addition to other mechanisms. It is therefore important to discuss the implications of this insight in strategic models of signaling.

2.8.1 Courtship Display and Quality Advertisement

A major challenge to evolutionary biologists, from the time Darwin (1871) posed the problem, is how to account for dramatic secondary sexual traits of males used in courtship display. These traits are maintained apparently by female preferences. How then are such preferences maintained? If females choose males with costly ornaments, their male offspring not only inherit the ornament but also carry its cost. A popular idea is that the ornament's cost, passed onto the female's offspring, is outweighed by the high viability of ornamented males. Grafen (1990) confirmed the logic of this argument in a game-theory model. Female preference for ornamented males can be evolutionarily stable, as long as the ornament is more costly for low-quality males than high-quality males.

An alternative explanation for the evolution of elaborate ornaments is that the preference for the ornament existed before the appearance of the ornament itself. The preference may have arisen for reasons unconnected with mate choice, or the preference simply may be the consequence of a sensory bias (Ryan et al. 1990). The preference or bias may be common or even fixed in the population when the male trait arises and would permit rapid spread of the trait even if there was no correlation between the trait and male quality. Once the trait arises, however, there could be a strong counterselection against the trait preference in females. For example, the trait preference may result in females mating with males that may be rejected for other reasons. We suggest that sensory biases can be responsible for the origin of a trait preference, but other mechanisms are needed to explain the preference's elaboration and maintenance.

One may consider, however, that (1) most signals vary along many dimensions; and (2) biases may arise continuously as a result of random drift in the recognition mechanism. Any adaptation in the female that tends to reduce her sensitivity to a signal trait along one dimension may be quickly countered by males exploiting a different direction of bias. An equilibrium between the trait and the preference may never be reached. Moreover, in the model of sender-receiver conflict discussed in section 2.7, increasing the receiver's resistance to signals actually causes signals to become more exaggerated, at least up to a moderate degree of conflict (for details, see Arak and Enquist 1995).

The difficulty of distinguishing between models that assume an adaptive origin of preferences (in the context of mate choice) and those that assume preferences arise because of perceptual biases become apparent when one considers the similar predictions of the models. For example, in both cases, one may expect to find a correlation between the development of the signal and some aspect of male quality. According to the adaptive model, this correlation causes female preference; however, according to the perceptual-bias model, this correlation is simply the outcome of selection of males once the signal is exaggerated because of perceptual biases.

Which explanation is most relevant to the evolution of courtship signals in nature? We are unlikely to answer this question until more information is obtained about the historical sequence of trait-preference evolution from a large number of species. Moreover, because the explanations are not exclusive, several mechanisms may be jointly important. We do not intend to persuade the reader that receiver bias is the only or most important mechanism that affects the evolution of courtship signals; however, we encourage an approach that combines strategic models with more realistic assumptions about the mechanisms of perception that underlie behavior.

Game-theory models show that the evolution of "honest" signaling is logically possible; however, these models cannot determine how quickly or easily honest equilibria can be reached in nature. It is possible, for instance, that honest signaling is continuously destabilized by the appearance of new exploitative signals. Krakauer and Johnstone (1995) recently used artificial neural networks to study the interactions that occur between sender and receiver as the signaling system approaches an equilibrium. They showed that if signals are costly, receivers obtain some honest information; however, exploitative signals, at the same time, continually arise and signalers gain short-term benefits. Thus, their model showed how honest signaling can coexist in a system with recurrent episodes of exploitation. Over time, therefore, signal form changes continuously and remains partially honest on average but never perfectly honest.

2.8.2 The Evolution of Warning Coloration

Warning coloration is one of the most spectacular and widespread examples of exaggerated signaling. The classic explanation for the existence of warning coloration is based on kin selection. According to this idea, warning coloration arises simultaneously in several related individuals. During avoidance learning, a predator samples and kills some individuals, which appear to be altruists, but the predator leaves the other individuals unharmed. Warning coloration essentially speeds the learning process, and fewer individuals must be sampled. As a consequence, there is decreased predation on the altruists' relatives, which carry copies of the same genes (Turner 1971; Harvey and Greenwood 1978; Harvey et al. 1982). We now describe a model developed by Leimar et al. (1986) that explains the evolution of warning coloration without relying on kinship. The model works even if the attacked individual always dies, so there are no direct benefits of conspicuous coloration either. In the model, it is assumed there is a bias in the predator's recognition mechanism, which develops as the predator learns to discriminate profitable from unprofitable prey. This example is one of the few studies of communication that combines strategic modeling with knowledge of the mechanisms that underlie behavior.

The details of the model are complex, but the essence of the model is conveyed as follows. The predator's problem is to discriminate between profitable and unprofitable prey. Consider a single dimension x in which the appearance of the unprofitable prey can differ from the profitable prey. When $x = 0$, both prey look alike; as x increases, discrimination becomes easier. At the same time, however, the unprofitable prey becomes more conspicuous and is more likely to be detected by the predator. Can conspicuous coloration (i.e. $x > 0$) be maintained in the population of unprofitable prey despite the higher detection rate of conspicuous coloration?

We now need to model how a predator generalizes and learns. Let $G(x)$ be the predator's generalization gradient expressed as the probability of attack as a function of x. Following Spence's classic model, we assume that such net gradients can be produced by combining excitatory $E(x)$ and inhibitory $I(x)$ gradients (Spence 1937; Rilling 1977). In the present case, the excitatory gradient is due to interactions with the profitable prey, and the inhibitory gradient is due to interactions with the unprofitable prey.

Assume now that a predator has a number of experiences with the profitable prey and then suddenly encounters a new potential, but unprofitable prey with appearance x. Initially, the net generalization gradient before any experiences with the unprofitable prey is equal to the excitatory gradient, and then changes occur as the inhibitory gradient increases. Figure 2.19a shows an example of how the net gradient G is modified as a result of a number of the predator's experiences with unprofitable prey with a particular value of x ($x_1 = 0.5$). The attack probability initially declines as x increases even beyond x_1 (the peak-shift effect). Thus, among those individual prey not yet encountered, those with a somewhat higher x than the unprofitable prey already attacked (x_1) are most likely to be left by the predator. This effect can offset the disadvantage of being more easily detected and can stabilize an aposematic strategy in the population in the absence of kinship. In fact, a number of solutions, varying from the adoption of a cryptic strategy to startling aposematism, are possible in the population (fig. 2.19b).

There is another potential strategy that allows aposematic coloration to evolve: Unprofitable prey take on an appearance that is so different from profitable prey that the predator simply ignores the unprofitable prey (Turner 1975). A typical reaction to novelty is avoidance or withdrawal. Certain new stimuli, especially ones that are big or loud, may elicit flight responses (e.g. Gray 1971; Archer 1976). Novel stimuli may also elicit exploratory behavior (Toates 1986), typically after some time has passed. An animal, however, will not carefully examine every stimulus for various reasons. Likelihood of exploration of a stimulus may partly depend on its similarity to other stimuli and the frequency of new stimuli. For instance, several investigators suggest that some predators are reluctant to sample certain novel prey (Coppinger 1969, 1970; Wicklund and Järvi 1982; Greenberg 1990; Schuler and Roper 1992). Consequently, potential prey can avoid being attacked by looking odd.

What should prey do to be ignored? In short, they should locate to an area in the predator's stimulus space that is far from profitable prey and other stimuli that elicit attraction. Alternatively the prey may take on the pattern of those

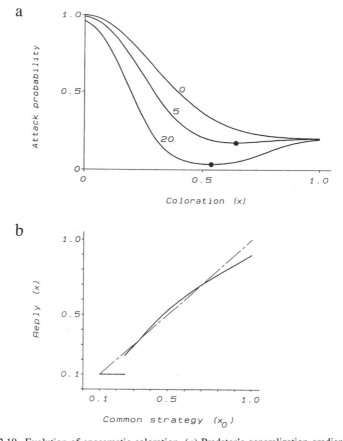

Figure 2.19. Evolution of aposematic coloration. (*a*) Predator's generalization gradient (attack probability) as a function of x (signal conspicuousness) developed during avoidance learning after no, five, and 20 encounters with unprofitable prey of $x = 0.5$ (dots indicate minimum attack probability). The gradient continues to decrease at 0.5, which means that individuals displaying an x greater than 0.5 experienced a lower probability of attack despite being more conspicuous; this is due to peak shift. (*b*) A game played among unprofitable prey. What should an individual look like in a population of unprofitable prey that adopts a particular pattern of coloration? The x-axis represents the appearance adopted by the majority of the population (x_0) and the y-axis represents the range of possible replies (in terms of x) to the population strategy. The best reply for each x_0 is represented by the solid line. Equilibrium points are the points of intersection of the solid and broken lines ($x = x_0$). In this particular example, there are three. The first point corresponds to a cryptic ESS ($x = 0.1$) in which both profitable and unprofitable prey are cryptic. When this is the case, the only option to avoid being attacked is to not stand out. The second point represents an unstable equilibrium (the population does not return to its initial state after a small perturbation). The third point represents the aposematic ESS in which a cost is taken in the form of an increased detection rate; however, increased detection is offset by the peak shift in the predator's generalization gradient (from Leimar et al. 1986).

stimuli that elicit avoidance reactions in the predator; in a broad sense, this kind of prey behavior is mimicry.

2.9 EVOLUTIONARY AESTHETICS: DID DARWIN GET IT RIGHT?

The greatest problem in Darwin's theory of sexual selection is often thought to be his failure to account for the origin of female preferences for extreme male traits, such as elaborate plumage and beautiful song. Darwin simply assumed that animals, like humans, possess a "sense of the beautiful." Darwin was acutely aware that discussion of an aesthetic sense invited criticism, but he insisted ample evidence existed that animals behave according to their own standards of beauty. Birds, for example, have "strong affections, acute perceptions, and a taste for the beautiful" (Darwin 1871). Oddly enough, Darwin did not attempt to seek an adaptive origin for the evolution of aesthetic sense. Instead, he came close to suggesting that the aesthetic sense is no more than the manifestation of basic properties of animal nervous systems: "The perception, if not the enjoyment, of musical cadences and of rhythm is probably common to all animals, and no doubt depends on the common physiological nature of their nervous systems" (Darwin 1871).

The position we adopted in this chapter is close to Darwin's view. We argued that signal form evolves in response to various response biases inherent in recognition systems. These biases often arise for historical reasons that are not connected with communication: They may be the by-products of receiver design for more general cognitive functions. If an animal shows stronger responses to certain stimuli that are outside the range of variation normally experienced, it is appropriate to consider these responses as aesthetic; these responses cannot be explained as purely adaptive. In this sense, however, to what extent is "aesthetic" an anthropomorphism? If the same underlying biases are responsible for the human aesthetic sense, we may feel that there is a stronger justification for speaking of an aesthetic sense in animals.

Many signals used by other species are judged to be beautiful by humans. Examples include the colors and symmetries of flowers, the patterns on butterflies' wings and coral reef fish, and the elaborate plumage and songs of birds. The nearly universal appeal of such signals to humans is surprising because these signals were selected to influence the nervous systems of other species. We seem to find most attractive those signals that are highly efficient in releasing behavior in animals; these signals can repel (e.g. warning colors) or attract (e.g. courtship displays) their intended recipients. The appeal of these signals to humans suggests that the human aesthetic sense may have evolved like sensory perception evolved vis-à-vis biological signals.

Clearly, humans have not been under selection to respond to flowers, butter-flies, or bird song in any particular way. Indeed, one of the defining character-istics of the human aesthetic sense is that it often projects disinterest in or detachment from the object under contemplation (Sheppard 1987). Yet, more generally, natural selection has probably acted strongly on those particular mental abilities that are concerned with recognizing, discriminating, and cate-gorizing the many complex stimuli that surround us, and so natural selec-tion has ultimately endowed human beings with the ability to comprehend and manipulate the world to their individual advantage. Whether butterflies, stamps, rocks and minerals, or a million other things are considered, humans are undoubtedly obsessive classifiers. In the end, the acquired library of inter-nal representations in the average human being's brain may be vast compared with that of many other animals. There is no reason, however, to suppose that, in the formation of such representations, the same kinds of biases have not been important.

For example, think of the way in which we come to recognize a particular kind of flower, say a tulip, and distinguish it from other kinds of flowers. With every experience, a tulip we see elicits a pattern of activation in our brain that is unique but nonetheless not too dissimilar from the patterns of activation that have been elicited by other tulips. We must assume, with repeated synaptic adjustments, the association between diverse patterns on the retina caused by tulips and a particular pattern of neuronal activation in the brain is strength-ened. This process occurs with other kinds of flowers. In this way, different kinds of flowers are represented internally as different regions in a multidimen-sional neuronal-activation space, where each dimension corresponds to some particular quality of flowers—such as shape and number of petals, color, length of stem, etc. (or combinations of features).

What then constitutes a beautiful tulip to such an experienced network? It is probably not the average example that is encountered in the garden: weather-beaten, petals missing, color fading, leaves browning. The most beautiful tulip possesses all those characteristics of tulipness that most clearly define it from other types of flowers: petal glossiness, upright form, pale green color, broad leaves, etc. We are most likely to admire a particular tulip for being a perfect example of its kind; in a sense, an ideal tulip conforms most closely to the prototypical representation. We have seen in our study of signal symmetry how preferences for average or prototypical forms may emerge when networks are asked to recognize a variety of stimuli as belonging to the same class or category. These preferences are likely to be relevant also in some aspects of the human aesthetic response to stimuli. For example, in a classic experiment, men were shown photographs of individual women's faces and composites of the same photographs superimposed on one another. When asked to judge the attractiveness of the different photographs, most subjects preferred the composites over any of the individuals' photographs (Daucher 1979, cited in Eibl-Eibesfeldt 1988).

However, this preference for stimuli that match a prototypical representation no more explains all aspects of the human aesthetic response than the responses of animals to a great diversity of signal forms in nature. Preferences for average values of a signaling trait do appear to be common in animals, but there are often biases that result in stronger responses to extreme or exaggerated characteristics as we have seen. Is there any evidence for such biases in humans? One obvious example is the ability of humans to recognize caricatures. Figure 2.20, from Brennan's (1985; see also Dewdney 1986) computer study of caricatures, demonstrates. The vector-average, or prototypical human face (fig 2.20a), was first generated by taking pictures of a large random sample of faces and calculating the average values along twenty dimensions of variation. Notice the sex, race, and age of the prototypical face is curiously ambiguous (the face is not even bad looking!). By comparison, figure 2.20b shows a faithful tracing of a photograph of former president of the United States Ronald Reagan. Brennan used a simple computer program (*Face Bender*) to draw a straight line between the prototypical face and Reagan's. She then extended the line further to some specified point in a multidimensional face-space. With this process, a series of faces was generated that differed from the prototypical human face in a way so that the face resembled Reagan's face but only more so. The resulting images are in fact caricatures of Reagan (figs. 2.20c and d). The caricatures are more quickly and easily recognized as likenesses of Reagan than his faithful portrait. These caricatures, in other words, act as supernormal stimuli: They elicit a stronger response in the part of our face center in the brain that is responsible for recognizing Reagan than the more accurate drawing.

This example illustrates a more general phenomenon that is probably common when humans are faced with a discrimination problem—namely, the operation of a bias for exaggerated characteristics. For example, if the task is to recognize women from men, these biases for exaggerated characteristics are likely to generate stronger responses to female forms that differ more strongly from the typical male form than from the average female form. Ultimately, these biases could lead to (the not uncommon) preference among men for women with larger-than-average breasts or a smaller-than-average waist-to-hip ratio; these body traits contrast most strongly with male body traits. Of course, the opposite is also true: Women may prefer men with narrow hips and broad shoulders (Ridley 1993).

If such recognition biases are common, we should expect them to be exploited by artists; their ability to manipulate stimuli far exceeds that of evolution. We suggest that in art and music there is ample evidence that the

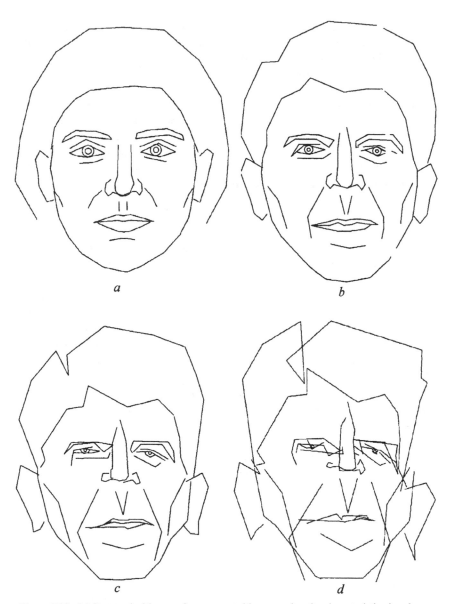

Figure 2.20. (*a*) Prototypical human face generated by averaging the characteristics in a large sample of humans of different races, ages, and sex. (*b*) Faithful tracing of a photograph of Ronald Reagan. (*c, d*) Caricatures of Reagan differ increasingly from the prototype but in the same direction as in *b* (from Brennan 1985).

operation of such biases have been at work: Everyday we confront many stimuli that are at least as extreme and as fanciful as the peacock's tail or the nightingale's song. Indeed, much art used for decoration shares those qualities of biologic signals: symmetries, repetitions, contrast between different elements, and exaggerated features that appear to be the outcome of a cultural process of ritualization (Gombrich 1984). Thus, we may be wrong to assume that human response to art is always based on intellectual capacity for critical evaluation (Freedberg 1989). The power of images and music to cause fear, stir revolt, evoke tears, arouse sexually, or produce empathy may be based on biases that are deep rooted in our nervous systems; connoisseurs of high art may prefer not to acknowledge these biases.

2.10 CONCLUSIONS

Throughout this chapter, we discussed the importance of cognitive mechanisms in the evolution of biological signals. Receiver mechanisms for decoding signals have a central function in communication because signals do not have the power to produce responses directly. Every signal influences receiver behavior because neuronal decoding mechanisms in the receiver associate the signal with a particular response or change the internal state. We presented evidence that these decoding mechanisms contain biases, and we attempted to develop a theory of how such biases may drive the evolution of signals. The biases we incorporated into our theory may represent constraints produced by the brain's construction, or these biases more likely represent an adaptive strategy for dealing with variable stimulation. For instance, the peak-shift effect observed by psychologists (from the perspectives of statistical decision and signal-detection theory) provide a good general strategy for discriminating between stimuli that are susceptible to natural variation. In particular cases, however, the evolutionary consequences of such biases are bizarre structures and behavior that could be maladaptive.

To understand how receiver biases may influence the evolution of signal form, we used computational models known as artificial neural networks. The rationale for use of such models is they reasonably reproduce the kinds of recognition biases found in animals by experimental psychologists. With simulation techniques, we used artificial neural networks to explore how signal form and receiver mechanisms may coevolve; these simulations resulted in some qualities of signals that approximate their biological counterparts: exaggeration, contrasting signal elements, and various types of symmetries common in ritualized signals.

Of course, the qualities of signal form that emerged from our models only crudely resembled the equivalent aspects of signals displayed by animals. For example, we have been unable, as yet, to evolve the intricate patterns seen on butterflies' wings or the remarkable plumes of birds of paradise! The crudeness of our models is partly the cause: the largest artificial retina used in our studies contained a mere 22 × 22 cells. Perhaps, more convincing signals (and therefore more persuasive arguments) could be generated with networks that allow finer resolutions. However, even if we could afford the computational resources and time necessary, we doubt a much finer level of analysis could be justified by anything but aesthetic desire. The primary value of such models is to show how, in principle, general features of signal form can evolve; with these models, we do not attempt to reconstruct for any particular species the historical journey that evolution has taken through signaling space. It is likely that future studies with simple networks will illuminate other general qualities of signals. For example, we are making some progress with our coworkers toward understanding patterns commonly observed in time-varying signals with use of recurrent networks.

We tried to show, where applicable, how various transmission constraints and strategic factors interact with receiver mechanisms to influence signal design. For example, we suggested that the need to recognize signals when viewed in different positions and orientations by the receiver may select for various symmetries in signals. The type of symmetry observed in natural signals (e.g. bilateral, radial, and spherical symmetry in flowers) may be a reflection of the manner in which signals are usually seen by the receiver (e.g. insect pollinators); this prediction could be tested with comparative studies. We also showed, by introducing a cost to signal production into models, that signal cost may eliminate or reduce the tendency for signals to become exaggerated as a result of receiver biases. Exaggeration of costly signals, however, is possible when signalers compete for responses from receivers or when the interests of signaler and receiver conflict. Thus, costly signals may evolve even when the receiver obtains no information from the signaler.

In summary, cognitive mechanisms serve a central function in the evolution of signal form. We encourage an investigative approach that incorporates the receiver mechanism into evolutionary arguments, with assumptions about transmission constraints and strategic factors. We believe this approach will provide a deeper understanding of the evolution of biological communication systems.

2.11 SUMMARY

Signals used by animals for communication are extremely diverse in form; explaining this diversity is a major challenge to biologists. In most recent models of the evolution of biological signaling systems, little attention has been paid to the receiver's mechanisms used for decoding signals. An approach that combines strategic modeling with a knowledge of the mechanisms underlying perception leads to a better understanding of the evolution of biological communication systems. Mechanisms are important because they contain biases that act as potent forces on the evolution of signal form. We showed there are theoretical justifications for the existence of recognition biases, and biases emerge naturally in models of recognition based on artificial neural networks. Artificial neural networks are useful because they enable simulation of the process of coevolution that occurs between senders and receivers in a signaling system. In the simulations presented, artificial signals evolved with qualities that correspond closely to their natural counterparts; for example, certain elements of signals became exaggerated, signals used for different purposes became distinct, and various symmetries in signals evolved. The consequences of strategic factors—such as conflicts between players and different assumptions about the costs of signaling—were also incorporated into the simulations. These simulations revealed that exaggerated signals are most likely to evolve in situations where there is a conflict of interest between players and when the cost of being conspicuous is relatively low. Finally, we compared biases in the perception of signals in animals to certain aspects of the human aesthetic response, and we speculated that ritualization in human culture arises from common properties in a nervous-system design that is shared with other animals.

Acknowledgments

We thank Reuven Dukas, Rufus Johnstone, Oren Hasson, Jerry Hogan, Don Hugie, Daniel Osario, and Risa Rosenberg for comments on an earlier draft of this chapter. Research was supported by grants form the Swedish Science Research Council and Marianne och Marcus Wallenbergs Stiftelse.

Literature Cited

Andersson, M. 1982. Female choice selects for extreme tail length in a widowbird. *Nature* 299:818–820.

———. 1994. *Sexual Selection: Monographs in Behavior and Ecology.* Princeton, NJ: Princeton University Press.

Arak, A., and M. Enquist. 1993. Hidden preferences and the evolution of signals. *Philosophical Transactions of the Royal Society of London,* ser. B 340:207–213.

———. 1995. Conflict, receiver bias, and the evolution of signal form. *Philosophical Transactions of the Royal Society of London,* ser. B 349:337–344.

Archer, J. 1976. The organisation of aggression and fear in vertebrates. In P.P.G. Bateson and P. Kloper, eds., *Perspectives in Ethology,* Vol. 2, 231–298. New York: Plenum.

Baerends, G. P. 1982. The herring gull and its eggs: General discussion. *Behaviour* 82:276–399.

Baerends, G. P., and J. P. Kruijt. 1973. Stimulus selection. In R.A. Hinde and J.

Stevenson-Hinde, eds., *Constraints on Learning: Limitations and Predispositions,* 23–50. New York: Academic.

Barlow, G. W. 1977. Modal action patterns. In T.A. Sebeok, ed., *How Animals Communicate,* 98–134. Bloomington: Indiana University Press.

Basolo, A. L. 1990. Female preference predates the evolution of the sword in swordtail fish. *Science* 250:808–810.

Bechtel, W., and A. Abrahamsen. 1991. *Connectionism and the Mind: An Introduction to Parallel Processing in Networks.* Oxford: Basil Blackwell.

Beer, C. G. 1975. Multiple function and gull display. In G. Baerends, C. Beer, and A. Manning, eds., *Function and Evolution in Behaviour,* 16–54. Oxford: Clarendon.

Berlyne, D. E. 1971. *Aesthetics and Psychobiology.* New York: Appleton.

Blurton Jones, N. G. 1968. Observations and experiments on causation of threat displays of the great tit (*Parus major*). *Animal Behaviour Monographs* 1:75–158.

Brennan, S. E. 1985. The caricature generator. *Leonardo* 18:170–178.

Brown, J. L. 1975. *The Evolution of Behavior.* New York: Norton.

Bruce, V., and P. R. Green. 1990. *Visual Perception: Physiology, Psychology, and Ecology.* Hove, England: Erlbaum.

Boyle, R. D., and R. C. Thomas. 1988. *Computer Vision: A First Course.* Oxford: Blackwell Scientific.

Caryl, P. G. 1979. Communication by agonistic displays: What can game theory contribute to ethology? *Behaviour* 68:136–139.

Caryl, P. G. 1981. Escalated fighting and the war of nerves: Game theory and animal combat. In P. P. G. Bateson and P. Klopfer, eds., *Perspectives in Ethology,* Vol. 4, 199–224. New York: Plenum.

Churchland, P. M. 1995. *The Engine of Reason, the Seat of the Soul: A Philosophical Journey into the Brain.* Cambridge: MIT Press.

Churchland, P. S., and T. J. Sejnowski. 1992. *The Computational Brain.* Cambridge: MIT Press.

Coppinger, R. P. 1969. The effect of experience and novelty on avian feeding behavior with reference to the evolution of warning coloration in butterflies. I. Reactions of wild-caught adult blue jays to novel insects. *Behaviour* 35:45–60.

———. 1970. The effect of experience and novelty on avian feeding behavior with reference to the evolution of warning coloration in butterflies. II. Reactions of naive birds to novel insects. *American Naturalist* 104:323–334.

Cott, H. B. 1957. *Adaptive Coloration in Animals.* London: Methuen.

Cottrell, G., and J. Metcalfe. 1991. EMPATH: Face, emotion, and gender recognition using holons. In R. Lippman, J. Moody, and D. Touretzky, eds., *Advances in Neural Information Processing Systems,* Vol. 3. San Mateo, Calif.: Morgan Kaufmann.

Cullen, J. M. 1966. Reduction of ambiguity through ritualization. *Philosophical Transactions of the Royal Society of London,* ser. B. 251:363–374.

Darwin, C. 1871. *The Descent of Man and Selection in Relation to Sex.* London: Murray.

———. 1872. *The Expression of Emotions in Man and Animals.* London: Murray.

Dawkins, M. S., and T. C. Guilford. 1995. An exaggerated preference for simple neural network models of evolution? *Proceedings of the Royal Society of London,* ser. B 261:357–360.

Dawkins, R., and J. R. Krebs. 1978. Animal signals: Information or manipulation? In J. R. Krebs and N. B. Davies, eds., *Behavioural Ecology: An Evolutionary Approach,* 282–309. Oxford: Blackwell Scientific.

Dewdney, A. K. 1986. Computer recreations: The complete computer caricaturist and a whimsical tour of face space. *Scientific American* 255:20–28.

Eibl-Eibesfeldt, I. 1975. *Ethology: The Biology of Behavior.* 2d ed. New York: Holt, Rinehart, and Winston.

———. 1988. The biological foundation of aesthetics. In I. Rentschler, B. Herzberger, and D. Epstein, eds., *Beauty and the Brain: Biological Aspects of Aesthetics,* 29–68. Basel, Switzerland: Birkhauser Verlag.

Enquist, M. 1985. Communication during aggressive interactions with particular reference to variation in choice of behaviour. *Animal Behaviour* 33:1152–1611.

Enquist, M., and A. Arak. 1993. Selection of exaggerated male traits by female aesthetic senses. *Nature* 361:446–448.

———. 1994. Symmetry, beauty, and evolution. *Nature* 372:169–172.

Enquist, M., and R. Johnstone. 1997. Generalisation and the evolution of symmetry preferences. *Proceedings of the Royal Society,* ser. B. 264:1345–1348.

Ehret, G., and H. C. Gerhardt. 1980. Auditory masking and effects of noise on responses of the green tree frog (*Hyla cinerea*) to synthetic mating calls. *Journal of Comparative Physiology,* ser. A, *Sensory, Neural, and Behavioral Physiology* 141:13–18.

Eysenck, M. W., and M. T. Keane. 1995. *Cognitive Psychology: A Students Handbook.* 3d ed. Hove: Erlbaum, Taylor, and Francis.

Fodor, J. A., and Z. W. Pylyshyn. 1988. Connectionism and cognitive architecture: A critical analysis. *Cognition* 28:3–71.

Freedberg, D. 1989. *The Power of Images: Studies in the History and Theory of Response.* Chicago: University of Chicago Press.

Futuyma, D. J. 1986. *Evolutionary Biology.* 2d ed. Sunderland, Mass.: Sinauer.

Garcia, J., and R. A. Koelling. 1966. Relation of a cue to consequence in avoidance learning. *Psychonomic Science* 4:123–124.

Gerhardt, H. C. 1976. Significance of two frequency bands in long distance vocal communication in the green tree frog. *Nature* 261:692–694.

———. 1978. Discrimination of intermediate sounds in a synthetic call continuum by female green tree frogs. *Science* 199:1089–1091.

———. 1983. Communication and the environment. In T. R. Halliday and P. J. B. Slater, eds., *Animal Behaviour,* Vol. 2, *Communication,* 82–113. Oxford: Blackwell Scientific.

Gerhardt, H. C., and J. A. Doherty. 1988. Acoustic communication in the gray tree frog, *Hyla versicolor:* Evolutionary and neurobiological implications. *Journal Comparative Physiology,* ser. A, *Sensory, Neural, and Behavioral Physiology* 162:261–278.

Gombrich, E. H. 1984. *The Sense of Order: A Study in the Psychology of Decorative Art.* 2d ed. London: Phaidon.

Grafen, A. 1990. Biological signals as handicaps. *Journal Theoretical Biology* 144:517–546.

Grafen, A., and R. A. Johnstone. 1993. Why we need ESS signalling theory. *Philosophical Transactions of the Royal Society of London,* ser. B 340:245–250.

Gray, J. A. 1971. *The Psychology of Fear and Stress.* London: Weidenfeld and Nicholson.

Green, D. M., and J. A. Swets. 1966. *Signal Detection Theory and Psychophysics.* New York: John Wiley and Sons.

Greenberg, R. 1990. Feeding neophobia and ecological plasticity: A test of the hypothesis with captive sparrows. *Animal Behaviour* 39:375–379.

Guilford, T. 1990. The evolution of aposematism. In D. L. Evans and J. O. Schmidt, eds., *Insect Defenses: Adaptive Mechanisms and Strategies of Prey and Predators,* 23–61. Albany: University of New York Press.

Guilford, T., and M. S. Dawkins. 1991. Receiver psychology and the evolution of animal signals. *Animal Behaviour* 42:1–14.

Hailman, J. P. 1967. The ontogeny of an instinct. *Behaviour* 15(suppl.):1–159.

———. 1977. *Optical Signals.* Bloomington: Indiana University Press.

Hanson, H. M. 1959. Effects of discrimination training on stimulus generalization. *Journal of Experimental Psychology: Learning* 58:321–333.

Harris, M. G., and G. W. Humphreys. 1994. Computational theories of vision. In A. M. Colman, ed., *Companion Encyclopaedia of Psychology,* Vol. 1. London: Routledge.

Harvey, P. H., and P. J. Greenwood. 1978. Anti-predator defence strategies: Some evolutionary problems. In J. R. Krebs and N. B. Davies, eds., *Behavioural Ecology: An Evolutionary Approach,* 129–151. Oxford: Blackwell Scientific.

Harvey, P. H., J. J. Bull, M. Pemberton, and R. J. Paxton. 1982. The evolution of aposematic coloration in distasteful prey: A family model. *American Naturalist* 119: 710–719.

Haykin, S. 1994. *Neural Networks: A Comprehensive Foundation.* New York: Macmillan.

Hinton, G. E. 1992. How neural networks learn from experience. *Scientific American* 267:105–109.

Hogan, J. A. 1994. Development of behaviour systems. In J. A. Hogan and J. J. Bolhuis, eds., *Causal Mechanisms of Behavioural Development,* 242–264. Cambridge: Cambridge University Press.

Hogan, J. A., J. P. Kruijt, and J. H. Frijlink. 1975. "Supernormality" in a learning situation. *Zreitschrift der Tierpsychologie* 38:212–218.

Hopkins, C. D. 1977. Electric communication. In T. A. Sebeok, ed., *How Animals Communicate,* 263–289. Bloomington: Indiana University Press.

———. 1983. Sensory mechanisms in animal communication. In T. R. Halliday and P. J. B. Slater, eds., *Animal Behaviour,* Vol. 2, *Communication,* 114–155. Oxford: Blackwell Scientific.

Hubel, D. H., and T. N. Wiesel. 1962. Receptive fields, binocular interaction, and functional architecture in the cat's visual cortex. *Journal of Physiology.* 160:106–154.

Hunter, M. L., and J. R. Krebs. 1979. Geographical variation in the song of the great tit (*Parus major*) in relation to ecological factors. *Journal of Animal Ecology* 48: 759–785.

Hurd, P. L., C. A. Wachtmeister, and M. Enquist. 1995. Darwin's principle of antithesis revisited: A role for perceptual biases in the evolution of intraspecific signals. *Proceedings of the Royal Society of London,* ser. B 259:201–205.

Huxley, J. S. 1938. Threat and warning coloration in birds with a general discussion of the biological function of colour. In *Proceedings of 8th International Ornithological Congress* 430–455.

———. 1966. Ritualization of behaviour in animals and men. *Philosophical Transactions of the Royal Society of London,* ser. B 251:249–271.

Johnstone, R. A. 1994. Female preference for symmetrical males as a by-product of selection for mate recognition. *Nature* 372:172–175.

Kaissling, K. E. 1971. Insect olfaction. In L.M. Beidler, ed., *Handbook of Sensory Physiology,* Vol. 4, 351–431. Berlin: Springer-Verlag.

Kalish, H. I., and N. Guttman, 1957. Stimulus generalisation after equal training on two stimuli. *Journal of Experimental Psychology* 53:139–144.

Koehler, O., and A. Zagarus. 1937. Beitraäge zum Brutverhalten des Halsbandregenpfeifers *(Charadrius h. hiaticula* L.) *Beitr. Fortpfl. biol.* 13:1–9.

Kraukaur D. C., and R. Johnstone. 1995. The evolution of exploitation and honesty in animal communication: A model using artificial neural networks. *Philosophical Transactions of the Royal Society of London,* ser. B. 348:355–361.

Krebs, J. R., and R. Dawkins. 1984. Animal signals: Mind-reading and manipulation. In J. R. Krebs and N. B. Davies, eds., *Behavioural Ecology: An Evolutionary Approach,* 2d ed., 380–402. Oxford: Blackwell Scientific.

Lehrer, M., G. A. Horridge, S. W. Zhang, and R. Gadagkar. 1995. Shape vision in bees: Innate preferences for flower-like patterns. *Philosophical Transactions of the Royal Society of London* B347:123–137.

Lehrman, D. S. 1970. Semantic and conceptual issues in the nature-nurture problem. In L. R. Aronson, E. Tobach, D. S. Lehrman, and J. S. Rosenblatt, eds., *Development and the Evolution of Behavior,* 17–52. San Francisco: W. H. Freeman.

Leimar, O., M. Enquist, and B. Sillén-Tullberg. 1986. Evolutionary stability of aposematic coloration and prey unprofitability: A theoretical analysis. *American Naturalist* 128:469–490.

Lekhy, S. R., and T. J. Sejnowski. 1988. Network model of shape-from-shading: Neural function arises from both receptive and projective fields. *Nature* London 333:452–454.

Lorenz, K. 1935. Der Kumpan in der Umwelt des Vogels. *Journal of Ornithology* 83:137–413.

———. 1978. *Vergleichende Verhaltensforschung.* Vienna: Springer-Verlag.

McClelland, J. L., and D. E. Rumelhart. 1985. Distributed memory and the representation of general and specific information. *Journal of Experimental Psychology, Learning* 114:159–188.

McClelland, J. L., D. E. Rumelhart, and the PDP Research Group. 1986. *Parallel Distributed Processing: Exploration in the Microstructure of Cognition,* Vol. 2, *Psychological and Biological Models.* Cambridge: MIT Press.

Mackintosh, N. J. 1974. *The Psychology of Animal Learning.* London: Academic.

Magnus, D. 1958. Experimentelle Untersuchung zur Bionomie und Ethologie des Kaisermantels *Argynnis paphia* Girard (Lep. Nymph.). *Zeitschrift der Tierpsychologie* 15:397–426.

Maynard Smith, J. 1974. The theory of games and the evolution of animal conflicts. *Journal of Theoretical Biology* 47:209–221.

————. 1982. *Evolution and the Theory of Games.* Cambridge: Cambridge University Press.

Møller, A. P. 1990. Fluctuating asymmetry in male sexual ornaments may reliably reveal male quality. *Animal Behaviour* 40:1185–1187.

————. 1992. Female swallow preference for symmetrical male sexual ornaments. *Nature* 357:238–240.

————. 1993. Female preference for apparently symmetrical male sexual ornaments in the barn swallow, *Hirundo rustica. Behavioral Ecological Sociobiology* 322:371–376.

————. 1995. Bumblebee preference for symmetrical flowers. *Proceedings of the National Academy of Science* 92:2288–2292.

Møller, A. P., and J. Höglund. 1993. Patterns of fluctuating asymmetry in avian feather ornaments: Implications for models of sexual selection. *Proceedings of the Royal Society of London,* ser. B 245:1–5.

Morris, D. 1957. "Typical intensity" and its relation to the problem of ritualization. *Behaviour* 11:1–12.

Morton, E. S. 1975. Ecological sources of selection on avian sounds. *American Naturalist* 109:17–34.

Pearce, J. M. 1987. *An Introduction to Animal Cognition.* Hillsdale, NJ: Lawrence Erlbaum Associates.

————. 1994. Discrimination and categorization. In N.J. Mackintosh, ed., *Animal Learning and Cognition,* 109–134. San Diego: Academic.

Perrett, K. A., K. A. May, and S. Yoshikawa. 1995. Facial shapes and judgements of female attractiveness. *Nature* 368:239–241.

Pinker, S., and A. Prince. 1988. On language and connectionism: Analysis of a parallel distributed processing model of language acquisition. *Cognition* 28:73–193.

Proctor, H. C. 1992. Sensory exploitation and the evolution of male mating behaviour: A cladistic test using water mites (Acari: Parasitengona). *Animal Behaviour* 44:745–752.

Radesäter, T., and H. Halldórsdóttir. 1993. Fluctuating asymmetry and forceps size in earwigs, *Forficula auricularia. Animal Behaviour* 45:626–628.

Ragge, D. R. 1965. *Grasshoppers, Crickets, and Cockroaches of the British Isles.* London: Frederick Warne and Co.

Razran, G. 1949. Stimulus generalisation of conditioned responses. *Psychological Bulletin* 46:337–365.

Rensch, B. 1957. Ästhetische Faktoren bei Farb- und Formbevorzugungen von Affen. *Zeitschrift der Tierpsychologie* 14:71–99.

————. 1958. Die wirksamkeit ästhetischer faktoren bei wirbeltieren. *Zeitschrift Tierpsychologie* 15:447–461.

Ridley, M. 1993. *The Red Queen: Sex and the Evolution of Human Nature.* London: Penguin. *Zeitschrift der Tierpsychologie.*

Rilling, M. 1977. Stimulus control and inhibitory processes. In W. K. Honig and J. E. R. Staddor, eds., *Handbook of Operant Behavior,* 432–480. Englewood Cliffs, N.J.: Prentice Hall.

Roper, T. J. 1983. Learning as a biological phenomenon. In T. R. Halliday and P. J. B. Slater, eds., *Genes, Development and Learning,* 178–212. Oxford: Blackwell Scientific.

Rumelhart, D. E., J. L. McClelland, and the PDP Research Group. 1986. *Parallel Distributed Processing: Exploration in the Microstructure of Cognition,* Vol. 1, *Foundations.* Cambridge: MIT Press.

Ryan, M. J. 1990. Sexual selection, sensory systems, and sensory exploitation. *Oxford Surveys in Evolutionary Biology* 7:157–195.

Ryan, M. J. 1994. Mechanisms underlying sexual selection. In L. A. Real, ed., *Behavioural Mechanisms in Evolutionary Ecology,* 190–215. Chicago: University of Chicago Press.

Ryan M. J., J. H. Fox, W. Wilczynski, and A. S. Rand. 1990. Sexual selection for sensory exploitation in the frog *Physalaemus pustulosus. Nature* 343:66–67.

Ryan, M. J., S. A. Perrill, and W. Wilczynski. 1992. Auditory tuning and call frequency predict population-based mating preferences in the cricket frog, *Acris crepitans. American Naturalist* 139:1370–1383.

Ryan, M. J., and A. Keddy-Hector. 1992. Directional patterns of female mate choice and the role of sensory biases. *American Naturalist* 139(suppl.):4–35.

Ryan, M. J., and A. S. Rand. 1993. Sexual selection and signal evolution: The ghost of biases past. *Philosophical Transactions of the Royal Society of London,* ser. B 340:187–195.

Schaller, F., and H. Schwalb. 1960. Attrappenversuche mit Larven und Imagines einheimischer Leuchtkäfer (Lampyrinae). *Zooligischer Anzeiger Supplement* 24:154–167.

Schenkel, R. 1956. Zur Deutung der Phasianiden balz. *Ornithologischer Beobachter* 53:182.

Schneider, D., B. C. Block, J. Boeckh, and E. Priesner. 1967. Die Reaktion der männlichen Seidenspinner auf Bombykol und seine Isomeren: Elekroantennogramm und Verhalten. *Zeitschrift für Vergleichende Physiologie* 54:192–209.

Schneirla, T. C. 1966. Behavioral development and comparative psychology. *Quarterly Review of Biology* 4:283–302.

Schuler, W., and T. J. Roper. 1992. Responses to warning coloration in avian predators. *Advances in the Study of Behaviour* 12:111–146.

Searcy, W. A. 1992. Song repertoire and mate choice in birds. *American Zoologist* 32:71–80.

Shanks, D. R. 1994. Human associative learning. In N. J. Mackintosh, ed., *Animal Learning and Cognition,* 335–374. San Diego: Academic.

Sheppard, A. 1987. *Aesthetics: An Introduction to the Philosophy of Art.* Oxford: Oxford University Press.

Shettleworth, S. J. 1972. Constraints on learning. *Advances in the Study of Behaviour* 4:1–68.

————. 1994. Biological approaches to learning. In N. J. Mackintosh, ed., *Animal Learning and Cognition,* 185–219. San Diego: Academic.

Smith, W. J. 1977. *The Behavior of Communicating.* Cambridge: Harvard University Press.

Spence, K. W. 1937. The differential response in animals to stimuli varying in a single dimension. *Psychological Review* 44:430–444.

Staddon, J. E. R. 1975. A note on the evolutionary significance of "supernormal" stimuli. *American Naturalist* 109:541–545.

Stewart, I., and M. Golubitsky. 1992. *Fearful Symmetry: Is God a Geometer?*

Sutherland, N. S. 1969. Shape discrimination in rat, octopus, and goldfish. *Journal of Comparative Physiological Physiology* 67:160–176.

Swaddle, J. P., and I. C. Cuthill. 1994. Preference for symmetric males by female zebra finches. *Nature* 367:165–166.

Ten Cate, C., and P. Bateson. 1988. Sexual selection: The evolution of conspicuous characteristics in birds by means of imprinting. *Evolution* 42:1355–1358.

Tinbergen, N. 1951. *The Study of Instinct.* Oxford: Clarendon.

———. 1959. Comparative studies of the behaviour of gulls (Laridae): A progress report. *Behaviour* 15:1–70.

Tinbergen, N., and A. C. Perdeck. 1950. On the stimulus situation releasing the begging response in the newly-hatched herring gull chick (*Larus a. argentatus* Pont.). *Behaviour* 3:1–38.

Toates, F. M. 1986. *Motivational Systems.* Cambridge: Cambridge University Press.

Trivers, R. L. 1985. *Social Evolution.* Menlo Park, Calif.: Benjamin/Cummings.

Turner, J. R. G. 1971. Studies of Müllerian mimicry and its evolution in burnet moths and heliconid butterflies. In R. Creed, ed., *Ecological Genetics and Evolution,* 224–260. Oxford: Blackwell Scientific.

Turner, J. R. G. 1975. A tale of two butterflies. *Natural History* 84:28–37.

Van Valen, L. 1962. A study of fluctuating asymmetry. *Evolution* 16:125–142.

Watson, P. J., and R. Thornhill. 1994. Fluctuating asymmetry and sexual selection. *Trends in Ecology and Evolution* 9:21–25.

Weary, D. M., T. C. Guilford, and R. G. Weisman. 1993. A product of discrimination learning may lead to female preferences for elaborate males. *Evolution* 47:333–336.

Weiss, S. J. 1972. Stimulus compounding in free-operant and classical conditioning. *Psychological Bulletin* 78:189–208.

Wells, K. D. 1988. The effect of social interactions on anuran vocal behavior. In B. Fritzch, M. J. Ryan, W. Wilczynski, and W. Walkowiak, eds., *The Evolution of the Amphibian Auditory System,* 433–454. New York: John Wiley and Sons.

Wicklund, C., and T. Järvi. 1982. Survival of distasteful insects after being attacked by naive birds: A reappraisal of the theory of aposematic coloration evolving through individual selection. *Evolution* 36:998–1002.

Wiley, R. H. 1983. The evolution of communication: Information and manipulation. In T. R. Halliday and P. J. B. Slater, eds., *Communication, Animal Behaviour,* Vol. 2, 156–189. Oxford: Blackwell Scientific.

Wiley, R. H., and D. G. Richards. 1983. Adaptations for acoustic communication in birds: Sound transmission and signal detection. In D. E. Kroodsma and E. H. Miller, eds., *Ecology and Evolution of Acoustic Communication in Birds,* Vol. 1, 131–181. New York: Academic.

Zahavi, A. 1975. Mate selection—a selection for a handicap. *Journal of Theoretical Biology* 53:205–214.

———. 1977. The cost of honesty (further remarks on the handicap principle). *Journal of Theoretical Biology* 67:603–605.

Zeller, A. C. 1980. Primate facial gestures: A study of communication. *International Journal of Human Communication* 13:565–606.

CHAPTER THREE

Constraints on Information Processing and Their Effects on Behavior

REUVEN DUKAS

3.1 INTRODUCTION

Biological traits reflect some balance between adaptation and constraint. Natural selection may act to optimize the value of a trait; however, phylogenetic, historical, genetic, developmental, and structural limitations may keep a trait's value far from some theoretical optimum (Gould and Lewontin 1979; Cheverton et al. 1985; Herrera 1992; Arnold 1994a, 1994b). For examples, humans cannot distinguish ultraviolet (UV) light from gray light with identical brightness—although ultraviolet light conveys biologically relevant information (e.g. Jacobs 1981)—and the best athlete cannot run faster than a threshold limit or sustain maximum running speed for an extended time because of physiological constraints on muscles (e.g. Peterson et al. 1990).

Early models in behavioral ecology were instrumental for identifying key ecological components that relate animal behavior to fitness (reviewed by Pyke et al. 1977; Krebs and Davies 1978). Such models provided tremendous insight owing to their relative simplicity and generality, but they mostly avoided incorporating explicit mechanisms and constraints on behavior. Such neglect of mechanisms, however, was quickly recognized (reviewed by Stephens and Krebs 1986). Consequently, much current research in behavioral ecology addresses questions on mechanisms and constraints (reviewed by Hughes 1990; Real 1994).

Behavioral ecologists wish to know how various behaviors affect survival and reproduction. By contrast, neurobiologists and psychologists concentrate on proximate mechanisms of behavior; research is done mostly under highly controlled laboratory conditions. Neurobiologists and psychologists have thus accumulated a broad mechanistic understanding, which is changing gradually to encompass ecological and evolutionary considerations (Staddon 1983; Kamil 1994; Anderson 1995). Given the superiority of neurobiology and psychology for studying mechanisms and behavioral ecology for examining the evolutionary ecology of behavior, integration of these disciplines seems appropriate. This will allow us to evaluate how constraints on information processing interact with ecologically relevant behaviors of animals in their natural settings.

As animals move through the environment, they encounter an enormous amount of information. Animals must decide what set of stimuli to attend to and how to integrate current sensory information with knowledge stored in memory in order to determine subsequent behavior. Several constraints on the processing of information may affect strongly such behavior. I combine theory from behavioral ecology and data from neurobiology and cognitive psychology to discuss first how optimality considerations and knowledge of central cognitive constraints can explain observed behavioral patterns, and second, how evolutionary and ecological theory can be used to examine mechanisms and constraints that underlie animal cognition. I will begin, however, with a brief evaluation of the importance of constraints in evolutionary ecology.

3.2 CONSTRAINTS: DEFINITION AND BIOLOGICAL FOUNDATIONS

Constraints limit the design or performance of a system. Different constraints may vary in magnitude or temporal effects. Some constraints, such as the laws of physics and chemistry, are fundamental and persist indefinitely—at least within a certain environmental medium. Other constraints are temporary; they may only delay the evolution of a trait. For instance, no genetic variation in a certain characteristic can be a temporary limitation that can be overcome with future mutations. This is especially true if the trait can vary relatively independently of other traits.

In many cases, however, various characteristics are just minute components in a large and complex system. The interdependency of elements that form the system is probably the most common general biological constraint. Interdependency is ubiquitous; it underlies processes of ontogenic development, in which each step must proceed from the one before it (Oster and Alberch 1982; Maynard Smith et al. 1985), and interdependency is fundamental in gene action. First, single morphological, physiological, and behavioral traits are determined with the expression of tens to hundreds of genes that form a complex system, which still is not well understood (McAdams and Shapiro 1995). Second, gene expression is typically pleiotropic, meaning that a certain gene affects several related or separate traits (reviewed by Arnold 1994a, 1994b). Interdependency thus implies that no specific trait can change without altering other traits and no changes in specific traits enhance fitness unless these changes are accompanied by changes in other traits. Such requirements could delay or even prevent adaptive changes.

For example, many birds and lizards perceive the solar spectrum between UV and red light, but the perceived region by mammals is narrower. In mammals, color vision is most developed in old-world primates, but even they do not perceive UV light (Jacobs 1981; Tovee 1995). Having UV receptors is the ancestral condition, but mammals lost these receptors during their nocturnal evolutionary history. In addition, many mammals, including primates, have lenses that absorb UV light and thus preclude perception of UV light. The evolution of UV-light perception implies at least three acquisitions: a lens that transmits UV light, UV-light receptors, and other receptors with oil droplets that absorb UV light to protect these receptors from photodamage (see Goldsmith 1990). Such a combination of acquisitions is much less likely to occur than a single alteration.

Many constraints have an absolute effect: They limit performance temporarily or permanently. Other constraints may be considered costs: Specific performance may be enhanced but only at a certain price. For example, under artificial selection, the life span of fruit flies can be extended, so there is no absolute limit on the length of life; however, female flies with increased life span show reduced reproduction early in life. In other words, there is a trade-off between reproductive and survival rates (e.g. Kirkwood and Rose 1987; Rose 1991; Stearns 1992).

Constraints characterize also synthetic products such as computer keyboards. The general design of a computer keyboard was copied from the mechanical typewriter. Theoretically, the optimal layout of a keyboard should reflect ease of use. For example, placing the ten most used letters in the home row allows the user to compose about 70% of English words without moving the fingers to other rows. Such an existing key layout, the DSK, does allow faster and more convenient typing. However, a major constraint underlying the design of mechanical typewriters is frequent jamming during fast typing. Thus, in the nineteenth century, typewriter jamming was minimized by spreading the keys of the most used letters (David 1985). The mechanical constraint now is long gone, but keyboards have not "evolved" toward the optimal design. Consequently (and ironically), a central component of the information superhighway, the computer keyboard, illustrates an extreme case of "phylogenetic" and social constraints. In short, the contribution of constraints must not be neglected when deciphering a system's design. Moreover, full understanding of such constraints may require a deep comprehension of a feature's evolutionary history as a component in a global system with various spatial and temporal scales.

How adaptation and constraint combine to produce a cognitive system is mostly unknown. A perfect brain, however, can be envisioned readily. Such a brain can (1) perceive all incoming environmental stimuli; (2) process simultaneously all the relevant information; (3) compute accurately and rapidly the variables necessary for determining behavior; (4) remember accurately all information required for future reference; and (5) quickly retrieve that specific information in the future when relevant. Clearly, brains do not meet all these criteria; various constraints, costs, and trade-offs shape the somewhat limited capacity of the brain. I will discuss a few aspects of that limited ability and the way it shapes behavior.

3.3 ATTENTION

3.3.1 Neurobiological Background

The recognition of complex patterns is too demanding computationally to be performed at maximal resolution across the entire visual field. Recognition of complex patterns is manageable, however, because pattern recognition occurs mainly within a restricted window of attention that can be rapidly shifted in position and changed in spatial scale (Van Essen et al. 1992; Crick 1994; Maunsell 1995). Visual attention can be directed to a certain spatial location; alternatively, visual attention can be less restricted spatially and directed instead along other dimensions or attributes such as color, size, or shape of objects. Concentrating attention on a certain attribute modulates neuronal activity in the brain areas that are specialized for processing that attribute. Consequently, accuracy in the associated detection or discrimination task is improved, particularly under near-threshold discriminability (Corbetta et al. 1990; Posner and Peterson 1990). Most who study attention refrain from explicit definitions; however, it is commonly believed that attention is related to the brain's limited capacity to process information simultaneously, and selective attention allows one to filter out irrelevant information (Desimone and Duncan 1995). Given that most attention studies focus on the visual domain, I will address only visual attention.

In the past decade, numerous studies evaluated the simultaneous effect of attention on neuronal activity and behavioral performance. Such studies measured electrophysiological activity of single neurons or overall activity of parts of the brain with one of a few brain imaging techniques. Research in Desimone's laboratory focused on the effect of selective attention on the activity of single cells in the visual cortex of rhesus monkeys. Spitzer et al. (1988) asked if neuronal responses to attended stimuli are affected by the amount of attention devoted to these stimuli. They examined neurons in area V4 of the extrastriate cortex. Area V4 is an intermediate station along the pathway from the primary visual cortex to the temporal lobe. This pathway is critically involved in object recognition. Area V4 is one of the first cortical areas along the pathway in which neuronal activity is gated by spatially directed selective attention.

Spitzer et al. trained two rhesus monkeys to maintain fixation on a small spot on a video monitor and discriminate the orientation or color of a stimulus presented in the receptive field of a neuron in V4. While the monkey held a bar and gazed at the fixation point, a sample stimulus appeared for 200 msec; after 400–600 msec, a test stimulus appeared for 200 msec at the same location. When the test stimulus was identical to the preceding sample (a matching trial), the subject was rewarded if it released the bar immediately; when the test stimulus differed from the sample (a nonmatching trial), the subject was rewarded only if it delayed release of the bar for at least 700 msec. The stimuli were small colored bars presented at two levels of difficulty. For the easy task, the nonmatching test stimuli differed from the samples by 90° of orientation or 77 nm in wavelength. For the difficult task, the nonmatching test stimuli differed from the samples by only 22.5° or 19 nm.

The overall performances of subjects on the easy and difficult tasks were 93% and 73% correct, respectively; these results confirmed that the difficult task was indeed more difficult for the subjects. To verify that subjects actually processed stimuli differently in the two levels of task difficulty, difficult nonmatching probe trials were inserted in 6% of the trials in sessions of the easy task. Accuracy in these difficult probe trials was much lower when they were presented in the easy session (48% correct) than when they were presented in the difficult session (78% correct). These behavioral results and further signal detection analyses suggest that the monkeys devoted more attention to the stimuli in the difficult session.

The simultaneous recordings from 98 neurons in V4 revealed that 81% responded more strongly to the sample stimulus when it was presented in the difficult session (fig. 3.1a). Moreover, 77% of 42 cells tested showed a narrowing of their tuning bandwidths in the difficult task, indicating an improvement in selectivity (fig. 3.1b). These results and further controls suggest that concentration of attention on a stimulus is correlated with enhanced response and sharpened selectivity of the neurons that process that stimulus (Spitzer et al. 1988).

In an earlier experiment with use of similar procedures in the same laboratory, Moran and Desimone (1985) asked how the brain limits processing of irrelevant stimuli. They trained monkeys to attend to stimuli at one location in the visual field and ignore stimuli at another. When both locations were in the receptive field of a cell in area V4, the cell's response to the unattended stimulus was reduced by more than half. This suggests that attention gates visual processing by filtering out irrelevant information in the receptive fields of single neurons.

Results of studies with human subjects using positron emission tomography (PET) agree with these findings. PET studies are based on indirect measurements of blood flow. There is strong evidence, however, that changes in blood flow do reflect local neuronal activity (Posner 1995). Corbetta et al. (1990) found that attending to a specific attribute of a stimulus in a visual detection task was correlated with increased local blood flow in the area of the visual

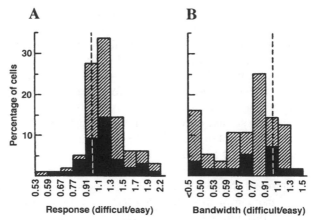

Figure 3.1. Effects of task difficulty on neuronal activity in area V4 of rhesus monkeys. (A) For each cell, the response to the stimulus in the difficult session was divided by the response in the easy session. A ratio of 1.0 (*dashed line*) indicates that task difficulty did not affect the response. Most cells showed larger responses in the difficult session (ratios > 1.0; $\chi^2 = 18$, $P < 0.01$). (B) Tuning bandwidth of neurons in the difficult session was divided by the bandwidth in the easy session. Most cells showed narrower bandwidths in the difficult session (ratios < 1.0; $\chi^2 = 7.5$, $P < 0.01$). *Hatched bars* = cells studied with orientation as the relevant stimulus dimension, *shaded bars* = cells studied with color (from Spitzer et al. 1988).

cortex that processes this attribute. Subjects' performances in the detection tasks were higher when subjects attended to a single attribute instead of three characteristics simultaneously. Similarly, in a study of spatial attention, Heinze et al. (1994) showed that attending to the left half of a stimulus array increased neuronal activation in the right posterior fusiform gyrus of the extrastriate visual cortex, and attending to the right half of a stimulus array produced activation in the left fusiform gyrus. Drevets et al. (1995) reported that local blood flow to unattended cortical areas decreased when subjects anticipated a stimulus that was processed by another cortical area.

These studies used monkeys or humans as subjects. Behavioral research with other species, however, strongly suggests that limitation on the amount of information that can be processed simultaneously and the employment of attentional mechanisms to cope with this constraint is ubiquitous. For example, D. Blough and P. Blough used pigeons that searched for targets on video monitors. Results from their experiments suggest, first, that pigeons attend selectively to specific locations or attributes while searching for targets, and second, that attending to a certain spatial location on the screen or a particular target type increases performance in detection tasks (P. Blough 1989, 1991; D. Blough 1993). Even honeybees appear to attend selectively to different visual attributes in a manner that enhances their foraging performance (Klos-terhalfen et al. 1978).

To summarize, the brain has a limited capacity to process information si-multaneously. Attentional mechanisms control the type and amount of infor-mation processed on the basis of the relevance of various sensory stimuli in time and space. Usually, only a limited set of stimuli is processed with a high degree of resolution. More information can be attended to simultaneously; however, in this case, the information is processed with less sensitivity.

All the studies mentioned used experimental paradigms of visual detection to study neurophysiological and behavioral aspects of attention in the labora-tory. However, the precise detection of or discrimination between various sig-nals is a universal task performed by many animals in nature. Thus, attentional constraints must influence key behaviors that determine animal fitness. Two classes of behavior I now will discuss concern foraging and the evasion of predators.

3.3.2 Attention and Prey Detection

Many animals must search for food or hosts that are inconspicuous. Under such conditions, attentional constraints may determine searching strategies and, consequently, the diet or preferred hosts of various species. For clarity, the discussion of attention and prey detection is divided into two sections. The first section considers optimal allocation of attention to spatial locations in the visual field. The second part concerns selective attention to distinct visual attributes or prey types.

Before proceeding, however, it is necessary to be explicit about the meaning of conspicuousness. Conspicuousness is typically defined as the degree of dis-similarity between an item and its surrounding background. One can systemat-ically measure conspicuousness of a target in specific background and lighting conditions (Endler 1984, 1993). Nonetheless, conspicuousness ultimately de-pends on the perceptual capacity of certain species. The resolving power, light sensitivity, and spectral sensitivity of eyes are among the most important fac-tors that determine conspicuousness (e.g. Land 1981; Endler 1991). In addi-tion, although vision dominates human perception, other species may use other sensory modalities to detect targets. Hence, the only right way to infer conspic-uousness is to test individuals of a species for what *they* perceive as conspicu-ous or cryptic.

3.3.2.1 Spatial Attention and Search Rate

The visual field of most animals is between 180° and 360°. As discussed in section 3.3.1, however, computational constraints prevent the simultaneous processing of information from the entire visual field. Hence, attention has to be focused on a small portion of the surrounding environment to allow a detailed extraction of information. Alternatively, attention can be distributed to a larger area, with a less sensitive extraction of information (LaBerge 1983; Eriksen and Yeh 1985; Desimone and Duncan 1995). The optimal allocation of attention depends on the visual task. If the task is sufficiently easy, a broader attentional scope increases the *effective* visual field without compromising performance. On the other hand, difficult detection tasks require maximal attention and thus a narrow attentional focus (see section 3.3.1). For example, one can attend simultaneously to a whole computer monitor and readily detect a large, 2×2-cm light gray target appearing at a randomly chosen location on a dark gray background. By contrast, if the task is to detect a one-pixel cryptic target, one must attend only to a very small portion of the monitor at a time and move the narrow attentional focus until spotting the target.

In the laboratory, researchers can define spatial focus by instructing subjects or measuring parameters such as eye movement. In the field, foragers usually pause and fix their gaze to several locations in succession before moving on in pursuit of prey (see Anderson 1981; O'Brian et al. 1990; Getty and Pulliam 1991, 1993). Hence, the rate of movement of foragers can provide a reasonable approximation of their attentional focus. Specifically, covering a certain area with highly focused attention must result in a low rate of movement, and searching the same area with broad attentional scope can be achieved quickly. It is optimal to focus attention for difficult search tasks and to broaden it for easy tasks. Thus, one would predict foragers to decrease rate of movement while searching for increasingly cryptic prey (see Gendron and Staddon 1983).

Gendron (1986) tested this prediction with bobwhite quail that foraged in the laboratory for several types of food pellets that varied in conspicuousness. Subjects' mean probabilities of finding the most conspicuous and most cryptic pellets were about 100% and 40%, respectively. Their search rates were approximately 25 m/min for the conspicuous food and 15 m/min for the cryptic food. Thus, the significant reduction in search rate for more cryptic food ($P < 0.001$) supports the notion that spatial focus of attention must be narrower for more difficult detection tasks.

The inference regarding spatial attention from Gendron's data is somewhat indirect. O'Brien and Showalter (1993) provided more direct evidence for the function of spatial attention in the search for food. They allowed arctic greyling *Thymallus articus* to forage for daphnias with two levels of conspicuousness in an experimental stream with four current speeds. The daphnias were cryptic when plant debris and other detritus were added to the water and conspicuous when no debris was added. Unlike cruise predators that move through a stationary medium in pursuit of prey, drift-feeding fish remain stationary while the medium moves. Consequently, drift-feeding fish are ideal subjects for the study of visual and attentional aspects of searching behavior in natural or seminatural conditions (see also Hughes and Dill 1990). O'Brien and Showalter videotaped the fish during the experiments and later determined the location-angle and -distance of each prey relative to the fish when the prey was first detected. The increase in water current, from 11.6 to 55.7 cm/sec, was associated with a decrease in location-angles and -distances ($P < 0.001$). Similarly, reduction in prey conspicuousness was associated with a decrease in location-angles and -distances ($P < 0.001$). These results strongly suggest that fish reduce the area of attentional focus when (1) search speed increases (water current increases); and (2) prey is more difficult to detect. Nonetheless, such data must be augmented with direct experimental evidence that demonstrates the fish indeed adjusted their attentional focus (and, alternatively, did not attend to the whole visual field but altered decisions about attacking prey).

3.3.2.2 Selective Attention

Although focusing attention to a specific spatial location can enhance performance, attending selectively to certain visual attributes may also be beneficial. To evaluate this issue, Dukas and Ellner (1993) added an attention factor to a basic prey model. In the model, attention was defined explicitly as the brain's limited capacity to process information simultaneously. The information processed was aspects of various *familiar* foods. Each food was distinctly different in appearance, and each food varied in conspicuousness, density, or energy content. As discussed, it is easy to detect a conspicuous item, but the detection of highly cryptic prey is difficult and attention-demanding.

On the basis of the neurophysiological and behavioral evidence detailed in section 3.3.1, the following functional form was chosen to describe the association between the probability of detecting food, P_d, and attention:

$$P_{d,i} \propto a_i^{1/k_i}, \tag{3.1}$$

where a_i is the fraction of a forager's attention devoted to prey type i ($0 \le a_i \le 1$, $\Sigma\, a_i \le 1$), and k_i is the conspicuousness index of prey type i, $k > 0$. A larger value of k means greater prey conspicuousness and, therefore, a higher probability of detection at any given attention level. Prey type i is cryptic if $k_i < 1$ in equation 3.1, and conspicuous if $k_i > 1$ (fig. 3.2).

We incorporated equation 3.1 into a modified prey model that takes into account effects of search rate on food detection (see Gendron and Staddon 1983):

$$R = \frac{S\sum_{i=1}^{m} D_i[1 - (S/M)^{k_i}]a_i^{1/k_i}e_i - (f + bS)}{1 + S\sum_{i=1}^{m} D_i[1 - (S/M)^{k_i}]a_i^{1/k_i}h_i}, \tag{3.2}$$

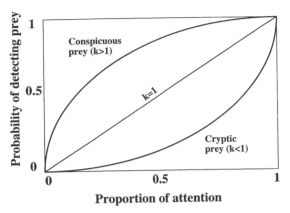

Figure 3.2. Probability of detecting an encountered prey as a function of attention (equation 3.1). Each curve represents prey with different conspicuousness (k) (modified from Dukas and Ellner 1993).

where R is the rate of energy gain, S is search rate (or the area searched per unit of time), D_i is the density of type i prey, M is the maximum possible search rate (at which the predator cannot detect any prey regardless of how conspicuous it is), e_i is the net energy gain from a type i item, f and b are species-specific positive constants for energy expenditure, and h_i is the handling time for a type i prey item (Dukas and Ellner 1993).

Equation 3.2 says that the forager's rate of energy intake depends not only on the density, energy content, and handling time of prey types, as in the basic prey model (e.g. Stephens and Krebs 1986, chap. 2). Two additional factors are (1) the forager's search rate (Gendron and Staddon 1983); and (2) the way the forager divides its attention between available prey types. Therefore, the forager's diet consists of those prey types it attends to (i.e. the types for which $a_i > 0$).

The incorporation of attentional limitations alters some basic predictions of the simple prey model. First, the simplest case to consider is where all prey types have the same parameter values (k, D, e, and h). Thus, each prey type has a distinct appearance, but all types are equally cryptic, abundant, and rewarding. Under such conditions, the basic prey model predicts that a forager should search for all available prey types. By contrast, predictions of the attention model (equation 3.2) depend on the conspicuousness of prey types. On one hand, when all potential prey types are equally cryptic ($k_i < 1$), the optimal strategy is to search for a single prey type only. The choice of the single cryptic prey type is arbitrary in this case because all prey types have identical parameter values. The forager may switch between the cryptic prey types over time, but the forager should not search for more than one type simultaneously.

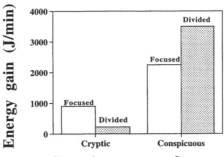

Conspicuousness of prey

Figure 3.3. Net rate of energy intake of a simulated predator encountering items of three prey types that were equally cryptic ($k = 0.5$) or equally conspicuous ($k = 2.0$). The predator searched for a single prey type (*white bars*, focused attention) or all three types simultaneously (*shaded bars*, divided attention). All three prey types had the same net energy contents (125.52 J/prey), prey densities (4/m²), and handling times (0.01 min/prey). Values of the constants for energy expenditure were f = 20 and b = 1.6 (see equation 3.2) (data from Dukas and Ellner 1993).

On the other hand, when all potential prey types are equally conspicuous ($k_i > 1$), the optimal strategy is to search for all prey types simultaneously. The above predictions are intuitively appealing because they agree with humans' typical behavioral decisions: We perform only a single difficult task but prefer to do a few simple things simultaneously when possible. For example, we don't attempt to read two books simultaneously, but we are able to walk and talk at the same time.

With use of simulations, the net rate of energy intake of a forager employing the optimal allocation of attention was compared with that of a forager using a feasible alternative. When all three prey types were cryptic ($k = 0.5$), the forager's net rate of energy intake was more than three times higher when the forager attended to one prey type instead of three types simultaneously (fig. 3.3). When all three prey types were relatively conspicuous ($k = 2.0$), the forager's net rate of energy intake was higher when the forager attended to all three prey types instead of a single type (fig. 3.3).

The only other case considered here is different conspicuousness values of prey types with the same energy contents, handling times, and densities. Because the basic prey model does not consider prey conspicuousness, the model's prediction in this case is the search for all available prey types. Once again, predictions of the attention model depend on the conspicuousness of prey types. When all prey types are cryptic ($k_i < 1$), the optimal strategy is to attend only to the prey type with the highest k value (fig. 3.4). When prey types are conspicuous ($k_i > 1$), the optimal strategy is to devote attention to all prey types. However, there is a distinct change in the optimal allocation

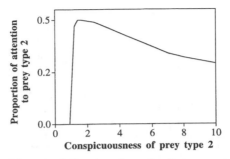

Figure 3.4. Example of the optimal allocation of attention for two prey types with different conspicuousness. Prey type 1 was fairly conspicuous ($k = 2$), and the conspicuousness of prey type 2 varied from very cryptic to very conspicuous. Proportion of attention to prey 2 was 0 when prey 2 was cryptic and 0.5 when its conspicuousness was identical to that of prey 1 ($k = 2$). If $k > 2$, less attention was devoted to prey 2 when it became increasingly more conspicuous (from Dukas and Ellner 1993).

of attention as prey becomes increasingly more conspicuous. When prey types have small conspicuousness values (for example, $1 < k_i < 2$), the optimal strategy is to allocate attention to all prey types and give more attention to the *most* conspicuous prey types; this is because the marginal increase in probability of detection with increasing attention is higher for more conspicuous prey. When prey types became more conspicuous ($k_i \gg 1$), however, the optimal strategy was to allocate more attention to the *less* conspicuous prey types (fig. 3.4). The marginal increase in probability of detection with increasing attention was higher for the less conspicuous prey.

This attention model takes into account the spatial focus of attention as it affects search rate and the selective attention to various attributes. Decreasing prey conspicuousness is indeed associated with a lower search rate as Gendron and Staddon (1983) originally proposed. However, attending selectively to a single attribute under difficult search conditions is still advantageous. In other words, substituting search rate (or stare duration) for selective attention (Guilford and Dawkins 1987) is inappropriate; this was shown theoretically in the model by Dukas and Ellner (1993) and empirically by Reid and Shettleworth (1992), Plaisted and Mackintosh (1995), and Langley et al. (1996).

Assumptions and predictions about the selective attention model have not yet been tested. Nevertheless, ethologists have paid some attention to perceptual constraints and their effects on searching behavior, but ethologists have worked in remarkable isolation from neurophysiologists and cognitive psychologists. L. Tinbergen (1960, 332) was the first biologist to appreciate what we now term attentional constraints and their effects on the consumption of prey by birds: "The intensity of predation depends to a great extent on the use of specific searching images. This implies that the birds perform a highly selective sieving operation on the visual stimuli reaching their retina. . . . There is some reason to believe that the birds can use only a limited number of different searching images at the same time." Several researchers attempted to demonstrate the existence of search images and contemplated the difficulties of distinguishing various factors that influence searching performance (e.g. Dawkins 1971a, 1971b; Lawrence and Allen 1983; Guilford and Dawkins 1987; Reid and Shettleworth 1992). Given the substantial controversy about and numerous definitions of search image, it is convenient to refrain from its use altogether. Instead, one can employ the somewhat less ambiguous vocabulary of neurobiologists and cognitive psychologists as I have.

One nicely designed experiment, which suggested the importance of limited attention for searching performance, was conducted by Pietrewicz and Kamil (1979). The beauty of this experiment's design was its combination of a highly controlled laboratory paradigm, typical of that of experimental psychologists, and a realistic imitation of a natural foraging problem faced by birds. Well-trained blue jays were supposed to detect images of two species of cryptic moth resting on tree bark. The moth *Catocala relicta* has white forewings with black and gray stripes; it usually rests head up on white birch. The moth *C. retecta* has gray forewings with a pattern of brown lines; it typically rests head down on trees such as oak, which have dark bark. Slides of the moths were projected in front of the jays; the jays were rewarded if they pecked ten times on the stimulus key during positive trials in which slides showed a moth on its background bark. During negative trials, slides of only tree bark were shown, and the jays were supposed to peck the advance key; the subsequent slide then appeared after a brief interval between trials. There were three types of sessions: runs with one moth, runs with the other moth, and nonruns. In a run session, all slides were of only one moth species; eight slides showed a moth on its corresponding bark, and eight slides showed only bark. In nonrun sessions, four slides of one moth, four slides of the other moth, and four slides each of the corresponding barks without a moth were shown.

During a run session, jays could fully attend to searching for one moth. However, during nonruns, they had to divide attention between searching for either of the two moths. Searching for a single moth resulted in close to a 100% accurate performance by the end of a session. By contrast, jays searching for the two species simultaneously showed only about 75% correct detection (fig. 3.5). This experiment suggested that attentional limitations have strong effects on the foraging success of predators that feed on cryptic prey, but this experiment did not allow separation of effects of attention from those of memory interference; separation of these effects will be discussed in section 3.5.2.1.

Recent experiments by Reid and Shettleworth (1992), Plaisted and Mackintosh (1995), and Langley (Langley et al. 1996; Langley 1996) provide additional support to the assertions that selective attention to prey attributes under-

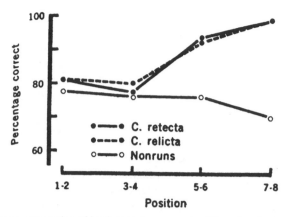

Figure 3.5. Average proportion of jays' correct responses to slides of moths in runs with a single moth (*C. retecta* or *C. relicta*) and nonruns, in which jays searched simultaneously for both moths. Figure shows performance as a function of the position of each of eight positive slides in a session. Performance was higher in runs than in nonruns ($P < 0.025$, ANOVA) (from Pietrewicz and Kamil 1979).

lie searching behavior, and that searching for cryptic prey involves focusing attention on some attribute, with a consequent decrease in attention devoted to other attributes. Nevertheless, a more systematic evaluation of the role of limited attention for prey detection is still necessary.

3.3.3 Attending Simultaneously to Prey and Predators

Animals foraging in nature must insure simultaneously that they eat but are not eaten. The optimal balance between feeding and antipredator demands implies usually that foragers must somehow divide their limited attention between the two difficult, concurrent tasks of finding cryptic food and detecting approaching predators (see Milinski and Heller 1978; Sih 1980, 1992; Lima and Dill 1990; Ydenberg, this vol. chap. 9). There are two ways of striking this balance. In the first way, searching for food requires a posture that impairs visual detection of distant predators; for example, the head may be close to the ground when searching for food and as high as possible when searching for predators. In this case, it is beneficial for the animal to alternate between bouts of searching for food or predators. Vigilance for predators during foraging has been studied extensively (Lima and Dill 1990; Ydenberg, this vol. chap. 9). In this case, however, analyses of selective visual attention are confounded by the visual field changes that occur when the head is raised and lowered.

In the second way, the head is in the same position for food searching and predator alertness, but the forager can alter its attentional focus. Thus, this

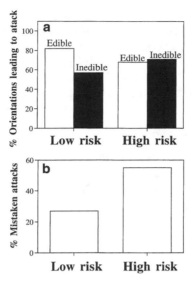

Figure 3.6. (*a*) Percentage of orientations to edible (*white bars*) and inedible (*black bars*) pellets that led to subsequent attack by juvenile salmon. Subjects oriented more often toward edible items only in the low-risk sessions ($P < .01$, Wilcoxon signed-ranks test). (*b*) Mistaken attacks on inedible pellets as a percentage of all attacks made by salmon. Twice as many mistaken attacks were made in the high-risk sessions ($P < .02$, Wilcoxon signed-ranks test) (redrawn from Metcalfe et al. 1987a).

case is more appropriate for evaluating the effects of limited attention on the simultaneous search for prey and predators. Metcalfe et al. (1987a) evaluated how increased predation risk affects the attention devoted to food by juvenile salmon *Salmo salar*. Specifically, they asked if a salmon's ability to distinguish between edible and inedible food varied with predation risk. These fish are sit-and-wait predators that attack prey passing in the water current. Since current speeds are often high, the salmon must initiate attacks immediately after detecting prey. In the experiments, edible food was small pellets that produce optimal growth. Larger food pellets, which were too big for the fish to swallow, were inedible. In low-risk sessions, no predator was visible. In high-risk sessions, a model predator was present in a nearby compartment for 30 sec immediately before the session began. In each session, a randomly chosen pellet type was placed in the current every 10 min.

In the low-risk sessions: (1) salmon were more likely to orient to edible pellets ($P < 0.01$); (2) more of these orientations led to subsequent attacks on food ($P < 0.01$; fig. 3.6*a*); and (3) the proportion of mistaken attacks was lower ($P < 0.01$; fig. 3.6*b*). It is well established that increased risk of predation alters optimal decisions about pursuing food (Ydenberg, this vol. chap. 9).

Indeed, Metcalfe et al. (1987a, 1987b) reported that salmon were slower to respond to approaching food and less likely to attack prey when under higher predation risk. In addition to this change in behavior, however, it appears that devoting more attention to a potential predator results in reduced attention directed toward food. Consequently, feeding efficiency is reduced (see also Milinski and Heller 1978; Milinski 1990; Krause and Godin 1996).

3.3.4 Further Thoughts on Attention

I have concentrated on the effects of limited attention on foraging and predator avoidance because, in these areas, more is known. Attentional constraints, however, probably influence many other behaviors as well. Moreover, although visual attention has been studied most, attentional constraints affect perception and processing of other kinds of sensory information. For example, humans cannot usually listen to more than one person simultaneously at a cocktail party, although we are within hearing range of several potentially interesting conversations. A hypothetical example in nature may be a decreased ability of females of various species to notice the sound of approaching predators while attending to courtship songs or calls of males (see Ryan 1994).

Many ecological consequences of limited attention are still in need of careful evaluation. Attentional constraints, through selection imposed by and on predators, have likely shaped morphology and visual appearance of many species. For instance, if two related species of cryptic prey share some visual attribute, a predator may search for both prey simultaneously by focusing attention on the shared trait. Divergence in appearance would force the predator to search for only a single prey simultaneously; this would reduce predation, at least in some circumstances (see Rausher 1978; Gilbert 1975, 228; Endler 1988, 1991; Dukas and Waser 1994).

I argued that it may be optimal for foragers to attend to only one cryptic food type at the same time. I did not discuss, however, the possibility of *sequentially* searching for two or more food types. In other words, limited attention explains why subjects prefer to conduct only a single difficult task at one time, but limited attention does not account for the tendency to have long runs of a single activity. The best explanation for such runs is that switching is costly (Dukas and Clark 1995a). There are at least two cognitive costs of switching. First, experiments on human subjects suggested that visual attention is not a high-speed switching mechanism; visual attention is a sustained state during which relevant information becomes available to influence behavior. More specifically, effective switching of attention in humans takes about half a second (Duncan et al. 1994). Second, conducting one activity may interfere with the memory of another. Such probable memory interference is discussed in section 3.5.2.1.

Throughout this discussion of attention, I accepted as a premise that attention is limited. We indeed know from experience and numerous studies that this is at least partly true. The argument given, however, to explain this constraint (which is a limited computational capacity to process all available sensory information simultaneously) is weak! Thus, one must pose the question Why does the brain have a limited capacity to process information simultaneously? In many cases, only a limited assortment of sensory information is relevant for determining subsequent behavior. So having some kind of filtering mechanism can allow one to focus on relevant information only. Filtering mechanisms are obviously advantageous. In other situations, however, processing more information simultaneously is beneficial. For instance, if limited attention causes reduced foraging success and increased predation, why have animals not evolved a larger attentional capacity? We do not yet have a satisfactory answer to this question.

A partial explanation of limited attention may be provided with a better understanding of another brain limitation I will discuss in the following section. Human studies suggested there is no rigid upper limit for attentional capacity. For example, in an emergency or when well-motivated experimental subjects are urged, more information may be processed simultaneously, at least for short periods. However, the consequence of such intense cognitive activity is fatigue. In other words, optimal attentional capacity may be at least partially determined with the nervous system's ability to sustain continuous processing of information over time. This capacity is apparently limited, and besides its interactions with attention, it has other direct effects on behavior.

3.4 SUSTAINED VIGILANCE

3.4.1 Background

Data from behavioral and neurophysiological research suggest that the central nervous system cannot sustain a high quality of information processing for a long time. Subjects that perform a continuous and difficult task, such as detecting highly cryptic targets, show a gradual reduction in performance, or vigilance decrement. Vigilance implies a general state of alertness that results in enhanced processing of information by the brain. Performing a difficult task reduces the quality of information processing and this reduction results in decreased performance over time (Mackworth 1969; Davis and Parasuraman 1982; Mckie 1977; Warm 1984). Throughout this section, I will use the psychological definition of vigilance, which is different from the more restrictive interpretation that is commonly used in behavioral ecology (Lima and Dill 1990; Ydenberg, this vol. chap. 9). In other words, I will use vigilance to mean a general state of enhanced capacity to process information about all

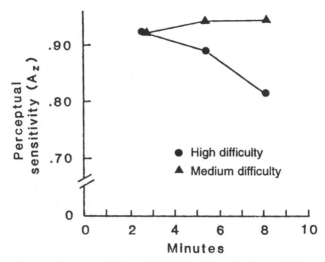

Figure 3.7. Vigilance decrement during a difficult detection task. Curves depict mean values of A_z, which is an index of perceptual sensitivity derived from receiver operating characteristic curves. The index is a corrected measure of sensitivity equivalent to the percentage of correct trials. Significant decrease in sensitivity occurred for the difficult task only ($P < .05$) (from Nuechterlein et al. 1983).

subjects, whether they are prey, competitors, or predators; vigilance is not just a state of alertness toward predators.

Behavioral evaluations of vigilance decrement have been conducted with humans only; however, neurophysiological animal studies suggest that animals other than humans possess similar mechanisms that control vigilance (e.g. Aston-Jones et al. 1984). An example from numerous experiments on sustained vigilance is depicted in figure 3.7. Human volunteers watched images of single-digit numbers (0–9) projected on a screen for 40 msec every second. Subjects were asked to detect the number 0, which appeared randomly with a probability of 0.25. In the easy task, images were moderately blurred, and in the difficult task, images were very blurred. Initial performance was identical in the easy and difficult tasks, but subjects showed a decrement in sensitivity only in the difficult task. In other words, in the difficult task only, the proportion of correct detections and false-positive responses decreased rapidly (Nuechterlein et al. 1983).

3.4.2 Sustained Vigilance and Temporal Patterns of Activity

Psychological studies on vigilance have focused on the proximate factors causing vigilance decrement or ways to reduce decrement for practical proposes (e.g. Parasuraman 1979; Wickens 1984; Parasuraman and Mouloua 1987). For animals in the field, sustaining vigilance seems highly relevant for determining temporal patterns of activity. Dukas and Clark (1995b) designed a set of models to predict, first, optimal allocation of time between foraging and rest, and second, optimal length of each foraging episode. The effects of vigilance decrement on foraging under the risk of predation was examined because foraging is a central animal activity in nature (e.g. Gilliam 1990; Ydenberg, this vol. chap. 9). The models, however, are relevant for the study of many other behaviors, including mate choice, nest building, and care of offspring.

In the first model, it was assumed that a forager alternates between short periods of activity and rest, and the forager's vigilance level, $v(t)$, declines and recovers during foraging and rest, respectively:

$$\frac{dv}{dt} = \begin{cases} -\alpha v & \text{while foraging} \\ \beta(1 - v) & \text{while resting,} \end{cases} \tag{3.3}$$

where α and β are positive constants. In this equation, α is the rate of vigilance decrement, which is positively associated with task difficulty (Parasuraman 1979; Neuchterlein et al. 1983; Parasuraman and Mouloua 1987), and β is the rate of vigilance recovery, which is determined by the quality of rest and sleep (see the following). It was assumed also that the forager devotes a portion θ of its time to foraging and $(1 - \theta)$ to rest. The variable θ was referred to as "foraging effort". The dynamics of vigilance level is thus

$$\frac{dv}{dt} = -\theta\alpha v + (1 - \theta)\beta(1 - v), \tag{3.4}$$

and the equilibrium level of vigilance (where $dv/dt = 0$) is

$$\bar{v}(\theta) = \frac{(1 - \theta)\beta}{\theta\alpha + (1 - \theta)\beta}. \tag{3.5}$$

Equation 3.5 implies that the equilibrium level of vigilance, $\bar{v}(\theta)$, is a decreasing function of foraging effort, θ. For a given foraging effort, $\bar{v}(\theta)$ is inversely related to the ratio α/β (fig. 3.8a). For example, assuming that β is constant, the equilibrium level of vigilance is lower for more difficult tasks, which have a higher α/β ratio.

If the forager allocates proportion θ of its time to foraging and its average rate of food intake is proportional to its current vigilance level, then the long-term average rate of food intake is

$$\bar{f}(\theta) = \lambda\theta\bar{v}(\theta), \tag{3.6}$$

where λ is a positive constant proportional to prey density. Increasing foraging effort has two effects. First, it increases the forager's rate of encountering prey, $\lambda\theta$. Second, it decreases the equilibrium level of vigilance, $\bar{v}(\theta)$. The

108   Reuven Dukas

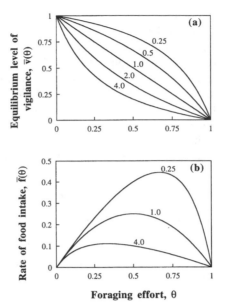

Figure 3.8. (*a*) Equilibrium level of vigilance, $\bar{v}(\theta)$, and (*b*) long-term average rate of food intake, $\bar{f}(\theta)$, as functions of the proportion of time allocated for foraging activity, θ, for a few values of α/β, which are the rates of decline and recovery of vigilance during foraging and rest, respectively. Assuming that β is constant, then a higher value of α/β can be perceived as depicting more difficult tasks; with this assumption, the highest curve in both figures depicts the easiest task, and the lowest curve depicts the most difficult activity (from Dukas and Clark 1995).

combined outcome is an initial increase in rate of food intake, $\bar{f}(\theta)$, followed by a decline. Both the rate and magnitude of increase and decrease of $\bar{f}(\theta)$ depend on the ratio α/β (fig. 3.8*b*). The optimal foraging effort, θ^*, is inversely related to the ratio α/β (fig. 3.8*b*; see Dukas and Clark (1995b) for the analytical derivations, which also incorporate effects of predation risk and metabolic costs).

Assuming again that β is constant, then our model predicts that foraging effort should decrease, and correspondingly, the proportion of time devoted for rest should increase when foraging becomes more difficult (i.e. the ratio α/β increases; see fig. 3.8b). This prediction agrees with data from empirical studies on humans (Horne and Minard 1985; Horne 1988), insects (Tobler 1983), and birds (Shaffery et al. 1985). In their bird study, Shaffery et al. (1985) fed experimental pairs of herring gull, *Larus argentatus,* a daily ration of 500 g of meat; control pairs continued foraging naturally. Therefore, the experimental pairs had an easy foraging task, and the control birds had a difficult task. Gulls that foraged naturally spent 40% more time sleeping compared to the fed subjects. In other words, increased task difficulty was associated

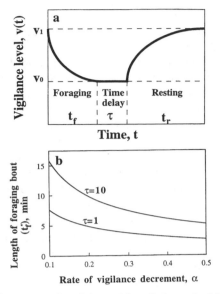

Figure 3.9. (*a*) Graphic representation of the model predicting optimal length of a foraging bout, t_f^*, where v_1 and v_0 are the levels of vigilance at the beginning and end of a foraging bout, respectively. (*b*) Optimal length of a foraging bout, t_f^*, as a function of the rate of vigilance decrement (α) for two values of the costs of switching (τ) between foraging and rest, and with $\beta = 0.012$/min. Given a certain β, the length of a foraging bout decreases (1) for more difficult activities, which have higher rates of vigilance decrement (α); and (2) for lower costs of switching (τ) (from Dukas and Clark 1995b).

with a decrease in overall activity time and an increase in rest. These results may indicate effects of vigilance decrement on activity patterns, but further experiments must control for a few alternative explanations, such as the quality and quantity of food eaten and differences in energy expenditure.

In the second model, (Dukas and Clark 1995b), the assumption that the forager alternates rapidly between short periods of activity and rest and thus maintains an equilibrium level of vigilance, $\bar{v}(\theta)$, was relaxed. The forager now initiates a period of foraging activity with a vigilance level v_1, which decreases over time until reaching v_0 by the end of the foraging period. After a time delay, τ (representing the cost of switching from foraging to resting), recuperation of vigilance begins (fig. 3.9*a*). Under this condition, the forager's daily activity comprises a series of bouts of foraging and rest. The optimal length of a foraging bout is a function of the cost of switching (τ) and of the rates of vigilance decrement (α) and recovery (β). Given a certain β, the length of a foraging bout decreases (1) for more difficult activities, which have higher rates of vigilance decrement; and (2) for lower costs of switching (fig. 3.9*b*).

Both predictions are intuitively appealing, but I know of no empirical data that addresses these issues. Of course, several other factors, such as food availability, digestive capacity, predation risk, and energy conservation, influence temporal patterns of activity (e.g. Wolf and Hainsworth 1977; Daan 1981; Diamond et al. 1986; Elgar et al. 1988; Horne 1988; Lima and Dill 1990). Nonetheless, vigilance decrement and recovery are additional important elements that deserve further experimental examination.

Although little is known about ecological aspects of rest and sleep (Horne 1988), experiments by Lendrem (1983, 1984) suggest that the quality of rest affects rate of vigilance recovery. Sleep in many birds involves a period of eye closure interrupted by "peeks" of eye opening. Peeking allows a bird to scan the sleeping site for predators. Lendrem monitored rates and durations of peeking in two related studies involving field observations of mallards (*Anas platyrhynchos*) and experimental manipulations of Barbary doves (*Streptopelia risoria*). Several factors associated with elevated predation risk increased peeking rates. For instance, decreasing doves' group sizes from six to one caused a twofold increase in individual peeking. While the advantage of peeking is obvious, the fact that it is always kept to some optimal minimum implies that it is costly. That cost of peeking is most likely a reduction in rate of vigilance recuperation; this suggestion can readily be evaluated empirically.

3.5 MEMORY

Animals rely heavily on neuronal representations of past and present experiences. It is tempting to assume that memory capacity is limited, but the surprising conclusion will be that no such general constraint is well established. First, I must clarify some confusing vocabulary. Both neurobiologists and cognitive psychologists usually distinguish between short-term and long-term memory. But many contemporary discussions of memory replace the term "short-term memory" with "working memory" (e.g. Baddeley 1986; Anderson 1990). Because working memory seems a more meaningful description of actual information processing, I will use this term as well. Note that although memory and attention are not independent entities, one can typically distinguish between the two. Memory is the capacity to store and retrieve information from the near or distant past. Attention is the capacity to process information simultaneously and the ability to process selectively relevant information.

Working memory typically comprises newly perceived sensory information and information activated from long-term memory. This combined knowledge in working memory is used to determine behavior. The newly acquired information in working memory may be incorporated into long-term memory. Long-term memory contains many items, but only a small number of representations may be activated and held in working memory at once. The distinction between working and long-term memory is strongly supported by data from research on amnesia in humans. Some amnestic patients can remember a short list of items for several minutes if they keep rehearsing; the portion of their working memory that holds newly acquired information is intact. These patients, however, cannot recall information from the more distant past and usually forget the events of daily life almost as fast as they occur (Squire 1986; Squire et al. 1993).

3.5.1 Working Memory

Detailed research on the capacity of working memory has been conducted mostly on humans (Wickens 1984; Baddeley 1986). Nevertheless, working memory is most likely essential for the behavior of other species (Spear and Riccio 1994; Anderson 1995, 182). For example, a bee foraging for pollen and nectar continuously updates her spatial position in relation to her nest (Dyer, this vol. chap. 5). The bee also keeps track of her orientation to a flower relative to her arrival direction, so she can depart in an appropriate direction to minimize revisitation of flowers (Pyke and Cartar 1992). While sampling other flowers, the bee evaluates average quality and quantity of reward for each species sampled (Bateson and Kacelnik, this vol. chap. 8). She might also include in her evaluation estimates of predation risk for each alternative (Ydenberg, this vol. chap. 9). All of this information must readily be available to the bee in working memory because that knowledge determines her present behavior.

Extensive research on constraints in working memory was stimulated by Miller (1956). Miller and his followers argued that working memory can hold only several items, and that this limited capacity strongly confines information processing and problem solving. It is easy to demonstrate that working memory has *some* limitation. For example, most people find it impossible to repeat accurately the following navigational information, which may be issued from the ground control to a pilot: "Change heading to 179 and speed to 240 knots when you reach latitude 47°21′, longitude 15°30′" (Wickens 1984). Another example, which can be readily tried with a friend, is a typical working-memory span task. Two people can read each other lists of random numbers (0–9) and try to repeat the list back immediately after hearing it. Different lengths of lists, from five to 10 numbers, can be used. Most people can repeat back a list of five numbers accurately but not a 10-number list because most people's working memory span for *random items* is limited to only about seven items (Miller 1956; Anderson 1990).

Clearly, there is one aspect of working memory that is limited; this can be shown in amnestic people who lack long-term memory and with span tasks, in which random or unfamiliar items must be held in working memory. Working memory, however, usually interacts with long-term memory, and research on trained human subjects demonstrates this interaction can effectively eliminate the apparent limitation of working-memory capacity (Ericsson and Kintsch 1995). For instance, I mentioned that *most* people can repeat accurately lists of only less than seven numbers, but extensive training can dramatically increase performance. Ericsson et al. (1980; see also Ericsson and Chase 1982) trained an undergraduate student with average memory abilities and intelligence on a task of working-memory span for about 1 h a day from 3 to 5 days a week for more than 1½ years. The student was read a list of random numbers at the rate of 1 digit/sec; immediately afterwards, the student had to repeat back the number list. During the course of 20 months of practice, the length of the repeated list steadily increased from the typical seven numbers to almost 80 numbers (fig. 3.10). On the basis of the student's descriptions of his techniques and further tests, Ericsson et al. (1980) concluded the student effectively used familiar information in long-term memory to bypass the limit of working memory.

Ericsson and Kintsch (1995) recently elaborated on the previous findings of Ericsson and his colleagues. First, Ericsson and Kintsch reviewed a large body of literature concerning various types of expertise such as reading comprehension, chess, and medical analysis. They showed that, in all cases, experts use an effective working memory, or "long-term working memory" with a capacity much higher than seven items. Second, they suggested that acquired memory skills accompany the development of expertise. Such skills allow the expert to store end-products of information processing in long-term memory but still keep these products accessible to working memory through retrieval cues. In short, common convention that working memory has a limited capacity is not incorrect, but humans readily can acquire skills to circumvent the capacity limitation of working memory (Ericsson and Kintsch 1995). In other words, effective working memory may not be limited!

Some laboratory experiments with monkeys and pigeons suggest these animals also possess a working memory with limited capacity (reviewed by Anderson 1995, 182). Nonetheless, such laboratory studies cannot indicate if working memory limits information processing and decision making in animals in their natural settings. Most likely, various species, like trained humans, possess long-term working memory (Ericsson and Kintsch 1995), which allows the management of large amounts of information. In other words, animals may possess domain-specific enhanced memory capacities (Sherry, this vol. chap. 7); such capacities could have evolved under certain ecological conditions, or they may develop given a subject's specific experience, as in the case of the human expert. Further empirical research is required to establish whether and how the limited capacity of working memory constrains information processing and decision making.

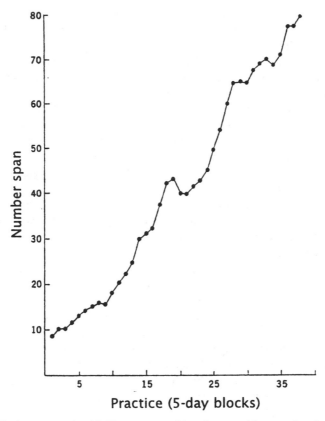

Figure 3.10. Average number (digit) span repeated by a human subject as a function of practice. Number span is defined as the length of the sequence that is correct 50% of the time. Each day represents about 1 h of practice, so the total 38 blocks represent about 190 h of practice (from Ericsson et al. 1980).

3.5.2 Long-Term Memory

Long-term memory apparently has no boundaries. First, an accurate memory may be maintained for a very long time, and second, there is no known limit to the amount of information that can be kept in long-term memory (e.g. Anderson 1990, 1995). Older people may vividly recall events that happened half a century ago, and people that learn more know more. In other words, learning allows one to increase the amount of information in memory rather than merely replace one item with another; replacement would occur if memory capacity was limited. As usual, more is known about human memory than other species' memories. Nonetheless, extensive research on the use of spatial memory by seed-caching birds supports the assertions regarding long-term memory. Several well-studied species can remember the location of numerous items and recall this information after extensive time periods (e.g. Balda and Kamil 1992; Sherry, this vol. chap. 7).

Although long-term memory has no known limits, it probably incurs a cost. Such cost can explain, for example, the positive correlation between dependency on cached food and spatial memory capacity in several groups of birds (Balda and Kamil 1989; Krebs et al. 1990; Sherry, this vol. chap. 7). In other words, if memory was cost-free, then all species would have a perfect spatial memory. This is not the case, and the memory capacity of a species probably represents some trade-off between benefits and costs. Currently, little is known about the cost of memory. However, research on genes suggests what factors may be involved.

First, encoding, maintaining, and transferring genetic information results in errors. Fewer errors can be made, but reducing error requires a substantial increase in expenditure and a lower speed of operation. Hence, some optimal level of accuracy is determined by the cost-benefit trade-off (Kirkwood et al. 1986). Second, because errors do occur, essential information must be redundant. An obvious example of redundancy is the double-strand structure of DNA. Single-strand damage, which is common, can readily be corrected with use of the redundant information in the complementary strand (e.g. Bernstein and Bernstein 1991). Theoretical research and empirical studies on humans with extensive brain damage suggest that, in brain functioning, there is substantial redundancy, which enables long-term reliability and insurance against errors with potentially fatal consequences (Glassman 1987; Bernstein and Bernstein 1991). How such redundancy is coordinated to insure maintenance of crucial neuronal representations if individual cells or a whole neural network fail is unknown. The exact cost of such redundancy is also not known. In short, there is a strong biological foundation for the notion that maintenance of accurate memories requires substantial resource expenditure. The mechanisms and types of resources involved have not been examined empirically. Better understanding of such mechanisms, however, could explain many aspects of memory, including the ecologically important phenomenon of interference, which I now will discuss.

3.5.2.1 Interference

Interference implies that learning or use of one item of information interferes with the later recall of another item. I use a functional definition of interference; this definition is more general than the one typically used by psychologists (Wickens 1984; Anderson 1995). Most importantly, my definition includes interference that results from the use of known information, not just new information; this is because I am interested in the effects of interference on real-life activity, which often involves switching between known tasks. For that reason, I also make no distinction between two proximate mechanisms of interference, retroactive interference (new information interferes with old) and proactive interference (old information interferes with new) (Spear and Riccio 1994). The magnitude of interference can determine if it is worthwhile to alternate between two or more tasks—such as foraging for two distinct food types. If it is worthwhile (or necessary) to alternate, the cost of switching caused by interference may also determine the frequency of rotation between tasks (see section 3.4.2 and fig. 3.9*b*).

The most convincing evidence for interference in animals other than humans comes from Stanton's (1983) observations of free-flying female *Collias* butterflies in Rocky Mountain meadows. There have been numerous laboratory studies on interference in animals (Spear and Riccio 1994; Anderson 1995; see the following), but Stanton's study is most intriguing because it was done under natural settings. The butterflies alternated between bouts of foraging for flower nectar and egg laying on leaves of legume species, which are the only suitable host. Stanton found that a female's probability of making a mistaken landing on nonlegume plants was 12% higher immediately after a foraging run. The best explanation for this observation is that foraging interferes with the memory of suitable hosts. Laverty (1994) also observed bumblebees (*Bombus fervidus*) that made independent decisions about switching between the relatively complex flowers of *Aconitum napellus* and *Impatiens capensis* in natural settings. The bumblebees showed an approximate 20% decrease in handling time for the first flower after switching.

Laboratory experiments on interference in butterflies and bees are inconclusive. Lewis (1986) and Woodward and Laverty (1992) used similar experimental paradigms to test for interference. Their subjects first learned how to handle flowers of one species; then subjects learned how to forage on a second species. Afterwards, subjects' performances on the first species were tested. In both studies, performance decreased immediately after the single switching event. In nature, however, subjects may often alternate between tasks. Thus, it is more relevant to evaluate performance after more than one switching event. In one of my bumblebee experiments (Dukas 1995), I allowed subjects to alternate six times between two foraging tasks. After the first switch, performance decreased initially, as in the experiments of Lewis (1986) and Woodward and Laverty (1992). However, performance of bees did not decline during subsequent switches (fig. 3.11). This finding suggests that, at least below some threshold of difficulty and with sufficient practice, subjects can alternate between tasks without experiencing the negative effects of interference. A crucial experiment, still required for a more complete evaluation of interfer-

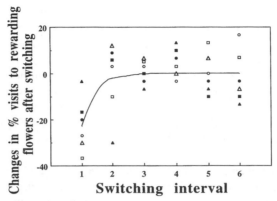

Figure 3.11. Changes in percentage of visits by bees to rewarding flowers during the first trial after each of six switching intervals. Rewarding flowers were yellow (*open symbols*) or pink (*filled symbols*) in one set and white in the other set. Each shape represents an individual bee. For example, a bee visited the following: first, an array of rewarding yellow flowers and non-rewarding pink flowers; second, an array of rewarding white flowers and nonrewarding green flowers; third, the original array of rewarding yellow and nonrewarding pink flowers (switching interval #1); fourth, the array of white and green flowers (switching interval #2); and so on. Interference was significant during the first switching interval ($P < 0.05$) but not during any of the subsequent intervals ($P > 0.1$) (from Dukas 1995).

ence, must examine the effects of the degree of task difficulty and prior switching experience.

In section 3.3, I argued that limited attention explains why animals typically do not conduct more than a single difficult task *simultaneously*. Subjects do not usually alternate rapidly between tasks probably because there are costs for switching. Memory interference is one of the most likely cognitive costs. Empirical results are still inconclusive, but it is relevant to ask how switching costs may affect performance. Effects of attention and memory may be separated experimentally. For example, one could compare a subject's capacity to alternate between searching for two target types in a memory experiment in which only one type is presented at a time. In an attention experiment, one could compare a subject's ability to search simultaneously for two target types.

Depending on the foraging task, the cost of switching may affect prey detection, prey handling, or both. Interference has the cost of reducing initial probability of detecting prey. In addition, when the probability of prey detection is reduced, the forager has less reliable information about the subsequent probability of finding prey at its present location. Therefore, interference after switching may cause the forager to spend more time searching at a spot where prey is not present (Dukas and Clark 1995a). The effects of interference on prey handling consist of the reduced probability of capturing detected prey and longer prey ingestion time (Croy and Hughes 1991). In our theoretical analysis (Dukas and Clark 1995a), we found that an initial 40% reduction in performance was sufficient to make specializing on a single prey optimal under a wide range of parameter values. For lower switching costs, the relative payoffs of the 'switch' versus 'specialize' strategies was more sensitive to parameters such as relative cost of forager movement, rate of relearning, and relative prey densities. We concluded that if the cost of switching is sufficiently high and a forager estimates that one prey type is more common, the forager should search for this prey only.

In summary, it is commonly agreed that reactivating memory of one task after time is devoted to another task results in reduced initial performance because of some kind of interference. Neither the cognitive mechanisms underlying interference nor the exact magnitude of interference are known. Nevertheless, memory interference is probably a crucial factor that determines the cost of switching between activities. Thus, memory interference influences ecological factors such as diet and temporal regulation of animal activity.

3.6 CONCLUSIONS AND FUTURE DIRECTIONS

Like any biological or artificial system, the brain has numerous constraints that limit its capacity to process information. I suggested that animals (1) can process only a limited amount of information at one time; (2) cannot sustain effective information processing for extended periods without rest; and (3) may be unable to reactivate memories of past experience. Ecological analysis of these constraints reveals they strongly affect many types of animal activity. The study of cognitive constraints helps us understand brain functioning and allows us to better understand various ecological interactions within and between species. Although issues such as diet and temporal patterns of behavior have concerned behavioral and evolutionary ecologists, the relationship of such issues to cognition has been largely neglected until recently.

Because of my cognitive limitations, I concentrated only on familiar central constraints from my research. For example, I did not discuss limitations of computation and representation although they are probably crucial for determining behavior (see Dyer, this vol. chap. 5; Bateson and Kacelnik, this vol. chap. 8). In any event, both empirical data and theory are highly suggestive. Nevertheless, a critical evaluation of the function of cognitive constraints in evolutionary ecology is lacking. I noted where additional experiments are needed to establish further the function of cognitive constraints in animal behavioral ecology (see also Dukas and Ellner 1993; Dukas and Clark 1995a, 1995b).

Our understanding of cognitive constraints is still superficial! Cognitive constraints have been inferred mostly from observed behavior. These observations and my ecological analyses are only a beginning. The neuronal mechanisms that underlie observed cognitive constraints remain to be understood. What constraints, costs, and tradeoffs at the level of (1) a single neuron; (2) a neural network; (3) an assembly of networks; and (4) the whole brain determine various cognitive capacities? We cannot yet answer this question despite the rapid progress in various disciplines devoted to brain research. Nevertheless, there is an enormous body of knowledge on neuronal mechanisms, and numerous modeling and experimental techniques are available for the study of neural networks (e.g. Gardner 1993; Wilson and McNaughton 1993; Posner and Dehaene 1994; Desimone and Duncan 1995; Kandel and Abel 1995; Singer and Gray 1995; Whittington et al. 1995). Information is probably sufficient for initiating evolutionary and ecological analyses of costs and tradeoffs in cognitive performance.

Ecologically oriented investigations could aid understanding of attention, sustained vigilance, and memory. *First,* the focus of spatial attention for the recognition of complex patterns is only about 1% of the information capacity of the optic nerve (Van Essen et al. 1992). What determines this limited capacity? The answer that information processing is computationally demanding (Van Essen et al. 1992) is simply unsatisfactory because unattended information could crucially alter a subject's fitness. Therefore, an acceptable answer must consider the fitness costs and benefits of increased attentional capacity, given fundamental limitations in the neuron and network. *Second,* the neuronal mechanisms that underlie vigilance decrement are unclear. Of course, sustaining a maximum degree of vigilance could enhance behavior and fitness. Apparently, however, there are fitness costs of sustained vigilance or fundamental neuronal constraints that result in vigilance decrement. Most likely, constraints and costs determine the ubiquitous phenomenon of vigilance decrement, which is probably closely associated to mechanisms of rest (see Horne 1988; Dukas and Clark 1995b).

Third, it is unclear what determines the limited capacity of working memory, how experience reduces that limitation in humans, and whether and how other species are constrained by a limited working memory. *Fourth,* statements regarding the unlimited capacity of long-term memory are somewhat puzzling. On one hand, a relatively small population of neurons can hold a very large number of distributed representations; a mechanism that functionally connects neurons representing one item (Singer 1994; Morton and Chiel 1994; Douglas and Martin 1995; Singer and Gray 1995; Whittington et al. 1995) and a system that retrieves appropriate representations on demand are all that are needed. On the other hand, the correlation between hippocampus size and the spatial-memory capacity in seed-storing birds (Sherry, this vol.

chap. 7) suggests that storage of more or longer-lasting memories requires more neurons. We must know more about the costs and constraints associated with storing more information within a single network.

In summary, I set two goals. The first and easier of the two was to show how current knowledge of cognitive limitations can be combined with ecological analyses of behavior. I attempted to demonstrate that such an integrative approach can provide new testable predictions regarding commonly observed, but poorly explained behavioral phenomena. The second goal was more challenging. Cognitive functions are by-products of a lengthy evolutionary process that is subject to numerous constraints. Therefore, an interdisciplinary research program that couples ecological thinking with detailed knowledge regarding neuronal action is required to understand cognitive design. However, there is great inherent complexity in the operation of individual neural networks and the whole brain. Therefore, one should not expect quick and easy answers to questions about the neuronal mechanisms that underlie cognitive constraints.

The complexity of the evolutionary ecology of cognitive mechanisms is obviously intimidating, but one should not be deterred. First, similar difficulties are faced by students of morphology, physiology, and genetics. Yet, research on complex systems in these fields is progressing rapidly (e.g. Cheverud 1982; Oster et al. 1988; Wake and Roth 1988; Peterson et al. 1990; Loomis and Sternberg 1995). Second, several simple models of the nervous system are available (Srinivasan et al. 1993; Ferrus and Canal 1994; Hall 1994; Thomas 1994; Wu et al. 1994; Grillner et al. 1995); these convenient animal models can be adopted for examining selective aspects, costs, and trade-offs associated with cognitive design.

3.7 SUMMARY

Various constraints limit the brain's capacity to process information. I integrated neurobiological and psychological knowledge of such constraints with theory from behavioral ecology to (1) evaluate how cognitive constraints affect ecologically relevant behaviors of animals in their natural settings; and (2) examine how evolutionary and ecological theory can aid the investigation of the fundamental neuronal constraints, costs, and trade-offs that shape cognitive systems. First, the brain has a limited capacity to process information simultaneously. Increasing the amount of information a subject attends to may decrease the quality of information processing and associated performance. Visual attention may be focused on a small spatial portion of the visual field or along various dimensions such as color and shape. If searching for cryptic prey, a narrower spatial focus of attention is optimal. A narrower focus of attention may be associated with a reduced search rate; this hypothesis is supported by experimental evidence. In addition, attending to a single visual attribute while searching for cryptic prey may be beneficial. As a result, only one of several available prey types may be searched for at one time; this phenomenon has been widely observed in numerous species. Limited attention also restricts the capacity of animals to monitor simultaneously predator behavior and search for food.

Second, animals cannot sustain a high quality of information processing for extended periods. The consequent decrement in vigilance reduces the performance of various behavioral tasks and may affect temporal patterns of activity and rest. Third, working memory has a very limited capacity; however, the constraint on animal behavior is unclear. It is possible that animals, like trained humans, can overcome this constraint. Fourth, it is commonly believed that long-term memory is unlimited, But it is likely that maintaining memories is costly. One probable cost results from interference between various items in memory. Interference and limited attention most likely cause animals to concentrate on a single activity at a time. Current theory and evidence reveal the potential importance of cognitive constraints on animal behavior in the field; however, current knowledge must be supplemented with additional rigorous experimental data. Moreover, further research is required to establish the neuronal foundations of cognitive constraints.

ACKNOWLEDGMENTS

I thank P. Bednekoff, T. Laverty, L. Gass, D. Sherry, and W. Roberts for comments on the manuscript. C. Clark and S. Ellner collaborated with me on some of the research reviewed. During writing, I was supported by the NSERC grant #83990 to C. Clark and provided with a stimulating working environment by the Centre for Biodiversity, University of British Columbia.

LITERATURE CITED

Anderson, J. R. 1990. *Cognitive Psychology and Its Implications.* New York: Freeman.
———. 1995. *Learning and Memory.* New York: Wiley.
Anderson, M. 1981. On optimal predator search. *Theoretical Population Biology* 19: 58–86.
Arnold, S. J. 1994a. Constraints on phenotypic evolution. In L. A. Real, ed., *Behavioral Mechanisms in Evolutionary Ecology,* 258–278. Chicago: University of Chicago Press.
———. 1994b. Multivariate inheritance and evolution: A review of concepts. In R. B. Boake, ed., *Quantitative Genetic Studies of Behavioral Evolution,* 17–48. Chicago: University of Chicago Press.
Aston-Jones, G., S. L. Foote, and F. E. Bloom. 1984. Anatomy and physiology of locus coeruleus neurons: Functional implications. In M. G. Ziegler and C. R. Lake, eds., *Norepinephrine,* 92–116. Baltimore: Williams and Wilkins.

Baddeley, A. 1986. *Working Memory.* Oxford: Oxford University Press.

Balda, R. P., and A. C. Kamil. 1989. A comparative study of cache recovery by three corvid species. *Animal Behaviour* 38:486–495.

Balda, R. P., and A. C. Kamil. 1992. Long-term spatial memory in Clark's nutcracker *Nucifraga columbiana. Animal Behaviour* 44:761–769.

Bernstein, C., and H. Bernstein. 1991. *Aging, Sex, and DNA Repair.* San Diego: Academic.

Blough, D. 1993. Reaction time drifts identify objects of attention in pigeon visual search. *Journal of Experimental Psychology: Animal Behavior Processes* 19:107–120.

Blough, P. 1989. Attentional priming and visual search in pigeons. *Journal of Experimental Psychology: Animal Behavior Processes* 15:358–365.

———. Selective attention and search images in pigeons. *Journal of Experimental Psychology: Animal Behavior Processes* 17:292–298.

Cherverud, J. M. 1982. Phenotypic, genetic, and environmental morphological integration in the cranium. *Evolution* 36:499–516.

Cheverton, J., A. Kacelnik, and J. R. Krebs. 1985. Optimal foraging: Constraints and currencies. In B. Holldobler and M. Lindauer, eds., *Experimental Behavioral Ecology,* 109–126. Stuttgart: Fischer-Verlag.

Corbetta, M., S. Miezin, G. L. Dobmeyer, G. L. Shulman, and S. E. Petersen. 1990. Attentional modulation of neural processing of shape, color, and velocity in humans. *Science* 248:1556–1559.

Crick, F. 1994. *The Astonishing Hypothesis: The Scientific Search for the Soul.* New York: Macmillan.

Croy, M. I., and R. N. Hughes. 1991. The role of learning and memory in the feeding behaviour of the fifteen-spined stickleback. *Animal Behaviour* 41:149–159.

Daan, S. 1981. Adaptive daily strategies in behavior. In J. Aschoff, ed., *Handbook of Behavioral Neurobiology,* 275–298. New York: Plenum.

David, P. A. 1985. Clio and the economics of QWERTY. *American Economical Review (proceedings)* 75:332–337.

Davis, D. R., and R. Parasuraman. 1982. *The Psychology of Vigilance.* New York: Academic.

Dawkins, M. 1971a. Perceptual changes in chicks: Another look at the "search image" concept. *Animal Behaviour* 19:566–574.

———. 1971b. Shifts of 'attention' in chicks during feeding. *Animal Behaviour* 19:575–582.

Desimone, R., and J. Duncan. 1995. Neural mechanisms of selective attention. *Annual Review of Neuroscience* 18:193–222.

Diamond, J. M., W. H. Karasov, D. Phan, and F. L. Carpenter. 1986. Digestive physiology as a determinant of foraging bout frequency in hummingbirds. *Nature* 320:62–63.

Douglas R. J., and K. A. C. Martin. 1995. Vibrations in the memory. *Nature* 373:563–564.

Drevets, W. C., B. Harold, T. O. Videen, A. Z. Snyder, J. R. Simpson, and M. E. Raichie. 1995. Blood flow changes in human somatosensory cortex during anticipated stimulation. *Nature* 373:249–252.

122   Reuven Dukas

Dukas, R. 1995. Transfer and interference in bumblebee learning. *Animal Behaviour* 49:1481–1490.

Dukas, R., and C. W. Clark. 1995a. Searching for cryptic prey: A dynamic model. *Ecology* 76:1320–1326.

———. 1995b. Sustained vigilance and animal performance. *Animal Behaviour* 49: 1259–1267.

Dukas, R., and S. Ellner. 1993. Information processing and prey detection. *Ecology* 74:1337–1346.

Dukas, R., and N. M. Waser. 1994. Categorization of food types enhances foraging performance of bumblebees. *Animal Behaviour* 48:1001–1006.

Duncan, J., R. Ward, and K. Shapiro. 1994. Direct measurement of attentional dwell time in human vision. *Nature* 369:313–315.

Elgar, M. A., M. D. Pagel, and P. H. Harvey. 1988. Sleep in mammals. *Animal Behaviour* 36:1407–1419.

Endler, J. A. 1984. Progressive background matching in moths and a quantitative measure of crypsis. *Biological Journal of the Linnean Society of London* 22:187–231.

———. 1988. Frequency-dependent predation, crypsis, and aposomatic coloration. *Philosophical Transactions of the Royal Society of London,* ser. B 319:505–523.

———. 1991. Interactions between predators and prey. In J. R. Krebs and N. B. Davies, eds., *Behavioural Ecology,* 169–196. London: Blackwell Scientific.

———. 1993. The color of light in forests and its implications. *Ecological Monographs* 63:1–27.

Ericsson, K. A., and W. G. Chase. 1982. Exceptional memory. *American Scientist* 70: 607–615.

Ericsson, K. A., and W. Kintsch. 1995. Long-term working memory. *Psychological Review* 102:211–245.

Ericsson, K. A. W. G. Chase, and S. Faloon. 1980. Acquisition of a memory skill. *Science* 208:1181–1182.

Eriksen, C. W., and Y. Y. Yeh. 1985. Allocation of attention in the visual field. *Journal of Experimental Psychology: Human Perception and Performance* 11:583–597.

Ferrus, A., and I. Canal. 1994. The behaving brain of a fly. *Trends in Neuroscience* 17:479–485.

Gardner, A., ed. 1993. *The Neurobiology of Neural Networks.* Cambridge: MIT Press.

Gendron, R. P. 1986. Searching for cryptic prey: Evidence for optimal search rates and the formation of search images in quail. *Animal Behaviour* 34:898–912.

Gendron, R. P., and J. E. R. Staddon. 1983. Searching for cryptic prey: The effects of search rate. *American Naturalist* 121:172–186.

Getty, T., and H. R. Pulliam. 1991. Random prey detection with pause-travel search. *American Naturalist* 138:1459–1477.

Getty, T., and H. R. Pulliam. 1993. Search and prey detection by foraging sparrows. *Ecology* 74:734–742.

Gilbert, E. L. 1975. Ecological consequences of a coevolved mutualism between butterflies and plants. In E. Gilbert and P. H. Raven, eds., *Coevolution of Animals and Plants,* 210–240. Austin: University of Texas Press.

Gilliam, J. F. 1990. Hunting by the hunted: Optimal prey detection by foragers under predation hazard. In R. N. Hughes, ed., *Behavioural Mechanisms of Food Selection,* 797–820. Berlin: Springer-Verlag.

Glassman, R. B. 1987. A hypothesis about redundancy and reliability in the brains of higher species: Analogies with genes, internal organs, and engineering systems. *Neuroscience and Behavioral Reviews* 11:275–285.

Goldsmith, T. H. 1990. Optimization, constraint, and history in the evolution of eyes. *Quarterly Review of Biology* 65:281–322.

Gould, S. J., and R. C. Lewontin. 1979. The spandrels of San Marco and the Panglossian paradigm: A critique of the adaptationist programme. *Proceedings of the Royal Society of London,* ser. B 205:581–598.

Grillner, S., T. Deliagina, O. Ekeberg, A. El Manira, R. H. Hill, A. Lansner, G. N. Orlovsky, and P. Wallen. 1995. Neural networks that co-ordinate locomotion and body orientation in lamprey. *Trends in Neuroscience* 18:270–279.

Guilford, T., and M. S. Dawkins. 1987. Search images not proven: A reappraisal of recent evidence. *Animal Behaviour* 35:1838–1845.

Hall, J. C. 1994. The mating of a fly. *Science* 264:1702–1714.

Heinze, H. J., G. R. Mangun, W. Burchert, H. Hinrichs, M. Scholz, T. F. Munte, A. Gos, M. Scherg, S. Johannes, H. Hundeshagen, M. S. Gazzaniga, and S. A. Hillyard. 1994. Combined spatial and temporal imaging of brain activity during visual selective attention in humans. *Nature* 372:543–546.

Herrera, C. M. 1992. Historical effects and sorting processes as explanations for contemporary ecological patterns: Character syndromes in Mediterranean woody plants. *American Naturalist* 140:421–446.

Horne, J. 1988. *Why We Sleep.* Oxford: Oxford University Press.

Horne, J., and A. Minard. 1985. Sleep and sleepiness following a behaviourally "active" day. *Ergonomics* 28:567–575.

Hughes, N. F., and L. M. Dill. 1990. Position choice by drift-feeding salmonids: A model and test for arctic greyling (*Thalmallus arcticus*) in subarctic mountain streams, interior Alaska. *Canadian Journal of Fish and Aquatic Sciences* 47:2039–2048.

Hughes, R. N., ed. 1990. *Behavioral Mechanisms of Food Selection.* Berlin: Springer-Verlag.

Jacobs, G. H. 1981. *Comparative Color Vision.* New York: Academic.

Kamil, A. C. 1994. A synthetic approach to the study of animal intelligence. In L. A. Real, ed., *Behavioral Mechanisms in Evolutionary Ecology,* 11–45. Chicago: University of Chicago Press.

Kandel, E., and T. Abel. 1995. Neuropeptides, adenyl cyclase, and memory storage. *Science* 268:825–826.

Kirkwood, T. B. L., and M. R. Rose. 1991. Evolution of senescence: Late survival sacrificed for reproduction. *Philosophical Transactions of the Royal Society of London,* ser. B 332:15–24.

Kirkwood, T. B. L., R. F. Rosenberger, and D. J. Galas, eds. 1986. *Accuracy in Molecular Processes.* London: Chapman and Hall.

Klosterhalfen, S., W. Fischer, and M. E. Bitterman. 1978. Modification of attention in honey bees. *Science* 201:1241–1243.

Krause, J., and J. G. J. Godin. 1996. Influence of prey foraging posture on flight behav-

124   Reuven Dukas

ior and predation risk: Predators take advantage of unwary prey. *Behavioral Ecology and Sociobiology* 7:264–271.

Krebs, J. R., and N. B. Davies, eds. 1978. *Behavioural Ecology.* Oxford: Blackwell Scientific.

Krebs, J. R., S. D. Healy, and S. J. Shettleworth. 1990. Spatial memory of Paridae: Comparison of a storing and non-storing species, the coal tit, *Parus ater,* and the great tit, *P. major. Animal Behaviour* 36:733–740.

LaBerge, D. 1983. Spatial extent of attention to letters and words. *Journal of Experimental Psychology: Human Perception and Performance* 9:371–379.

Land, M. F. 1981. Optics and vision in invertebrates. In H. Autrum, ed., *Handbook of Sensory Physiology, Vol. VII/6B, Invertebrate Visual Centers,* 472–592. Berlin: Springer-Verlag.

Langley, C. M. 1996. Search images: Selective attention to specific visual features of prey. *Journal of Experimental Psychology: Animal Behavior Processes* 22:152–163.

Langley, C. M., D. A. Riley, A. B. Bond, and N. Goel. 1996. Visual search for natural grains in pigeons (*Columba livia*): Search images and selective attention. *Journal of Experimental Psychology: Animal Behavior Processes* 22:139–151.

Laverty, T. M. 1994. Costs to foraging bumblebees of switching plant species. *Canadian Journal of Zoology* 72:43–47.

Lawrence, E. S., and J. A. Allen. 1983. On the term "search image." *Oikos* 40:313–314.

Lendrem, D. W. 1983. Sleeping and vigilance in birds. I. Field observations of the mallard (*Anas platyrhynchos*). *Animal Behaviour* 31:532–538.

———. Sleeping and vigilance in birds. II. An experimental study of the Barbary doves (*Streptopelia risoria*). *Animal Behaviour* 32:243–248.

Lewis, C. A. 1986. Memory constraints and flower choice in *Pieris rapae. Science* 232:863–865.

Lima, S. L., and L. M. Dill. 1990. Behavioral decisions made under the risk of predation: A review and prospectus. *Canadian Journal of Zoology* 68:619–640.

Loomis, W. F., and P. W. Sternberg. 1995. Genetic networks. *Science* 269:649.

Mackworth, J. F. 1969. *Vigilance and Habituation: A Neurophysiological Approach.* Harmondsworth, England: Penguin.

Maunsell, J. H. R. 1995. The brain's visual world: Representation of visual targets in cerebral cortex. *Science* 270:764–769.

Maynard Smith, J. R., R. Burian, S. Kauffman, P. Alberch, J. Campbell, B. Goodwin, R. Lande, D. Raup, and L. Wolpert. 1985. Developmental constraints and evolution. *Quarterly Review of Biology* 60:265–287.

McAdams, H. H., and L. Shapiro. 1995. Circuit simulation of genetic networks. *Science* 269:650–656.

Mckie, R. R., ed. 1977. *Vigilance: Theory, Operational Performance, and Physiological Correlates.* New York: Plenum.

Metcalfe, N. B., F. A. Huntingford, and J. E. Thorpe. 1987a. The influence of predation risk on the feeding motivation and foraging strategy of juvenile Atlantic salmon. *Animal Behaviour* 35:901–911.

———. 1987b. Predation risk impairs diet selection in juvenile salmon. *Animal Behaviour* 35:931–933.

Milinski, M. 1990. Information overload and food selection. In R. N. Hughes, ed., *Behavioral Mechanisms of Food Selection,* 721–737. Berlin: Springer-Verlag.

Milinski, M., and R. Heller. 1978. Influence of a predator on the optimal foraging behaviour of sticklebacks (*Gasterosteus aculeatus* L.). *Nature* 275:642–644.

Miller, G. A. 1956. The magical number seven, plus or minus two: Some limits on our capacity for processing information. *Psychological Review* 63:81–97.

Moran, J., and R. Desimone. 1985. Selective attention gates visual processing in the extrastriate cortex. *Science* 229:782–784.

Morton, D. W., and H. J. Chiel. 1994. Neural architecture for adaptive behavior. *Trends in Neuroscience* 10:413–420.

Nuechterlein, K. H., R. Parasuraman, and Q. Jiang. 1983. Visual sustained attention: Image degradation produces rapid sensitivity decrement over time. *Science* 220:327–329.

O'Brian, W. J., H. I. Browman, and B. I. Evans. 1990. Search strategies of foraging animals. *American Scientist* 78:152–160.

O'Brian, W. J., and J. J. Showalter. 1993. Effects of current velocity and suspended debris on the drift feeding of arctic greylings. *Transactions of the American Fisheries Society* 122:609–615.

Oster, G. F., and P. Alberch. 1982. Evolution and bifurcation of developmental programs. *Evolution* 36:444–459.

Oster, G. F., N. Shubin, J. D. Murray, and P. Alberch. 1988. Evolution and morphogenetic rules: The shape of the vertebrate limb in ontogeny and phylogeny. *Evolution* 42:862–884.

Parasuraman, R. 1979. Memory load and event rate control sensitivity decrements in sustained attention. *Science* 205:924–927.

Parasuraman, R., and M. Mouloua. 1987. Interaction of signal discriminability and task type in vigilance decrement. *Perception* 41:17–22.

Peterson, C. C., K. A. Nagy, and J. Diamond. 1990. Sustained metabolic scope. *Proceedings of the National Academy of Sciences of the United States of America* 87:2324–2328.

Pietrewicz, A., and A. C. Kamil. 1979. Search image formation in the blue jay (*Cyanocitta cristata*). *Science* 204:1332–1333.

Plaisted, K. C., and M. J. Mackintosh. 1995. Visual search for cryptic stimuli in pigeons: Implications for the search image and search rate hypotheses. *Animal Behaviour* 50:1219–1232.

Posner, M. I. 1995. Modulation by instruction. *Nature* 373:198–199.

Posner, M. I., and S. Dehaene. 1994. Attentional networks. *Trends in Neuroscience* 17:75–79.

Posner, M. I. and S. E. Petersen, 1990. The attention system of the human brain. *Annual Review of Psychology* 13:25–42.

Pyke, G. H., and R. V. Cartar. 1992. The flight directionality of bumblebees: Do they remember where they came from? *Oikos* 65:321–327.

Pyke, G. H., R. H. Pulliam, and E. L. Charnov. 1977. Optimal foraging: A selective review of theory and tests. *Quarterly Review of Biology* 52:137–154.

Rausher, M. D. 1978. Search image for leaf shape in a butterfly. *Science* 200:1071–1073.

Real, L. A., ed. 1994. *Behavioral Mechanisms in Evolutionary Ecology.* Chicago: University of Chicago Press.

Reid, P. J., and S. J. Shettleworth. 1992. Detection of cryptic prey: Search image or search rate? *Journal of Experimental Psychology: Animal Behavior Processes* 18: 273–286.

Rose, M. R. 1991. *Evolutionary Biology of Aging.* New York: Oxford University Press.

Ryan, M. J. 1994. Mechanisms underlying sexual selection. In L. A. Real, ed., *Behavioral Mechanisms in Evolutionary Ecology,* 190–215. Chicago: University of Chicago Press.

Shaffery, J. P., N. J. Ball, and C. J. Amlaner. 1985. Manipulating daytime sleep in herring gulls (*Larus argentatus*). *Animal Behaviour* 33:566–572.

Sih, A. 1980. Optimal behavior: Can foragers balance two conflicting demands? *Science* 210:1041–1043.

———. 1992. Prey uncertainty and the balancing of antipredator and feeding needs. *American Naturalist* 139:1052–1069.

Singer, W. 1994. Putative functions of temporal correlations in neocortical processing. In C. Koch and J. L. Davis, eds., *Large-Scale Neuronal Theories of the Brain,* 201–237. Cambridge: MIT Press.

Singer, W., and C. M. Gray. 1995. Visual feature integration and the temporal correlation hypothesis. *Annual Review of Neuroscience* 18:555–586.

Spear, E. N., and D. C. Riccio. 1994. *Memory: Phenomena and Principles.* Boston: Allyn and Bacon.

Spitzer, H., R. Desimone, and J. Moran. 1988. Increased attention enhances both behavioral and neuronal performance. *Science* 240:338–340.

Squire, L. R. 1986. Mechanisms of memory. *Science* 232:1612–1619.

Squire, L. R., B. Knowlton, and G. Musen. 1993. The structure and organization of memory. *Annual Review of Psychology* 44:453–495.

Srinivasan, M. V., S. W. Zhang, and B. Rolfe. 1993. Is pattern vision in insects mediated by 'cortical' processing? *Nature* 362:539–540.

Staddon, J. E. R. 1983. *Adaptive Behavior and Learning.* Cambridge: Cambridge University Press.

Stanton, M. L. 1983. Short-term learning and the searching accuracy of egg-laying butterflies. *Animal Behaviour* 31:33–40.

Stearns, S. 1992. *The Evolution of Life Histories.* Oxford: Oxford University Press.

Stephens, D. W., and J. Krebs. 1986. *Foraging Theory.* Princeton, N.J.: Princeton University Press.

Thomas, J. H. 1994. The mind of a worm. *Science* 264:1698–1699.

Tinbergen, L. 1960. The natural control of insects on pinewoods. I. Factors influencing the intensity of predation by songbirds. *Archives Neerlandaises de Zoologie* 13:265–343.

Tobler, I. 1983. Effects of forced locomotion in the rest-activity cycle of the cockroach. *Behavioral and Brain Sciences* 8:351–360.

Tovee, M. J. 1995. Ultraviolet photoreceptors in the animal kingdom: Their distribution and function. *Trends in Ecology and Evolution* 10:455–460.

Van Essen, D. C., C. H. Anderson, and D. J. Felleman. 1992. Information processing in the primate visual system: An integrated systems perspective. *Nature* 255:419–423.

Wake, D. B., and G. Roth, eds. 1988. *Complex Organismal Functions: Integration and Evolution in Vertebrates.* New York: Wiley.

Warm, J. S., ed. 1984. *Sustained Attention in Human Performance.* New York: Wiley.

Whittington, M. A., R. D. Traub, and J. G. R. Jeffreys. 1995. Synchronized oscillations in interneuron networks driven by metabotropic glutamate receptor activation. *Science* 373:612–615.

Wickens, C. D. 1984. *Engineering Psychology and Human Performance.* Columbus, Ohio: Bell and Howell.

Wilson, M. A., and B. L. McNaughton. 1993. Dynamics of the hippocampal ensemble code for space. *Science* 261:1055–1058.

Wolf, L. L., and F. R. Hainsworth. 1977. Temporal patterning of feeding by hummingbirds. *Animal Behaviour* 25:976–989.

Woodward, G., and T. M. Laverty. 1992. Recall of flower handling skills by bumblebees: A test of Darwin's interference hypothesis. *Animal Behaviour* 44:1045–1051.

Wu, J. Y., L. B. Cohen, and C. X. Falk. 1994. Neuronal activity during different behaviors in *Aplysia:* A distributed organization. *Science* 263:820–822.

CHAPTER FOUR

Evolutionary Ecology of Learning

REUVEN DUKAS

4.1 INTRODUCTION

The neurobiology, psychology, and ecology literature on learning is large and diverse. Clearly, it cannot be reviewed in a single chapter or by a single person. Hence, I will focus on only two major issues. The first is an analysis of learning as a type of phenotypic plasticity, and the second is an evaluation of the interaction between life history and learning. Even with discussion limited to these two topics, I will examine only central ideas and present selected examples from the extensive list available.

Phenotypic plasticity implies a change in the expressed phenotype of a genotype as a function of the environment. This means that individuals with identical genotypes that experience different environments may have considerably different, ecologically important phenotypic traits (Schmalhausen 1949; Stearns 1989; Scheiner 1993). Learning is typically defined as a change in behavior as a result of experience. If behavior is considered the phenotypic trait of interest and experience is considered interaction with the environment, learning can readily be seen as a kind of phenotypic plasticity (see Via 1987; West-Eberhard 1989; Stephens 1991; Moran 1992; Scheiner 1993). Thus, the rich literature on phenotypic plasticity can be used to evaluate evolutionary and ecological aspects of learning. Such analysis constitutes the first part of this chapter. In this part, only general ecological considerations will be highlighted. In the second part of the chapter, I will address specific life-history and behavioral traits that could have had differential effects on the evolution of various learning capacities.

4.2 PHENOTYPIC PLASTICITY

Evolutionary studies usually focus on the association between canalized gene expression and the resulting phenotypes. The effects of environmentally induced variation, or phenotypic plasticity, have been acknowledged and considered, but only recently has phenotypic plasticity been examined closely as an adaptive and evolved trait (Via and Lande 1985; Lively 1986; Stearns 1989). The simplest phenotypic plasticity is exemplified by a phenotype that is determined once during development and remains unchanged (Lively 1986; Moran

1992; Scheiner 1993). Specific examples of irreversible plasticity include alternative morphologies associated with predator defense (fig. 4.1), alternative wing patterns on butterflies (Shapiro 1976; Kingsolver and Wiernasz 1991), and winged and wingless forms in insects (Roff 1986). Note that the magnitude of plasticity in irreversible phenotypes is somewhat limited: An individual has only a single opportunity for a phenotypic decision. This is significantly different from canalized gene expression, and so the initial focus of students of plasticity on this simplest case is well justified. Students of animal behavior and learning, however, may find the very definition of irreversible plasticity a bit odd, given that irreversible plasticity looks rigid *relative* to most behaviors.

For example, Greene (1989) described irreversible plasticity in the geometrid moth, *Nemoria arizonaria* (Grote). Many moth caterpillars have cryptic coloration, which helps reduce detection by predators. There are two generations of *N. arizonaria* per year, and caterpillars of each generation live on visually distinct substrates—oak staminate flowers in the spring and oak leaves in the summer. If there was only a single caterpillar morph determined with canalized gene expression, then either one or both generations would not have been well camouflaged on their respective substrates. In this moth species, however, input from the environment early in development informs a caterpillar about its substrate; that information determines the visual appearance of the caterpillar. The appearance of the spring generation caterpillars mimics that of the flowers on which they feed, and the appearance of the summer generation caterpillars, which feeds on leaves, mimics that of twigs. The caterpillar morph is plastic: Two eggs with identical genotypes placed on two different substrates develop into distinct morphs. Each morph's appearance mimics that of the respective surroundings, and thus the caterpillar's probability of detection by visually oriented predators is reduced.

Phenotypic plasticity can be considered as an adaptation to environmental variation in time and space. Although the type of variation matters, I will concentrate on temporal variation. When the environment varies over time, irreversible plasticity may be favored if there is a high correlation between the state of the environment during trait development and the state of the environment at the time of selection. Temporal variation in the environment must be small relative to lifetime so that the irreversible trait can match the environment for much of the individual's life (Levins 1968; Moran 1992; Scheiner 1993). Two additional key factors that determine the value of irreversible plasticity are, first, the fitness benefits and costs of possessing the right and wrong irreversible traits, respectively. Because of imperfect information or environmental change between the times of trait development and maturity, the wrong irreversible phenotype, *a,* may be expressed in environment B and offer lower fitness than the right phenotype, *b.* Another factor that influences the value of irreversible plasticity is the cost of maintaining a certain

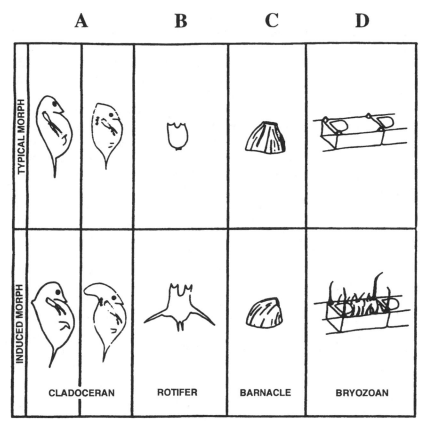

Figure 4.1. Examples of plastic morphologies associated with predator defense in four aquatic invertebrate groups (from Clark and Harvell, 1992). Usually, the typical morph (*top row*) develops in environments with no predation, and the induced morph (*bottom row*) is an irreversible plastic response to predation. The typical morph has a higher fitness than the induced morph in environments without predators because the typical morph invests less in growth, has a shorter maturation time, or feeds more efficiently. Conversely, the induced morph has a higher fitness than the typical morph in environments with predators because the induced morph's predation mortality is lower. (*a*) Cladocerans produce offspring with larger crests (helmets) and neck teeth in response to aquatic cues associated with predation (Havel 1985). (*b*) Rotifers produce spined progeny in response to cues from predatory rotifers (Stemberger and Gilbert 1987). (*c*) During early development, barnacles exposed to a predatory gastropod develop asymmetrically. The induced morph has lower fitness than the typical morph in environments without predatory gastropods; however, the induced morph has higher fitness with predators (Lively 1986). (*d*) Bryozoans rapidly develop spines on edge zooids when exposed to cues from predatory nudibranch (Harvell 1986).

plastic response. This includes, for example, the extra enzymatic machinery required for determining and developing a plastic phenotype (Moran 1992; Scheiner 1993).

Studies on irreversible plasticity are illuminating because the relative simplicity of irreversible plasticity allows theoretical and empirical tractability. Many plastic traits, however, are reversible; this means that an individual can change a certain phenotypic trait frequently in response to some environmental change. An example of reversible plasticity is changes in pupil diameter in response to changes in light intensity (Gomulkiewicz and Kirkpatrick 1992). We do not have a canalized genetically determined pupil size, nor do we have a single chance during development to determine irreversible pupil size for life. Instead, an environmental variable, light intensity, continuously determines pupil diameter within certain boundaries. This reversible plastic trait is clearly adaptive because changes in pupil diameter help maintain clear vision under a variety of light conditions.

Reversible plasticity implies that an individual maintains throughout life the sensory capacity to recognize a certain environmental cue and the mechanical ability to alter some aspect of its phenotype in response to that cue. This phenotypic change is reversible and may last from seconds to seasons (Levins 1968; Scheiner 1993). Unlike irreversible plasticity, reversible plasticity can allow adaptation to temporal changes when the proximate cues during development do not forecast the ultimate state of the environment during an individual's life, or when the environment changes frequently within lifetime. I will focus on reversible behavioral traits; even so, reversible phenotypic plasticity may occur in other kinds of traits such as physiological and morphological traits. Examples of reversible physiological plasticity include secretion of sweat, immune responses, and adaptations to high altitudes (Hochachka and Somero 1984; Clark 1995); a ubiquitous instance of reversible morphological plasticity is the change in muscle size as a result of exercise (Komi 1992).

4.3 LEARNING AS A TYPE OF PHENOTYPIC PLASTICITY

Students of phenotypic plasticity have only considered a narrow range of reversible traits such as the plastic pupil size mentioned (Levins 1968; Gomulkiewicz and Kirkpatrick 1992; Scheiner 1993). Many reversible traits, however, are more complex. Complexity accompanies the gradual transition from simple forms of reversible plasticity to various types of learning. For example, many times reversible plasticity involves interactions among several traits. Even if an individual possesses the capacity to respond to an event every time it is perceived, that individual may decide to react only if a set of conditions are met. For instance, a host-seeking rule may be activated only if a female possesses developed eggs; even if she has eggs, she may postpone laying and remain sheltered if predators are present. In general, a certain reversible response usually interacts with a subject's internal and external environment and with other reversible rules. In addition, a reversible response may be repeated numerous times throughout life, and this repetition allows an individual to alter its subsequent responses on the basis of experience.

Many times the strength of association between a stimulus and an event that affects fitness is less than 100%. Under some conditions, a repeated response to a cue does not produce the expected change because the cue is not associated with the event. In this instance, it is beneficial to stop responding to the cue (i.e. habituate) after a sequence of failures. For instance, a male fruit fly that unsuccessfully courts an already mated female stops courting after about 30 min; he then does not initiate additional courting for a few hours (Hall 1994). Note that such kind of habituation may require a rudimentary memory, which allows a subject to record a series of executions of a behavioral rule as failing to produce the desired change. Habituation is typically considered a simple form of learning.

Compared with habituation, more developed types of learning allow an individual to record and use experience for determining behavior throughout life (see Staddon 1983; Dudai 1989; Alcock 1993). Specifically, learning may be defined as the acquisition of or change in memory that allows a subject to alter its subsequent responses to certain stimuli. Basic types of reversible plasticity only permit a genetically determined association between environmental stimuli and responses. More complex kinds of reversible plasticity, including learning, also allow modification of the plastic response throughout life. Many instances of learning allow an individual to choose the stimulus to respond to or the response to a certain stimulus on the basis of experience. These choices can be updated and altered throughout life. I do not attempt to place various behaviors along the continuum of reversible plasticity, which includes anything from very basic reversible traits to sophisticated types of learning. In fact, probably most reversible responses show some complexity that does not fit the simple case typically discussed in the literature on phenotypic plasticity. Instead, I aim to integrate two relatively distinct bodies of literature that deal with plasticity into a unified discipline.

A common form of learning implies that an individual records its experience of responding to a stimulus. On the basis of that memory, the individual then adjusts or modifies its response to that same stimulus. Instead of having a simple reversible plastic rule that always indicates the same response to a certain stimulus, an individual has a more sophisticated plastic rule that enables a change in response on the basis of experience. An example for such learning is the timing of egg laying by great tits in Switzerland (Nager and van Noordwijk 1995). Breeding success of these great tits is considerably higher if the nestling stage coincides with peak caterpillar abundance. The females, however, must decide when to lay eggs a few weeks beforehand. They must use the best cue available earlier in the spring to predict the future peak of caterpillar abundance. The great tits use temperature in early spring as a cue for determining the time of egg laying despite the fact that early spring temperature correlates only weakly with the later peak caterpillar abundance over a large geographical area. The females, however, can improve on their initial guess: The correlation between early spring temperature and peak caterpillar abundance within a restricted locality is much higher than the correlation over a large area. This disparity is caused by considerable differences between the dates of peak caterpillar abundance in different localities. The date of peak abundance is influenced by factors such as vegetation type; location along, direction of, and steepness of a mountain slope; and proximity to bodies of water. Hence, the great tits apparently record the peak date of caterpillar abundance as it relates to the nestling feeding period; the great tits then adjust the time of egg laying *in the same locality* during subsequent years to increase synchronization (see fig. 8 in Nager and van Noordwijk 1995). In other words, learning allows the birds to strengthen the association between a cue (early spring temperature) and the environmental event that affects fitness (peak caterpillar abundance in a specific location) over time.

This example of learning can be related readily to the description of very basic reversible plasticity. Other classes of learning, however, are somewhat different. First, a subject may begin activity with a period of sampling. During that period, the subject may respond to several types of resources, compare its performance in each case, and then restrict its response to the one associated with the highest payoff. One field example is from an experiment by Heinrich (1979). Inexperienced bumblebees (*Bombus vagance*) visited an average of five species of rewarding and nonrewarding flowers on their first two foraging trips. Later, however, the bees gradually reduced the number of flower species visited. By the seventh foraging trip, most bees visited only jewelweed (*Impatiens biflora*), which was the most rewarding species (fig. 4.2). Another class of learning allows a subject to create representations in its memory of certain aspects of the environment. Examples include learning a food source location, song learning, and individual imprinting on parents (Lorenz 1965; Dyer, this vol. chap. 6; Beecher, this vol. chap. 5; Sherry, this vol. chap. 7).

In summary, the literature on phenotypic plasticity has focused on cases in which some predetermined phenotypic trait is produced in response to a certain predetermined environmental stimulus. For example, the fur of snowshoe hares always changes to white in response to shorter days in the fall and switches back to brown when days become longer in the spring. By contrast, the literature on animal learning addresses cases of reversible plasticity in which an animal creates representations in memory of some aspects of the environment for future use. With learning, there is an additional level of re-

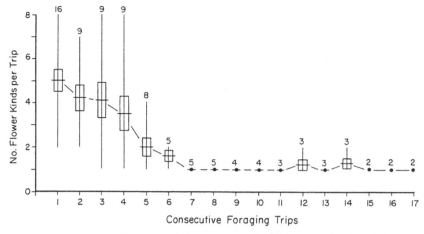

Figure 4.2. Number of different flower kinds visited by bumblebees as a function of the number of consecutive foraging trips. Only one bee foraged at a time. Numbers on graph indicate sample sizes. Boxes show standard errors (SE). Vertical lines depict ranges (from Heinrich 1979).

versibility, which allows an individual to change either the stimulus it responds to or the response to that same stimulus. For example, bumblebees can readily be trained to (1) collect nectar only from yellow flowers within a grid of rewarding yellow and nonrewarding blue flowers; (2) change and collect nectar from blue flowers if these become rewarding and yellow flowers become nonrewarding; and (3) change and collect pollen from the blue flowers if these change to offer ample pollen but no nectar.

4.4 ECOLOGICAL FACTORS FAVORING THE EVOLUTION OF LEARNING

4.4.1 Comparing Types of Plasticity

All types of phenotypic plasticity, including learning, can be seen as adaptations to some pattern of environmental variation. This has been emphasized in most writings on plasticity and learning. There is still no consensus however, on the exact ecological conditions and patterns of variation that give learning a fitness advantage over some feasible null models. One cause for this general disagreement is that various authors focus on different properties of learning or on distinct null models. In sections 4.2 and 4.3, I distinguished three types of phenotypic plasticity. This distinction only reflects qualitative differences along a multidimensional continuum, but this distinction is nevertheless useful for understanding ecological correlates of various types of plasticity and relating the literature on phenotypic plasticity to that on animal learning. Hence, I will use this distinction to compare irreversible plasticity to a canalized phenotype, simple forms of reversible plasticity to irreversible plasticity, and learning to simple forms of reversible plasticity.

Compared to a canalized phenotype, irreversible plasticity results in higher fitness when there is environmental variation between generations but little variation within a generation (e.g. Moran 1992; Scheiner 1993). In cases of simple reversible plasticity, the environment may vary frequently within a generation as long as the change between the alternate phenotypes happens sufficiently quickly. If there is a long lag between the environmental and phenotypic changes, then reversible plasticity may not be advantageous (Padilla and Adolph 1996). In both forms of plasticity, there must be a close long-term association from generation to generation between an environmental event (E_1) that affects fitness and the stimulus (S_1) that predicts or represents that event (i.e. a large value of $Prob[E_1 | S_1]$). A reliable stimulus is required for an individual to create the right phenotype in its respective environment.

Learning, as an advanced form of reversible plasticity, also allows variation within a generation as long as the information about the change is acquired sufficiently faster than the environmental change occurs. Learning rate must be adequately higher than the rate of environmental change. In contrast to the simpler forms of plasticity mentioned, learning does not require a close association from generation to generation between a certain stimulus and an environmental event. This is because an individual can learn about various stimuli and their associations to environmental events even if these associations are temporary. Moreover, a learning strategy does not even require persistence within a generation; all a learning strategy requires is that the rate of learning is sufficiently faster than the rate of environmental change. For example, each spring, an emerging bee can search only for certain yellow flowers if there has been an association between such yellow flowers and high floral reward for numerous generations. By contrast, if there has not been such a long-term relation, an emerging bee should sample various flower types and choose the type providing highest rewards in her current location in time and space. The bee then can choose to visit pink flowers after her initial sampling if she finds these flowers to be the most rewarding; later, however, if the bee learns about environmental change, she may switch to other, more rewarding flowers.

With use of the distinction of plasticity types, it is easy to recognize that Stephens's (1991) model of learning is similar to models of irreversible plasticity (section 4.2). In Stephens's model, the environment contains two resources, one variable and one stable. The variable resource can be in good or bad states, yielding g or b units of fitness per time period, respectively. States vary at certain probabilities between time periods within and between generations, where each generation consists of two time periods only. The stable resource always yields s units of fitness per period, where $g > s > b$. A plastic individual experiences the variable resource during the first time period; during its second and last time period, that individual makes a single choice between the variable and stable resource. For example, if the individual experiences the variable resource in a bad state at $t = 1$, it switches to the stable resource at $t = 2$. Conversely, the two nonplastic strategies always use the stable or the variable resource. Note that in this model, learning is actually irreversible plasticity because the plastic strategy has a period of sensing (and experiencing) the environment before a decision is made about its irreversible phenotype. Indeed, as in other models of irreversible plasticity, Stephens also emphasized the high correlation between the environmental state during the period of assessment and subsequent period of life. Most instances of learning, however, do not *require* predictability within a generation. The only requirement is that the learning rate be sufficiently higher than the rate of environmental change.

4.4.2 Factors Determining the Value of Learning

Some parameters that are involved in many cases of learning affect its fitness value. These parameters are probable initial costs, learning rate, rate of change in value of information acquired, and expected life span. Some types of learning necessitate an initial sampling period when experience is gained about available alternatives, some of them inferior (e.g. Stephens and Krebs 1986; Krebs and Inman 1994). The sampling by bumblebees depicted in figure 4.2 illustrates that cost. During the sampling period, bees visited flowers with a reward rate less than 10% of that offered by jewelweed, which bees visited almost exclusively later. In addition, unskilled animals with a trainable generalized motor system incur an initial cost when sampling alternatives; these animals also incur a cost during a subsequent period of low performance when motor skills are being acquired for the preferred task.

Laverty and Plowright (1988) nicely demonstrated learning costs in their comparison of flower handling by specialist and generalist bumblebees. *Bombus consobrinus,* the only known specialist bumblebee, is a Eurasian species that feeds primarily on monkshood (*Aconitum* spp.). All other species of bumblebee, such as *B. fervidus* and *B. pennsylvanicus* studied by Laverty and Plowright, are generalists that feed on a wide variety of flowers. Monkshood has morphologically complex flowers with concealed nectar. During a typical visit to a flower, a bumblebee enters from the front, crawls over the reproductive structures, and moves up under the helmet to probe the nectar petals (fig. 4.3) (Heinrich 1976; Laverty 1980). Laverty and Plowright tested worker bumblebees with no previous foraging experience. All specialist foragers probed under the helmet and discovered the nectar on the first or second visit. Con-

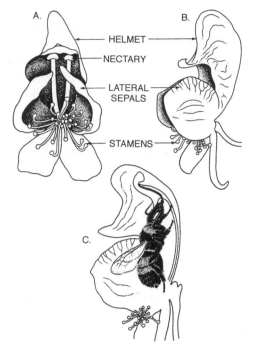

Figure 4.3. Flower of *Aconitum variegatum* in (*a*) frontal view, (*b*) side view, and (*c*) with a worker bumblebee inserting her tongue into a nectar petal (from Laverty and Plowright 1988).

versely, the generalists searched for nectar primarily around the stamens and sepal margins, and only about half of these bees located nectar on the first foraging trip. Moreover, while all specialist bees visited monkshood flowers continuously after initial foraging, 71% of generalist bees stopped visiting flowers at least once after failing to locate the nectar. This observation suggests that naive generalist foragers that sample a new resource might underestimate its intrinsic value. Overall, the initial handling time per flower for the specialists was one quarter of that for the generalists (fig. 4.4). The specialist bees took an average of nine visits or 2 min to reach the performance criterion of 80% correct visits. The generalist bees reached the same performance criterion after 54 visits or 17 min.

Although initially costly, performance with learning can increase over time. The rate of the increase, or learning rate, depends on a set of internal and external factors. When a subject acquires only a memory of environmental features, learning rate may be very high. For example, learning of prominent landmarks surrounding the nest or feeding location can be rapid because only the creation of neural "snapshots" of an area is required (e.g. Cartwright and

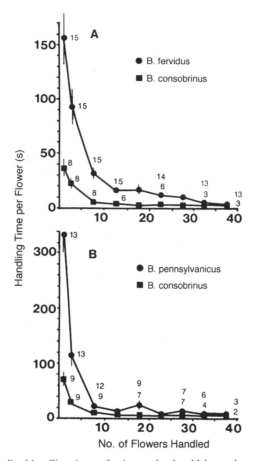

Figure 4.4. Standardized handling times of naive worker bumblebees when visiting flowers of (a) *Aconitum napellus* and (b) *A. variegatum*. Each data point depicts a mean (±1 SE) calculated over an interval of five consecutive visits for the generalists (*dots*) and the specialists (*squares*). Specialists reached asymptotic performance much faster than the generalists (*P* < 0.005, Mann-Whitney *U* test) (from Laverty and Plowright 1988).

Collett 1983; Dyer, this vol. chap. 6). In other cases, however, learning rate is much lower because of internal or external limitations. First, if a subject has to acquire a complex motor skill, extensive practice may be needed to achieve maximal speed, accuracy, and coordination (Newell 1991; Sale 1992; Stamps 1995). Practice is necessary for humans to acquire complex skills such as biking or skiing; however, practice can also be observed in the case of flower handling by bumblebees (fig. 4.4). Second, when a subject begins an activity by sampling available alternatives, the time of that sampling, or learning period, depends on the following parameters: (1) the number of distinct alternatives; (2) the magnitude of differences in the alternatives' mean qualities; (3) spatial and temporal variance around these means; (4) differences in appearance between alternatives; and (5) the subject's perceptual ability to detect differences in both quality and appearance (Staddon 1983; Stephens and Krebs 1986; Dukas and Real 1993; Krebs and Inman 1994; Bateson and Kacelnik, this vol. chap. 8).

Yet, as many of us know, information acquired through learning may be relevant for only a limited time. For instance, computer software we learned to use only a few years ago is most likely obsolete now. We must acquire knowledge of new software despite its impending obsolescence. In a natural setting, research on honeybee foraging provides an example. Visscher and Seeley (1982) monitored recruitment dances of honeybees housed in an observation hive and used that information to produce daily maps of food patches used by the colony. Their results suggest that the colony adjusted its distribution of foragers among new and old food patches daily. Such redistribution seemed to be determined by the relative richness of various sources and poor weather conditions, which increased the relative profitability of patches close to the hive. The reduction rate of the value of knowledge acquired may be caused by environmental changes or internal changes in the subject. For instance, ontogenetic or seasonal changes that result in changes in diet, habitat, reproductive state, or social status may make specific resources undesirable. In other words, the change is not in the environment but in the way it is perceived and exploited by the subject.

4.4.3 A Few Examples

To better understand the evolutionary ecology of learning, it helps to examine cases in which learning does not occur (or has not been demonstrated). In many situations, learning may not enhance performance. For instance, many insects can learn a wide variety of tasks regarding spatial orientation, food types, hazardous locations, competitors, and mates (e.g. Corning et al. 1973; Papaj and Lewis 1993; Craig 1994). Yet there are many aspects of insect life that do not involve learning. For example, specialized worker honeybees remove from the hive any bee emitting the "death pheromone" (Visscher 1983). This pheromone is released only after the death of a bee; in this situation, there is a perfect association between the cue (death pheromone) and the environmental event (dead bee). Worker bees do not need to learn this response; hence, a basic reversible plastic rule is sufficient in this case. Similarly, many ground-nesting birds respond to an egg outside the nest by rolling the egg back inside (Tinbergen 1951). There is a certain probability for an egg to roll out of the nest and an obvious fitness advantage to returning that egg. In this case, however, a simple behavioral rule is sufficient because no additional information is likely to be gained with learning in most natural conditions.

In another general example, many animals have a restricted food source, host, or habitat, and these animals appear to make no attempt to learn about alternative resources. One reason for the absence of learning is that a certain species possesses a whole set of unique, complex adaptations to a specific resource. These adaptations may include seasonal or daily timing of activity, sensory capabilities, mouth parts, physiologic processes, or the capacity to find mates or evade predators (e.g. Ehrlich and Raven 1964; Smiley 1978; Bernays and Graham 1988; Minckley et al. 1994; Singer 1994). Therefore, learning about alternative resources may not occur if this learning does not increase fitness. The magnitude of instances in which learning does not occur is difficult to quantify because (1) researchers usually prefer to report and elaborate on positive results; and (2) negative results may reflect experimental failure rather than absence of learning. Parmesan et al. (1995) provided a detailed report on the absence of learning in the Edith's checkerspot butterfly (*Euphydryas editha*); they suggested that this butterfly does not learn to search for its host plant, although this search could be adaptive in a recently altered environment. Similarly, Potting et al. (1997) have recently documented absence of learning in the parasitoid wasp *Cotesia flaripes*.

Another example for lack of learning involves bacteria. Although bacteria are relatively simple organisms, they possess a large set of plastic rules that allow them to move toward gradients of nutrients and away from harmful compounds. Chemotaxis enables bacteria to track resources in environments that vary in time and space, but the behavioral rule used in bacterial chemotaxis is very limited. A bacterium compares the quality of a resource, such as an amino acid, at times t and $t + 1$. If a resource concentration is higher at $t + 1$, the bacterium continues to move in the same straight line; if the resource concentration is lower at $t + 1$, the bacterium tumbles briefly and then sets off in a new random direction. Such tumbling increases the probability that the bacterium returns to its original location at time t (Koshland 1980). The bacterium, however, does not *learn* its location at times t and $t + 1$; therefore, the bacterium cannot move back to its precise location at t when necessary. Why does the bacterium not learn? There are two possible answers. First, bacterial environments do not have sufficiently distinct features that make spatial representation feasible. Second, and more likely, bacteria do not have the basic sensory capacity for spatial representation (see Staddon 1983, 98). For both possibilities, there is no expected benefit from use of a spatial-learning algorithm.

The example of bacteria illustrates how one can evaluate expected benefits of learning in a system in which learning does not occur. I will now discuss a system in which learning occurs in some species but, perhaps, not closely related species. Such sets of species are ideal for a comparative study of learning and its ecological correlates. Many organisms acquire food, mates, or hosts without learning about the spatial features of their environment. A solitary wasp that finds her host and then digs a cavity in which she deposits the parasitized victim does not have to learn spatial features. By contrast, a wasp that uses the same cavity or an abandoned cavity of another species to deposit collected hosts on successive trips would benefit from spatial learning. Similarly, a wasp that digs a nest and then initiates a search for a host must acquire some spatial knowledge of her environment to relocate her nest. Under certain environmental conditions, it is advantageous to use a multiple-cell nest or first dig a nest. If a wasp uses abandoned cavities that are at low density, use of the same cavity a few times could increase fitness. Similarly, if nest digging takes considerable time, reuse of the same cavity may increase overall reproduction. If digging is time consuming, digging should be done before finding a host because the host should not be unattended for a few hours while the wasp digs (see Evans 1953; Evans and Eberhard 1970). The only unlearned plastic rule that may serve as an alternative to spatial learning is scent marking of the nest. Scent marking is probably effective only for finding a relatively close nest and may limit host searching. Furthermore, scent marking makes the nest easier to locate by parasites and predators. Thus, even a relatively ineffective spatial learning—perhaps some primitive form of path integration or pictorial representation of prominent features surrounding the nest (see Dyer, this vol. chap. 6)—may increase a wasp's probability of relocating her nest.

In short, I expect to find a positive correlation between a wasp's dependency on a nest she must return to and her capacity for spatial learning. This prediction is analogous to the positive correlation between dependency on cached food and spatial memory in several groups of birds (Sherry, this vol. chap. 7). Species of spider wasps (Pompilidae) vary greatly in their nesting habits in the manner I described (Evans 1953). Hence, these wasps are good subjects for testing the prediction presented.

The cost of being naive depends on the nature of the task to be learned. Spending a few minutes hovering around the nest to learn surrounding spatial features probably has negligible cost; however, naivete about feeding can be very costly. Sullivan (1988a, 1988b, 1989) detailed various aspects of inexperience in young yellow-eyed juncos (*Junco phaeonotus*). Newly independent juveniles were much less efficient at capturing and handling prey than adults. For example, the time to handle a small mealworm was 6.5 and 3.5 sec for young and mature juncos, respectively. While young juncos spent over 90% of the daytime foraging, adults foraged for less than 30% of the day at the same site and during the same season. Daily mortality rates from starvation or predation during that period were 3.85% and 0.11% for young and adult juncos, respectively. Hence, in this species, the initial cost of learning was high. Species of birds and other groups of animals vary greatly in their feeding habits and the time required to learn foraging techniques (see Davies 1976; Davies and Green 1976; Greenberg 1983). Thus, evaluating costs of learning and the ecological correlates among species with little to extensive learning is feasible. The study of Laverty and Plowright (1988) described in section 4.4.2 (figs 4.3, 4.4) provides this kind of evaluation.

I listed there a few ecological parameters that determine learning rate. Learning rate is sometimes limited because an individual must distinguish and evaluate several variable alternatives. Variations in appearance or quality are external factors that limit learning rate. For example, one must first determine that perceived variations reflect a true pattern instead of a sampling bias. Learning rate is also determined with internal factors such as the ability to detect differences between alternatives, accurately record new information with each encounter, and change motor performance to achieve a particular behavior. For example, the primary sense in humans and birds is vision, but many mammals and insects rely most heavily on smell. Consequently, honey-bees seem to learn faster when artificial flowers must be discriminated by odor instead of color (Menzel 1985).

I know of no systematic study that examines internal limitations on the learning of cognitive or motor tasks. We know from experience and various animal studies that learning takes time (e.g. Wooler et al. 1990; Pyle et al. 1991; Dukas and Visscher 1994). We also know that inexperience is costly; this was depicted dramatically in Sullivan's (1988a) study, which suggested that only about 50% of the juvenile juncos that reached independence survived to the end of the summer (only six weeks later), and most of the mortality could be attributed to reduced foraging experience and lowered vigilance. So there was a very strong selection pressure on increased learning rate! Thus, slow learning implies the existence of fundamental neuronal constraints in addition to the external issues mentioned. Internal constraints and related factors that determine learning rate require further study.

Even if learning is fast, the effective learning rate may be low because of a reduction in the value of information gained. The actual benefit of learning is a product of (1) increased performance as a result of experience; and (2) lowered performance as a result of simultaneous environmental changes that reduce the fitness value of the task learned. Such environmental changes can be measured by monitoring the relative quality of alternatives available to a subject over time. This measurement, however, would probably require a coordinated team to simultaneously record the qualities of resources over a subject's range of activity. A system such as bees foraging on flowers seems especially attractive and tractable (see Heinrich 1979; Thomson 1982; Visscher and Seeley 1982; Rathcke 1988). Of course, some knowledge acquired through learning may remain relevant for a long time, at least in relation to a subject's expected life span. For instance, use of a distinct tree as a landmark for nest location is probably satisfactory for animals because the expected rate of change in tree location is low.

Finally, a given task may be learned only by an organism that lives long enough to benefit from the resulting increased fitness. In other words, there should be a positive correlation between capacity to learn complex tasks and expected life span. Nevertheless, this does not mean that short-lived animals should not learn! However, they may learn only tasks that permit enhanced performance early in relation to their life span. Moreover, theoretically the *proportion* of the lifetime used for learning may not differ between short- and long-lived species. The prevalent assertion that short-lived animals should not learn is most likely incorrect. Ample evidence that short-lived species do indeed learn is accumulating rapidly (Corning et al. 1973; Papaj and Lewis 1993); however, long-term studies on learning in such species are scant (but see Dukas and Visscher 1994).

4.4.4 Genetic and Phenotypic Constraints

In evaluating the evolution of plastic behavioral responses including learning, I have focused on costs and benefits; however, various constraints may have dominant effects as well (see also Dukas, this vol. chap. 3, section 3.2). The ultimate genetic limitation is simply no genetic variation for a certain characteristic, be it sensory capacity, ability to record information in memory, cognitive potential to make a behavioral decision, or mechanical ability to pursue such a decision (see Atchley and Hall 1991; Falconer 1989; Arnold 1994a, 1994b). One probable case of learning constrained by no genetic variation comes from extensive research on cowbird (*Molothrus ater*) parasitism on North American birds (reviewed by Rothstein 1990). Rothstein (1986) compared Eastern phoebes' (*Sayornis phoebe*) responses to partial clutch reduction and cowbird parasitism. Partial clutch reduction is the predation of one or more but not all eggs in a clutch. Phoebes often respond to partial clutch reduction by deserting the clutch and investing in a new full-sized clutch. On the other hand, Phoebes tolerate parasitic cowbirds' eggs despite the severe fitness loss. Partial clutch reduction has always occurred at a small frequency of 10% or more; however, cowbird parasitism has been a strong selective pressure for only the last several decades. The different responses to cowbird parasitism and partial clutch reduction suggest that there is no genetic variation in the response to cowbird parasitism (Rothstein 1986).

A related and ubiquitous genetic constraint, discussed in chapter 3, is that a single trait is typically only a component within a complex system. Hence, having one capacity without having another may not increase fitness; for example, the ability to integrate information to make a decision is useless if one is unable to express that decision.

Another common constraint results from genetic correlations. Each behavioral response is controlled by a large number of genes that affect various aspects of stimuli detection, integration of information, and the actual response. Thus, most behaviors are subject to either pleiotropy (in which individual genes exert their effects on more than one behavior) or linkage disequilibrium (in which there is some statistical association between alleles at different loci that affect different behaviors) (Arnold 1994a). For example, Arnold (1981) found a high, positive genetic correlation between the predatory reaction of garter snakes to slugs and leeches. Selecting to recognize or avoid one of the two prey types could affect or be affected by the predatory reaction to the other. Significant negative or positive genetic correlations have been documented in numerous studies (Arnold 1994b). Such correlations could constrain the operation or evolution of plastic traits, including learning. Nonetheless, although genetic correlations typically are perceived as constraints, they could sometimes have a positive effect on the direction or evolution of some plastic trait. For instance, learning in one domain may be more likely to evolve if a related learning capacity already exists (Papaj 1986).

4.5 COMPONENTS OF FITNESS, LIFE HISTORY, AND LEARNING

4.5.1 Measuring Learning

In this discussion of learning, I have not been explicit about the contribution of learning to various aspects of fitness or about probable differential effects that specific fitness components have had on the evolution of learning. In this section, I will address these issues. Behavioral ecologists usually distinguish four classes of behavior that affect fitness: foraging, escaping predation, reproduction, and social interactions. There are some reasons to believe that the magnitude of learning and the selection pressure on the evolution of certain learning capacities vary among different classes of behavior. Before discussing these behaviors, however, I will consider two central issues regarding the measurement of simple and complex learning capacities.

The ultimate way to demonstrate that a certain behavioral trait affects a specific learning capacity is to compare closely related species, one with and the other without the trait. Such research, with the appropriate controls, is laborious but feasible. To date, elaborate studies of this sort have been conducted only for the examination of food caching and spatial learning and memory in a few groups of birds (Sherry, this vol. chap. 7). Spatial learning, as discussed by Sherry and Dyer (this vol. chap. 6), is a specific and well-defined domain that is ideal for multispecies comparisons. Some other tasks, such as the capacity to evaluate alternative resources, might be measurable as well (Bateson and Kacelnik, this vol. chap. 8; Ydenberg, this vol. chap. 9; Dugatkin and Sih, this vol. chap. 10). Analyses, however, are less clear for more complex learning capacities—ones that are sometimes hard to label, define, and quantify. I mostly will use the term "complex cognitive capacity" to refer to a subject's ability to accumulate and manipulate successfully a large amount of information for the solving of complex problems and the making of intricate decisions. Other words used to describe such capacity are plasticity, cognition, intelligence, and expertise (Lark et al. 1980; Greenberg 1983; Fischler and Firschein 1987; Heinrich 1989; Cheney and Seyfarth 1990; Byrne 1995; Lenat 1995; Balda et al. 1996). These terms, including the one I will use, can readily be criticized for being insufficiently explicit or plainly wrong. I prefer to postpone further terminological arguments until additional cognitive understanding allows us more refined analyses.

To conduct a relevant test regarding such enhanced cognition, a task must be found that is sufficiently related to the specific ability in question yet general enough for testing in species with different habits. Too general a task, however, may prevent the "smart" species from performing well, and too specific a task may not relate to one or both species. The problem is not much different when determining intelligence quotients in humans (see Bryne 1995). Humans sometimes show domain-specific enhanced performance (Lark et al. 1980; Singley and Anderson 1989; Cosmides and Tobby 1992), which strongly suggests that comparing sophisticated cognitive abilities in different species is difficult. Nevertheless, carefully controlled experiments in birds suggest that empirical evaluation of complex learning or cognitive capacities is feasible (Greenberg 1983, 1990, 1992; Balda et al. 1996). The two keys for success of evaluation are first, a wise choice of closely related species for comparison and second, extensive experimentation—ideally by more than one laboratory—to substantiate empirical results and eliminate various alternative factors.

A different line of research involves multispecies comparisons at the level of the whole brain or a brain component (Jerison 1973; Harvey and Krebs 1990). Such studies may be much easier to carry out than experimentation, but study outcome remains questionable as long as these studies are not augmented with direct empirical evidence of a certain cognitive capacity. First, brain and body size are somewhat correlated. Therefore, the method used to remove effects of body size may bias results, because (1) body size is variable, and very small sample sizes are employed; or (2) body size is not independent of the behavioral trait of interest. For example, species with different diets also greatly differ in the size of their digestive systems; these differences could affect body size. A certain diet may select for larger body sizes, which may result in lower ratios of brain to body size (see Harvey and Krebs 1990; Byrne 1995).

Second, the association that exists between a specific learning capacity and the size of a particular brain component is not clear. The possession of a certain learning ability does not have to alter the size of the whole brain, and a certain learning ability may not even change the size of a specific brain part. After all, there are many additional dimensions, such as cell structure and physiological processes, neurotransmitter function, and complex interactions within and between neural networks that determine cognitive capacity. Moreover, learning and cognitive function in general involve parallel neural processes that are distributed in various brain parts (e.g. Sergent et al. 1992; Gardner 1993; Simmers et al. 1995). This distribution of neural processes can make it difficult to relate a certain learning ability to a change in the size of a specific brain part (see Finlay and Darlington 1995). For example, although much of the comparative study of spatial memory has focused on the hippocampus, head direction neurons are located outside the hippocampus (Sherry, this vol. chap. 7, section 7.4). Nevertheless, the comparative study of spatial memory and the hippocampus discussed in chapter 7 and recent intriguing comparative date on a positive correlation between feeding innovations and forebrain size in birds (Lefebvre et al. 1997) suggest that we can learn about brain and cognition from the comparative approach.

Given the uncertain nature of comparative studies of brain size and the scarcity of empirical evidence on learning and natural history, I will concentrate on theoretical considerations and ideas for future experimental investigation. Examples of learning given should be considered only as preliminary studies that necessitate further evaluation. Each section will focus on one class of behavior at a time, and if possible, I will assume that other behavioral and life-history traits do not interact with the behavior examined. Alternatively, if interaction is inevitable, I will evaluate its relevance for conceptual understanding or future experimental studies. Although I will concentrate on the description of behavior, the general ecological conditions for the evolution of learning outlined in section 4.4 must still be met; however, for brevity, these conditions will not be reiterated here.

4.5.2 Foraging

Because of the similarities, searching for and choosing food, hosts, and other resources are considered together. The simplest thing to learn about food is an easy way to find it. Parasitic wasps do this by learning to associate their host with the more prominent odor of the host's substrate (Vet and Dicke 1992; Turlings et al. 1990, 1993). Practically, odor recognition enables the wasp to increase the frequency of perceiving its host. Of course, a forager may use another sense, such as vision, to increase the frequency of finding food or another resource (e.g. Traynier 1984). Food locations may be learned in space or time. For a solitary species with no parental care, spatial learning is relevant if a subject uses a daily shelter far from its food source or if a species forages on a few patches of one or more types of food with a distinct spatial distribution. Spatial learning can enhance fitness under these conditions if food sources or potential shelters are sufficiently far apart so that direct visual or odor cues are ineffective. The behavior of some butterflies of the genus *Heliconius* (*H. charitonia* and *H. erato*) seems to reflect these conditions. Individuals spend the night at a constant roosting site and depart in the morning for regular feeding or egg-laying grounds up to 1.2 km away. Indeed, the butterflies appear to use visual landmarks for orientation (Mallet 1986; Mallet et al. 1987). Butterflies of the Heliconiini tribe have been studied extensively by Gilbert (1975, 1991) and colleagues, and these butterflies appear to be an ideal group for a comparative study of learning and its ecological correlates.

The steps just discussed are relevant even for species that feed on a single type of food: learning may be as important for specialists as for generalists. Moreover, given the difficulties involved in distinguishing between the specialist and generalist (Fox and Morrow 1981), I would not emphasize the distinction in the analyses of learning. For example, Papaj (1986) described the pipevine swallowtail butterfly (*Battus philenor*) as an *extreme specialist* that uses only species of the genus *Aristolochia* as hosts. However, interest in that butterfly's learning arose because of its *generalist* nature; female butterflies in eastern Texas learn to search for the broad-leaved *A. reticulata* or the narrow-leaved *A. serpentaria* (Rausher 1978; Papaj 1986). In addition, a species that strictly uses a single host may still distinguish between a few intraspecific host morphs. Practically, one can classify that species as a generalist.

Nonetheless, a broader diet could be correlated with better learning ability. If various items in the diet vary in profitability or abundance, learning the preferred type in current local conditions may increase a subject's foraging performance. Such learning is more elaborate than merely representing spatial locations in memory because a subject must sample items of a few food types with their variable appearances and qualities. On the basis of neural representations of the complex information on food type quality and density, the subject has to rank the food types (see Bateson and Kacelnik, this vol. chap. 8; Ydenberg, this vol. chap. 9). Of course, a generalist species does not *have to* show such sophisticated capacity to choose the best food type in time and space; however, a generalist species that possesses that capacity must have better learning ability than a generalist that merely develops a feeding preference for the first type it encounters. Once again, I de-emphasize the ecological distinction between the specialist and generalist. Instead, I highlight the functional aspect of how information is used to determine behavior. Indeed, many species appear to search for the first host type they encounter, which may be the one they emerged from or the one closest to their emergence site (see Prokopy et al. 1982; Tranier 1984). Such species seem to show little or no prior exploration and sampling of alternatives. By contrast, bumblebees do spend considerable time evaluating alternatives (see fig. 4.2). Hence, an empirical study that evaluates further such interspecific differences in choice behavior and the ecological and learning correlates is feasible.

Extreme generalization, in which subjects sample and add novel food types to their diet, is associated with aversion learning, at least in some taxa. In many species, when an animal encounters a new potential food, the animal first inspects the food with the senses of vision, smell, and finally taste. If the food passes that inspection, the animal eats only a small amount of the food. Sickness sometime afterward would be associated with the novel food, and that food would be remembered as one to exclude from the future diet. More generally, food-related sickness may even cause the exclusion of an already familiar food from the diet (Garcia and Koelling 1966; Rozin and Kalat 1971).

Lindquist and Hay (1995) nicely demonstrated the ecological importance of aversion learning in the omnivorous pinfish *Lagodon rhomboides*. Low doses of the secondary metabolite didemnin from adults and larvae of the Caribbean tunicate *Trididemnum solidum* induced vomiting in the fish, and they rapidly learned to avoid food pellets that contained didemnin (fig. 4.5). Moreover, it is likely that learning in the fish would have been faster if the poisonous pellets had an aposomatic color (Gittleman and Harvey 1980; Guilford and Dawkins 1991; Enquist and Arak, this vol. chap. 2). By contrast, the particle-feeding anemone *Aiptasia pallida* did not learn to avoid the poisonous food. Throughout the experiment, the anemones regurgitated about 80% of the food pellets; they showed an innate response that allowed them to avoid a large proportion of the noxious pellets, but they did not show an increase in avoidance. For a month, the anemones were fed the chemical equivalent of fifteen tunicate larvae per day, which amounted to less than 2% of their daily food intake. The small amount of didemnin consumed was sufficient to reduce adult growth by 82% and production of daughter clones by 44%. Significant differences in production of daughter clones occurred after only 4 days. This study shows that even a small amount of noxious food can have strong negative effects on generalist foragers (Lindquist and Hay 1995). Hence, rapid aversion learning that results in nearly 100% rejection could have substantial positive consequences on fitness. Such aversion learning is advantageous over an innate vomiting response to didemnins because (1) individuals do not have to waste time pursuing and handling poisonous items; (2) vomiting may involve rejection of a mixture of high-quality food with the

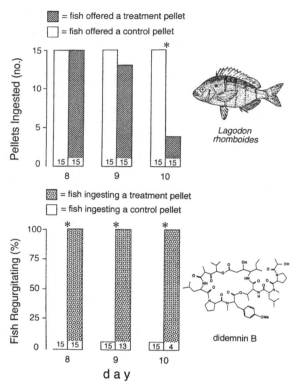

Figure 4.5. *Top:* Number of treatment (*hatched bars*) and control (*open bars*) fish ingesting a food pellet with and without didemnins, respectively. *Bottom:* Percent of these fish regurgitating an ingested pellet within two hours. Either type of pellet was offered only on days 8–10. Numbers at the bases of each bar are sample sizes. For each day, a significant difference between the number of treatment and control fish ingesting or regurgitating pellets is indicated with an asterisk ($P < 0.001$, Fisher exact test). The structure of the secondary compound didemnin B is also depicted (from Lindquist and Hay 1995).

noxious items; and (3) some digestion of the damaging poison may occur before regurgitation. In addition, a generalist forager that feeds on numerous items may not be able to possess a genetically determined set of rules that allows the forager to reject every harmful compound.

Aversion learning is a fairly specific mechanism that may be absent in species that never encounter novel poisonous food. Such species may have an exclusively nontoxic diet, or they may feed on nonpoisonous food types. For example, I expect broadly generalist birds that feed on seed and fruit, but not nectar-feeding birds, to possess aversion learning. This prediction is based on the assumption that nectar is almost never poisonous, but some seeds and fruits are often inedible to a certain taxon (see Herrera 1982).

The last two issues discussed in this section (choosing among food types and aversion learning) involve specific learning capacities that allow locating, catching, or handling food. Extensive spatial learning is required by foragers that must cover a large area to locate patches of food or by species that cache food for future consumption. I will discuss the former instance only (the latter is discussed by Sherry, this vol. chap. 7). Eisenberg and Wilson (1978) convincingly argued that relative brain size is larger in fruit- than insect-feeding bats because locating fruit trees in space and time requires an enlarged brain capacity for storing and processing information. The larger brains of insectivorous bats that exploit distinct microhabitats compared with those of insectivores that capture prey by flying provides added support to the proposition of Eisenberg and Wilson. Further support comes from additional studies that report similar results for two other orders of mammals. In rodents and primates, herbivores have smaller relative brain sizes than species that do not feed on leaves. Once again, smaller brain size seems to correlate with the smaller home range of leaf-eating mammals (Clutton-Brock and Harvey 1980; Mace et al. 1981). Nevertheless, as argued earlier (section 4.5.1), comparative measures of brain size should only be used as inspiration for rigorous experimental examination. Certainly, there is more to the brain than mere relative size.

Many species of birds and mammals develop and refine complex techniques for pursuing and manipulating food over a period of months and even years (e.g. Desrochers 1992a). Such species may show more elaborate learning capacities compared with those of closely related species that have simpler foraging habits. A central complication of testing such predictions is that one must control for expected life span and the period of parental care because both are predicted to allow increased learning (see sections 4.5.4 and 4.5.6 on reproduction and life span). As mentioned in section 4.5.1, designing a test for evaluating and comparing complex cognitive skills in different species is difficult. The very definition of "complex tasks" is neither obvious nor easy to quantify.

4.5.3 Evading Predation

An individual that fails to respond to an approaching predator faces a high mortality risk. Hence, an innate escape response, even to a stimulus with a low association to predation, may be beneficial. A learning alternative may not be feasible because the cost of failing to recognize a cue associated with predation could be death. Compared with learning about food, mates, and other resources, animals may retain more innate responses to predators.

Nevertheless, a few aspects of predation may be learned. First, subjects may initially overreact to various cues that may be associated with predation and learn to ignore the neutral cues over time. The cost of overreaction is that associated with stopping a current activity and fleeing; however, such cost is typically much smaller than that of failing to recognize a predator in time (Bednekoff et al. unpublished manuscript). Second, after successfully evading a predator, a subject may record information that could lower the probability of another attack. Perhaps the most obvious thing a subject can learn is the exact location of attack. Once again, on the basis of a single encounter, it is advantageous to overreact and learn, even if the knowledge acquired has a low probability of being useful in the future. Usually, the cost of approaching the location of a previous attack cautiously or bypassing the location altogether is small compared with the probable benefit of reducing mortality risk.

The results of many laboratory learning studies by psychologists, which involve electric shock as punishment, may be interpreted as related to learning about mortality factors, although the relevance of such research to natural settings is not obvious. An example of a field study is Craig's (1994) demonstration that stingless bees (*Trigona fluviventris*) can learn to avoid spiders' orb webs after a successful escape. Finally, subjects can acquire knowledge about novel types of hazards through observation of other individuals (Curio 1993; Maloney and McLean 1995). Sometimes the observation can be as subtle as detection of an alarm pheromone in association with a novel animal. Chivers and Smith (1994) nicely demonstrated that predator-naive fathead minnows (*Pimephales promelas*) recognized as potential predators a natural predator (northern pike, *Esox lucius*) or a nonpredator (goldfish, *Carassius auratus*) after the minnows had seen these fish paired with an alarm substance (fig. 4.6). The alarm substance, released by fish after physical injury caused by predation, is possessed by members of the superorder Ostariophysi.

Note that rapid learning about a probable mortality factor is similar to aversion learning described in the section on foraging (4.5.2). In both cases, overreaction is the rule because the assumed benefit is reduced probability of mortality; the most likely cost of overreaction is merely reduced foraging efficiency because of wasted energy and time. Such experiences, like the exclusion of a certain food from the diet after coincidentally getting sick from a stomach virus, are shared by many people (e.g. Garcia et al. 1966; Seligman and Hager 1972).

4.5.4 Reproduction

4.5.4.1 Finding and Choosing Mates

The discussion on reproduction is divided into two parts that deal with mate choice and parental care. My discussion on mate choice will be short because it is examined by Dugatkin and Sih (this vol. chap. 10). Also, conceptually, there are many similarities between demands on learning when searching for

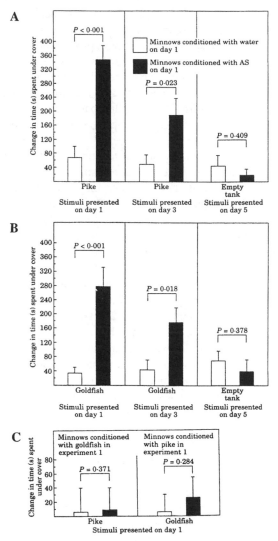

Figure 4.6. The mean (+ SE) increase in time spent under cover by minnows initially exposed to the sight of either (*a*) a pike or (*b*) a goldfish associated with either control water (*white bars*) or alarm substance (*dark bars*). Exposure to either fish and alarm substance on day 1 resulted in a strong antipredator response on day 3 compared with the two controls of water or an empty tank. (*c*) Antipredator response was specific for the fish species associated with the alarm substance; minnows that learned that goldfish are potential predators did not show antipredator response to pike and vice versa (from Chivers and Smith 1994). *AS* = alarm substance.

and choosing food or mates. This is especially true for species in which mating occurs at or near the food or host substrate. For example, just as female parasitoid wasps benefit from learning about the more prominent odor of their host's substrate, males may benefit from learning this same odor to increase their probability of finding females. Of course, these learning benefits would apply only to species in which the females are still receptive during egg laying. Similarly, males may benefit from learning the spatial location of a few distinct nest aggregates and food patches where receptive females may be found (Eickwort and Ginsberg 1980).

Males may learn more than just spatial location. Male bees searching for females approach a variety of objects that are similar to bees in size, shape, and color (e.g. Barrows et al. 1975). Males may also encounter numerous females that reject their copulation attempts. The capacity to learn which objects or individual females are not worth courting could allow males to use their limited time more effectively for pursuing receptive females. Indeed, male *Lasioglossum zephyrum* are able to distinguish between individual females. Barrows (1975) presented males with moist filter papers impregnated with a female odor and recorded the rate of copulation attempts addressed toward a small dark dot during trials of 5 min. The males received another filter paper with the same female odor 1–2 h later. Immediately afterward, the males were exposed to the odor of another female. On average, a male's response to the new female odor was higher than that to the old female odor (fig. 4.7). This experiment suggests that males discriminated among individuals or at least among a number of classes of females with different odors. Of course, if the number of classes is sufficiently large, class discrimination functionally becomes true individual discrimination (Barrows et al. 1975; Michener and Smith 1987). Further experiments with *L. zephyrum* revealed that a female's odor is genetically determined (Greenberg 1979; Smith 1983).

For species with large home ranges such as territorial mammals, locating females may be a challenge for males. Males are probably familiar with several neighboring males and females through some direct encounter but probably more often through detection of odors and vocalization (Leyhausen 1956; Macdonald 1983; Halpin 1986; Kerby and Mcdonald 1988). Males may possess a large-scale spatial representation of the location of other individuals and their qualities as competitors or mates. Obviously, these kinds of species also forage within a large home range and typically have extended parental care; therefore, separating causes of increased learning capacities may be difficult. Nevertheless, in such species, if selection for increased spatial learning acts on males only, males will experience better spatial learning than females. Experimental evaluations of this prediction are inconclusive (Sherry, this vol. chap. 7).

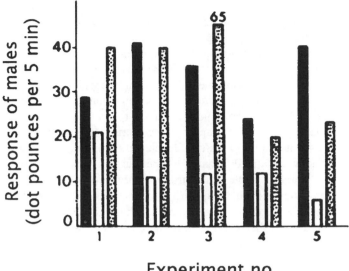

Figure 4.7. Results from one series of experiments that tested the responses of male *Lasioglos-sum zephryum* to (1) odor of a female (*black bars*); (2) second presentations of the same female's odor 1–2 h later (*open bars*); and (3) odor of a different female immediately after the second presentation (*stippled bars*). The males reduced courtship attempts addressed to the first female but showed larger responses toward a new female's odor (total $n = 14$; Wilcoxon test, $P < 0.005$) (from Barrows 1975). These results were substantiated by Greenberg (1979) and Smith (1983).

Of course, males must only mate with females of the same species, and recognition of the other gender and copulation behavior need not be learned. Indeed, for most species, copulation behavior and recognition of intraspecific mates are innate. In many birds and mammals, however, species and mate recognition and species-specific calls or songs are learned through observation or imitation (Lorenz 1965; Nottebohm 1972; Bateson 1979, 1981; Beecher, this vol. chap. 5). Birds other than oscine song birds, as well as various insects, use song efficiently without imitation learning (see Kroodsma 1987; Andersson 1994). Thus species recognition and communication can be highly effective without learning. Hence the adaptive significance of such learning compared with an innate alternative must still be elucidated.

Current verbal and formal models emphasize sexual selection as the main force in the evolution of song learning (see Nottebohm 1972; Aoki 1989; Kroodsma and Byers 1991). Such models, however, do not critically evaluate how and why learning is advantageous over some unlearned alternative (section 4.4.1). Genetic correlations with other learning capacities may have a crucial function in the evolution of song learning and species recognition (see section 4.4.4). A combination of positive genetic correlations with the currently suggested, somewhat less plausible explanations may provide a scenario for the evolution of song learning. For example, phenotype matching implies that an individual learns the phenotypes of relatives or itself. A related example is the learning of local environmental cues. Kin recognition with phenotype or environmental matching is widely used among animals (see various chap. in Fletcher and Michener 1987 and Hepper 1991; Ode et al. 1995). Given this learning capacity, it may be simpler to reconstruct the evolution of learned songs used for mate choice or male-male interactions because one can ask how learning capacity that already exists for other purposes has been elaborated to include song learning. Other than further theory, promising approaches are detailed evaluations of (1) learned song components in birds other than oscine song birds (such as the suboscines, parrots, and hummingbirds) (see Kroodsma 1983, 1987); and (2) the function of song learning in oscines with a single song type.

4.5.4.2 Parental Care

Parental care has probably been a central factor in the evolution of learning. Apparently, selection on parents and, more importantly, the increased learning opportunity for young under parent supervision contributed substantially to increasing the function of learning as a basic trait that affects fitness. A primitive type of parental care is the deposition of eggs in a nest with food. Many variations of this behavior occur in many animal groups (Clutton-Brock 1991), but I will focus again on hymenopterans because they are well studied and ideal for comparative studies of learning (Evans 1958; Wilson 1971; Michener 1974; Evans and Eberhard 1970; Dukas and Real 1991). A female that lays eggs on a host does not have to possess spatial learning. If a female, however, invests time and energy finding a cavity or digging her nest, she may benefit from learning the nest location in space. The female does not have to learn the spatial location of the nest if she finds the host and then digs a nest that is used only once. However, digging a nest first or reusing the same nest a few times requires spatial learning. Thus, even minimal parental care selects for enhanced spatial learning. This prediction may be tested with solitary wasps (see details in section 4.4.3) or species of other groups, such as fish or beetles (see Clutton-Brock 1991, chap. 2).

The basic parental care just depicted is termed "mass provisioning" because all the food needed for larval development is deposited with the egg. Progressive provisioning is a slight modification on this elementary pattern. In this case, a female remains inside the nest until her egg hatches; she then provides the larva with increasing amounts of food throughout its growth (Evans 1958). Locating the nest during the feeding stage is crucial for the mother's fitness because failure to find the nest means larval starvation. By contrast, with mass provisioning, failure to find the nest means a smaller fitness loss caused by the time and energy needed to locate another cavity or dig one. Hence, progressive provisioning could select for a more refined spatial learning. This prediction may be tested in groups of wasps such as the sphecid tribe Bembicini or the family Eumenidae (which includes potter wasps). In both groups, closely related species express either mass or progressive provisioning (Evans 1957, 1958, 1966).

Many solitary hymenopterans live in aggregations of thousands of nests, and the nests are a few centimeters or less apart (Michener 1974; Batra 1984). To find her nest within the large aggregation, a female must have precise spatial memory. At close proximity, a female may locate her nest on the basis of an individually recognized odor; there is some evidence in support of this proposition, but it has not been evaluated rigorously (Holldobler and Michener 1980). Similarly, in nest aggregations of birds, parents may have to learn to recognize their eggs or offspring among those of their neighbors. For related reasons, young individuals in colonies may gain by learning to identify their parents. Both capacities are known in a variety of species (Beer 1969, 1979; Falls 1982; Beecher et al. 1985; Beecher 1991).

In hymenopterans, advanced stages of parental care involve interactions between a mother and her mature daughters, which I will discuss in the section on sociality (4.5.5). For solitary hymenopterans, parental care does not include a stage of protection that offspring use for learning; however, such a stage is prevalent in birds and mammals. Functionally, parental care of offspring allows young to gain experience without paying the full costs of learning (Mayr 1974; Johnston 1982). Knowledge can be acquired by following the parents or direct exploration of novel tasks. Parents can reduce the cost of learning by (1) providing food and other resources; (2) leading the young to various resources including a shelter; and (3) alerting and protecting against predators and other hazards. Reducing the cost of learning allows animals to begin life with an open cognitive system that can acquire information with parental guidance. Lengthening the period of parental care allows young to accumulate more knowledge that can be used during adulthood. In young, an elaborate cognitive system and extensive knowledge probably implies that learning can continue throughout adulthood. With the machinery to learn and an already substantial experience, young adults can continue to improve their performance in various activities without devoting explicit time to learning only. Extended periods of learning, which may span years, seem to be associated with complex cognitive capacities.

4.5.5 Sociality

In social species, adults spend a substantial portion of life in groups larger than a male-female pair. Sociality exists in a few groups of insects, birds, and mammals (Wilson 1971; Alexander 1974; Rubenstein and Wrangham 1986). Group living, especially one that includes long-term interactions among the same individuals, may select for increased cognitive capacity. This hypothesis has been presented in greatest detail for primates (e.g. Cheney et al. 1986; Byrne and Whiten 1988; Cheney and Seyfarth 1990; Byrne 1995), and in general form for bees and birds (e.g. Menzel et al. 1974; Heinrich 1984; Dukas and Real 1991; Balda et al. 1996). A group member may have to process a large and complex set of signals used to communicate states and events related to food, predators, social status, and sexual availability. For example, honeybees use an intricate chemical "language" with about 36 pheromones, a dance language, and numerous additional behavioral signals (von Frisch 1967; Seeley 1985; Winston and Slessor 1992). The overall number of signals produced and processed by social bees is probably larger than that of solitary bees; however, this assertion has not yet been tested. If social bees' signal number is greater, the bees probably require a more elaborate system of information processing than less social bees or solitary bees. This elaborate system may or may not require enhanced learning capacities.

In social groups, individuals can benefit from recognizing other members with whom they interact for extended periods. Individual recognition is prevalent among social mammals and birds, but the data for social hymenopterans is unclear. The most likely candidates to express individual recognition in hymenopterans are species of the genus *Polistes* or other taxa in which a few females share a nest with use of some dominance hierarchy (see various chap. in Fletcher and Michener 1987 and Hepper 1991). Note, however, that individual recognition is common in solitary species as well (see sections 4.5.4.1 and 4.5.4.2 on mate choice and parental care). The need to interact with many individuals in a complex social system could have selected for more elaborate cognitive capacity. This cognitive capacity, for example, could allow a subject to use transitive inference to deduce the social status of two individuals on the basis of their interactions with a third subject (e.g. Cheney and Seyfarth 1990). In another example, the capacity to deceive other individuals may require higher intelligence (Byrne and Whiten 1988; Cheney and Seyfarth 1990). Once again, however, transitive inference and deception exist in solitary species as well (Cheney and Seyfarth 1990, 82, 186).

Finally, a larger quantity and better quality of information may be generated and transmitted within and between generations in larger social groups. This may select for increased cognitive capacity that enables efficient processing, storage, and exchange of information. Although this is clearly the case for humans, data from other mammals and birds is inconclusive. Song learning is probably the only undisputed case of imitation in nonhuman species (see Beecher, this vol. chap 5), but several other cases of imitation are likely to be confirmed with further experimentation, especially in nonhuman primates. Many other proposed cases of imitation, such as opening of milk bottles by birds and potato washing by monkeys, have been questioned recently by numerous researchers (Galef 1976, 1988; Sherry and Galef 1984; Cheney and Seyfarth 1990; Byrne 1995). Instead, social facilitation, most commonly of offspring by their parents, seems to be the common norm. For example, mother cats provide their kittens with prey in increasingly more challenging situations; this allows the kittens to acquire knowledge about catching and handling prey. The kittens, however, do not *mimic* their mother (Caro 1980; Martin and Bateson 1988). Obviously, parental care in solitary species is a sufficient condition for imitation and social facilitation as they are known in nonhuman animals.

There is considerable disagreement among researchers about definitions and evidence that are needed to confirm imitation and related phenomena (e.g. Galef 1976, 1988; Visalberghi and Fragaszy 1990). Nevertheless, the functional issue is whether learning and cognition are facilitated in social groups regardless of the exact mechanism involved. Currently, no evidence exists for increased production and transmission of information in social nonhuman groups; therefore, enhanced cognitive ability in members of social groups cannot yet be justified on this basis. Nevertheless, further experimental data may alter this preliminary conclusion. The critical empirical study required is an evaluation of the amounts of information used for intraspecific communication in solitary and social species or social species that live in distinct group sizes. Relatively objective methods for quantifying information can be used for such an investigation (see Beecher 1989, 1991).

The interplay between sociality and cognition can be tested with two methods. One method mentioned throughout this chapter is comparison of closely related species. Taxa that may be used for study include hymenopterans (Dukas and Real 1991), birds such as corvids (Balda et al. 1996), and mammals such as ground squirrels (Armitage 1981; Murie and Michener 1984); all characterized by a variety of social organizations. A selected group of primates may also be used for comparison. The other method involves experimental manipulation. Cats, other carnivores, and perhaps other mammals and birds can dramatically switch from solitary to social life when concentrated amounts of food are provided by humans (Macdonald 1983; several chap. in Turner and Bateson 1988). An experiment can be conducted in which one group of subjects lives as a coherent group and another lives in separate territories. Interactions of individuals and some cognitive abilities in either group can be compared. It would be especially interesting to test for the development of social interactions within and between generations. Such experimental manipulation could enrich our knowledge of the transition from solitary to social life and its associated increased cognitive demands, if any. Nonetheless, a manipulative approach cannot replace comparisons between species that have been separated for a long time.

4.5.6 Life Span

Life span has already been mentioned a few times in this chapter as a factor that could select for increased learning and cognitive capacity. Given the potential importance of life span, however, it deserves further examination. First, the *proportion* of life devoted to learning should not depend on absolute life span, be it 7 days or 70 years (Dukas and Visscher 1994), but the *total* amount of knowledge gained could be much higher for a longer-lived species. The fundamental ability to learn should not depend on life span; however, increasing the duration of life could select for an enlarged capacity to acquire and manipulate larger amounts of and more complex information.

Longer life span has another central effect. Most environments undergo seasonal changes. Species with life spans equal to or shorter than a certain season do not experience seasonal change, but longer-lived species may experience a few seasonal environments. Some species may consume only a single or a few resources throughout the year or hibernate in the less favorable season. Species, however, that consume a large number of seasonal items may benefit from learning various aspects of the temporal and spatial distribution, quality, and handling of each type. In effect, living longer may result in a more generalist species; as I discussed in section 4.5.2, increased generalization may select for increased learning capacity. Note that I use the term "resources" rather than "food" in order to include water, various nutrients, and shelters.

Increased life span is also a necessary condition for the interaction of parents with offspring. This clearly happens in hymenopterans, in which interactions between a longer-lived mother and her mature daughters exemplifies an evolutionary path to sociality (Evans 1958; Wilson 1971; Michener 1974). Even for long-lived taxa, an extended period of parental care is optimal only if parents have a high probability to be alive by the end of their offsprings' dependence period. Similarly, offspring investment in an extended period of parental care devoted to learning is optimal only if life expectancy is long enough to use the knowledge that is acquired during the prolonged juvenile stage. An enlarged capacity to learn allows a subject to continue accumulating extensive experience during adulthood (see section 4.5.4.2) and may result in increased reproductive performance throughout life.

Substantial evidence from numerous bird studies indeed suggests that subjects show a gradual increase in reproductive performance with experience

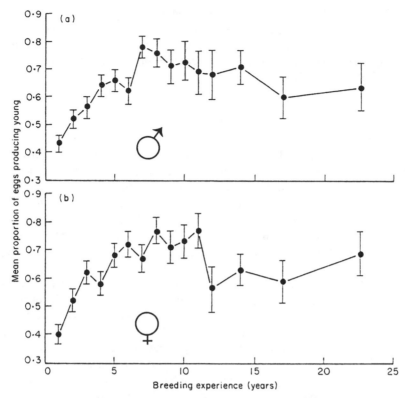

Figure 4.8. Mean (± SE) proportion of eggs that develop into free-flying young as a function of the breeding experience of their (a) male and (b) female short-tailed shearwater parents (*Puffinus tenuirostris*). The shearwater lays only one egg each year; lost or unsuccessful eggs are not replaced. Breeding success rose significantly for up to 8 years ($P < 0.001$). Very old individuals showed lower breeding success because of senescence (from Wooler et al. 1990).

(fig. 4.8). The increase in performance is depicted by short-lived song birds and long-lived species such as sea birds (Nol and Smith 1987; Newton 1988; Reid 1988; Wooller et al. 1990; Pyle et al. 1991; Desrochers 1992a; Black and Owen 1995). The gradual accumulation of information from experience over several years by the long-lived species suggests development of expertise that is similar to that in humans (see Lark et al. 1980; Anderson 1990). An alternative explanation for increase in performance with age is the elevated effort devoted to foraging and breeding by individuals that work harder and expand more resources at the cost of an increased mortality rate with advancing age. Indeed, Pugesek (1981; see also Pugesek and Diem 1990) repeatedly suggested that his observations on California gulls (*Larus californicus*) are better explained by this alternative. His results, however, were questioned by others (Nur 1984) and have not been repeated in other species (see Reid 1988). Moreover, there is direct evidence for the effects of experience, and not age alone, on factors that influence reproductive performance such as foraging, timing of breeding, nest location, and defense of eggs and offspring (Reid 1988; Desrochers 1992b; Nager and van Noordwijk 1995).

4.5.7 Synthesis

In the introduction to this section on components of fitness, natural history, and learning, I indicated that various life-history factors may have provided different selection pressures on different learning and cognitive capacities. Here I will integrate central arguments expressed in the preceding subsections. This integration, however, should be regarded as preliminary, given the scarcity of actual comparative data on life history and learning. The capacity to learn is ubiquitous among animals, including tiny and short-lived ones such as fruit flies and parasitoid wasps. Basic functions of learning include locating various resources in space and time, kin recognition, and distinguishing among items that strongly affect fitness and those with little or no effect (which should be ignored or assigned lower priority). An elaboration on these basic functions alone can produce a sophisticated cognitive system that processes a vast amount of information efficiently. The lengthening of life span may select for investment in enhanced cognitive capacities because larger amounts of information can be stored and used by longer-lived animals.

The effect of sociality has been highlighted in many discussions on the evolution of complex cognition, or intelligence (e.g. most authors in Byrne and Whiten 1988). Nonetheless, this effect of sociality is not well defended with available data. First, complex intraspecific interactions are expressed by many solitary species. These are interactions (1) between males and females; (2) among members of the same sex (in which interactions are more common among males competing for access to females); (3) between parents and offspring; and (4) among young in the same cohort. Therefore, sociality does not uniquely select for the learning of individual identification. Clearly, solitary subjects can benefit from the capacity to individually recognize potential mates and rivals with whom they are likely to interact in the future (Falls and Brooks 1975; Fenton 1985; Halpin 1986; Michener and Smith 1987; Godard 1991; Schwagmeyer 1995). Similarly, most other sophisticated capacities described for social species, such as transitive inference and deception, are as useful for solitary species; indeed, these capacities have been observed in solitary species (Cheney and Seyfarth 1990, 82, 186).

There is, however, one aspect of sociality that may provide greater advantage. Extended parental care requires the parent to support its young during the time it takes the young to reach independence. Therefore, the probability of parental survival must be high. Alternatively, or additionally, other members in a social group can support dependent or partially dependent young in case of parental death. Hence, extended parental care devoted to offspring learning can be less risky for members of a social group because parental death does not cause the loss of an already extensive investment in still dependent offspring. Essentially, this is an extension of the argument presented by Clark and Dukas (1994) for solitary and social bees. In other words, if sociality includes support of dependent immature offspring by other group members, parental care may be longer than that in either solitary species or social groups in which helpless young die if their parents die.

In the previous paragraph, I already hint to my central conclusion: A long period of parental care coupled with extended life span are probably the most crucial factors in the evolution of advanced learning and cognitive capacities. First, parental care selects for parents with more refined cognitive abilities. Second, parental protection and guidance allows young to learn under relative security. Thus, young can possess an elaborate machinery that records and processes information during the dependent period. That same cognitive machinery, enriched with extensive experience, can continue to be used by individuals after they reach maturity. As adults, subjects may not spend extensive time on pure learning, but they can still acquire information and improve performance in their central life activities (see Fig. 4.8). Economists refer to such learning as "learning while doing," (as opposed to time devoted exclusively to learning with little or no actual productivity) (e.g. Arrow 1962).

Admittedly, highly social species do seem more intelligent than solitary species. This opinion may be substantiated with empirical data in the future. However, probable substantiation should be taken cautiously because, as members of a highly social species, we project a strong bias. Are some social species indeed more intelligent, or is our definition of intelligence innately biased toward the social domain? Perhaps some long-lived solitary species of birds and mammals are as intelligent but this intelligence is expressed in domains we still do not appreciate. Perhaps future impartial studies will classify and quantify numerous specific intelligence traits that allow various animals to survive and reproduce in their particular or even peculiar natural settings.

4.6 LEARNING ABOUT LEARNING: CONCLUSIONS AND PROSPECTS

Learning can certainly be classified as a central trait that affects animal fitness. Given the importance of learning, it is surprising that there have been only a few rigorous examinations of learning by evolutionary and behavioral ecologists (e.g. Mayr 1974; Johnston 1982; Shettleworth 1984; Papaj and Prokopy

1989; Stephens 1991; Krebs and Inman 1994). I hope ideas presented in this chapter will inspire the further investigation needed. In the first part of this chapter, I attempted to integrate studies on phenotypic plasticity with those on animal learning. Research on phenotypic plasticity has focused on plastic traits that allow a specific phenotypic change in response to a certain environmental variation. More complex types of plasticity, including learning, allow an animal to modify its responses to environmental events on the basis of experience. This modification adds an additional level of flexibility, which can allow animals richer interactions with their physical and biotic environments. Key factors that determine the value of learning are the initial cost, rate of learning, and rate of reduction in value of knowledge acquired.

Basic learning abilities are used in every aspect of life by many animals, including short-lived and small insects; however, the transition to possessing more elaborate learning capacities seems to be associated with parental care and longer life span. These allow young to acquire extensive knowledge under parental protection and guidance and continue the accumulation of relevant experience throughout a long life. Various life-history traits require specific learning capacities. Empirically measuring learning in closely related species that differ in such habits will help uncover the various stages in the evolution of learning. I explicitly suggested such conceivable future experimental directions.

Theoretically, additional models, including quantitative genetic models, are needed to include learning within a broader conceptual framework of phenotypic plasticity. Learning may be perceived as cognitive growth; this is somewhat analogous to physical growth, which is the increase in body size with age. Hence, models of growth (e.g. Kirkpatrick 1988; Lynch and Arnold 1988) may be modified to address issues that concern learning. Topics requiring further evaluation include optimal values of and constraints on (1) environmental variables learned; (2) the effect of a single experience on learning; (3) the type, resolution, and amount of knowledge stored in memory; (4) the length of time a certain memory is maintained; and (5) the magnitude of change in a response given a certain experience. Clearly, we have much more to learn about learning!

4.7 SUMMARY

In this chapter, I addressed two aspects of learning. First, learning within the ecological framework of phenotypic plasticity was discussed. Studies of phenotypic plasticity have focused on a narrow range of plastic traits in which there is a consistent phenotypic response to some environmental variable. Such research has not evaluated complex types of reversible plasticity, such as learning—in which animals record experience in memory and alter their responses to the environment throughout life. I assessed learning in relation to simpler forms of plasticity. Simple forms of reversible plasticity are effective as long as an identical phenotypic response is a sufficient adaptation to some frequent environmental change. An example is changes in pupil diameter in response to variable light intensities. Conversely, learning is favored if feedback from interactions with an environmental event can be used to improve the phenotypic response to subsequent occurrences of that event.

The second issue addressed was that various life-history traits could select for different learning abilities. I examined the relative effects of aspects of foraging, predation, mate choice, parental care, sociality, and life span on learning. Rich learning capacities already appear even in short-lived solitary species. However, the transition toward more complex learning appears to be associated with parental care and a longer life span. I concluded by stressing the potential of comparative experimental studies, detailed in the section on life history and learning. Such empirical research could reveal critical steps in the evolution of learning. I also outlined further promising theoretical directions that would help place learning within a broader framework of phenotypic plasticity.

ACKNOWLEDGMENTS

My decade of learning and writing about learning has benefitted from discussions with or comments from P. Bednekoff, C. Clark, D. Cohen, F. Dyer, L. Gass, T. Getty, D. Grunbaum, V. Mirmovich, S. Otto, D. Papaj, L. Real and B. Robinson. I have been supported by the NSERC grant #83990 to C. Clark and the spiritual hospitality of British Columbia's wilderness.

LITERATURE CITED

Alcock, J. 1993. *Animal Behavior.* 5th ed. Sunderland, Mass: Sinauer.

Alexander, R. C. 1974. The evolution of social behavior. *Annual Review of Ecology and Systematics* 5:325–383.

Anderson, J. R. 1990. *Cognitive Psychology and Its Implications.* New York: Freeman.

Andersson, M. 1994. *Sexual Selection.* Princeton, N.J.: Princeton University Press.

Aoki, K. 1989. A sexual-selection model for the evolution of imitative learning of song in polygynous birds. *American Naturalist* 134:599–612.

Armitage, K. B. 1981. Sociality as a life history tactic of ground squirrels. *Oecologia* 48:36–49.

Arnold, S. J. 1981. Behavioral variation in natural populations. I. Phenotypic, genetic, and environmental correlations between chemoreceptive responses to prey in the garter snake, *Thamnophis elegans. Evolution* 35:489–509.

———. 1994a. Constraints on phenotypic evolution. In L. A. Real, ed., *Behavioral Mechanisms in Evolutionary Ecology,* 258–278. Chicago: University of Chicago Press.

————. 1994b. Multivariate inheritance and evolution: A review of concepts. In R. B. Boake, ed., *Quantitative Genetic Studies of Behavioral Evolution,* 17–48. Chicago: University of Chicago Press.

Atchley, W. R., and B. K. Hall. 1991. A model for development and evolution of complex morphological structures. *Biological Reviews* 66:101–157.

Balda, R. P., A. C. Kamil, and P. A. Bednekoff. 1996. Predicting cognitive capacities from natural histories: Examples from four species of corvids. *Current Ornithology.* 16:33–66.

Barrows, E. M. 1975. Individually distinctive odors in an invertebrate. *Behavioral Biology* 15:57–64.

Barrows, E. M., W. J. Bell, and C. D. Michener. 1975. Individual odor differences and their social functions in insects. *Proceedings of the National Academy of Sciences of the United States of America* 72:2824–2828.

Bateson, P. 1979. How do sensitive periods arise and what are they for? *Animal Behaviour* 27:470–486.

————. 1981. Control of sensitivity to the environment during development. In K. Immelmann, G. W. Barlow, L. Petronovich, and M. Main, eds., *Behavioral Development,* 432–453. Cambridge: Cambridge University Press.

Batra, S. W. T. 1984. Solitary bees. *Scientific American* 250:86–93.

Beecher, M. D. 1989. Signaling systems for individual recognition: An information theory approach. *Animal Behaviour* 38:248–261.

————. 1991. Success and failures of parent-offspring recognition in animals. In P. G. Hepper, ed., *Kin Recognition,* 94–124. Cambridge: Cambridge University Press.

Beecher, M. D., P. K. Stoddard, and P. Loesche. 1985. Recognition of parents' voices by young cliff swallows. *Auk* 102:118–146.

Beer, C. G. 1969. Laughing gull chicks: Recognition of their parents' voices. *Science* 166:1030–1032.

————. 1979. Vocal communication between laughing gull parents and their chicks. *Behaviour* 70:118–146.

Bernays, E., and M. Graham. 1988. On the evolution of host specificity in phytophagous arthropods. *Ecology* 69:886–892.

Black, J. M., and M. Owen. 1995. Reproductive performance and assortive pairing in relation to age in barnacle geese. *Journal of Animal Ecology* 64:234–244.

Byrne, R. 1995. *The Thinking Ape.* Oxford: Oxford University Press.

Byrne, R., and A. Whiten, eds. 1988. *Machiavellian Intelligence.* Oxford: Oxford University Press.

Caro, T. M. 1980. Effects of the mother, object play, and adult experience on predation in cats. *Behavioural and Neural Biology* 29:29–51.

Cartwright, B. A., and T. S. Collett. 1983. Landmark learning in bees. *Journal of Comparative Physiology* 151:521–543.

Cheney, D. L., and R. M. Seyfarth. 1990. *How Monkeys See the World.* Chicago: University of Chicago Press.

Cheney, D., R. Seyfarth, and B. Smuts. 1986. Social relationships and social cognition in nonhuman primates. *Science* 234:1361–1366.

Chivers, D. P., and J. F. Smith. 1994. Fathead minnows, *Pimephales promelas,* acquire predator recognition when alarm substance is associated with the sight of unfamiliar fish. *Animal Behaviour* 48:597–605.

Clark, C. W., and R. Dukas. 1994. Balancing foraging and antipredator demands: An advantage of sociality. *American Naturalist* 144:542–548.

Clark, C. W., and C. D. Harvell. 1992. Inducible defenses and the allocation of resources: A minimal model. *American Naturalist* 139:521–539.

Clark, W. R. 1995. *The Double-Edged Sword of Immunity.* Oxford: Oxford University Press.

Clutton-Brock, T. H. 1991. *The Evolution of Parental Care.* Princeton, N.J.: Princeton University Press.

Clutton-Brock, T. H., and P. H. Harvey. 1980. Primates, brains, and ecology. *Journal of Zoology* 190:309–323.

Corning, W. C., J. A. Dyal, and A. O. D. Willow. 1972. *Invertebrate Learning.* New York: Plenum.

Cosmides, L., and J. Tooby. 1992. Cognitive adaptations for social exchange. In J. H. Barkow, L. Cosmides, and J. Tooby, eds., *The Adapted Mind: Evolutionary Psychology and the Generation of Culture,* 163–228. Oxford: Oxford University Press.

Craig, C. L. 1994. Limits to learning: Effects of predator pattern and colour on perception and avoidance-learning by prey. *Animal Behaviour* 47:1087–1099.

Curio, E. 1993. Proximate and developmental aspects of antipredator behavior. *Advances in the Study of Behavior* 22:135–238.

Davies, N. B. 1976. Parental care and the transition to independent feeding in the young spotted flycatcher *(Muscicapa striata). Behaviour* 59:280–295.

Davies, N. B., and R. E. Green. 1976. The development and ecological significance of feeding techniques in the reed warbler *(Acrocephalus scripaceus). Animal Behaviour* 24:213–229.

Desrochers, A. 1992a. Age and foraging success in European blackbirds: Variation between and within individuals. *Animal Behaviour* 43:885–894.

———. 1992b. Age-related differences in reproduction by European blackbirds: Restraint or constraint? *Ecology* 73:1128–1131.

Dudai, Y. 1989. *The Neurobiology of Memory: Concepts, Findings, Trends,* Oxford: Oxford University Press.

Dukas, R., and L. Real. 1991. Learning foraging tasks by bees: A comparison between social and solitary species. *Animal Behaviour* 42:269–276.

———. 1993. Effects of recent experience on foraging decisions by bumblebees. *Oecologia* 94:244–246.

Dukas, R., and P. K. Visscher. 1994. Lifetime learning by foraging honeybees. *Animal Behaviour* 48:1007–1012.

Ehrlich, P. R., and P. H. Raven. 1964. Butterflies and plants: A study in coevolution. *Evolution* 18:586–608.

Eickwort, G. C., and H. S. Ginsberg. 1980. Foraging and mating behavior in Apoidea. *Annual Review of Entomology* 25:421–446.

Eisenberg, J. F., and D. E. Wilson. 1978. Relative brain size and feeding strategies in the Chiroptera. *Evolution* 32:740–751.

Evans, H. E. 1953. Comparative ethology and the systematics of spider wasps. *Systematic Zoology* 2:155–172.

168   Reuven Dukas

————. 1957. *Comparative Ethology of Digger Wasps of the Genus* Bembix. Ithaca, N.Y.: Cornell University Press.

————. 1958. The evolution of social life in wasps. *Proceedings of the International Congress of Entomology* 2:449–457.

————. 1966. *The Comparative Ethology and Evolution of the Sand Wasps,* Cambridge, Mass.: Harvard University Press.

Evans, H. E., and J. W. Eberhard. 1970. *The Wasps.* Ann Arbor: University of Michigan Press.

Falconer, D. S. 1989. *Introduction to Quantitative Genetics* 3d ed. New York: Wiley.

Falls, J. B. 1982. Individual recognition by sounds in birds. In D. E. Kroodsma and E. H. Miller, eds., *Acoustic Communication in Birds,* 237–278. New York: Academic.

Falls, J. B., and R. J. Brooks. 1975. Individual recognition by song in white-throated sparrows. II. Effects of location. *Canadian Journal of Zoology* 53:1412–1520.

Fenton, M. B. 1985. *Communication in the Chiroptera.* Bloomington: Indiana University Press.

Finlay, B. L., and R. B. Darlington. 1995. Linked regularities in the development and evolution of mammalian brains. *Science* 268:1578–1584.

Fischler, M. A., and O. Firschein. 1987. *Intelligence: The Eye, the Brain, and the Computer.* Reading, Mass.: Addison-Wesley.

Fletcher, J. C., and C. D. Michener, eds. 1987. *Kin Recognition in Animals.* New York: Wiley.

Fox, L. R., and P. A. Morrow. 1981. Specialization: Species property or local phenomenon? *Science* 211:887–893.

Frisch, K. von. 1967. *The Dance Language and Orientation of Bees.* Cambridge, Mass.: Harvard University Press.

Galef, B. G. 1976. Social transmission of acquired behavior: A discussion of tradition and social learning in vertebrates. *Advances in the Study of Behavior* 6:77–100.

————. 1988. Imitation in animals: History, definition, and interpretation of data from the psychological laboratory. In T. R. Zentall and B. G. Galef, eds., *Social Learning: Psychological and Biological Perspectives,* 3–28. Hillsdale, N.J.: Erlbaum.

Garcia, J., and R. A. Koelling. 1966. Relation of cue to consequence in avoidance learning. *Psychonomic Science* 4:123–124.

Gardner, A., ed. 1993. *The Neurobiology of Neural Networks.* Cambridge, Mass.: MIT Press.

Gilbert, L. E. 1975. Ecological consequences of a coevolved mutualism between butterflies and plants. In E. Gilbert and P. H. Raven, eds., *Coevolution of Animals and Plants,* 210–240. Austin: University of Texas Press.

————. 1991. Biodiversity of a Central American *Heliconius* community: Pattern, process, and problems. In P. W. Price, T. M. Lewinsohn, G. W. Fernandes, and W. W. Benson, eds., *Plant-Animal Interactions,* 403–430. New York: Wiley.

Gittleman, J. L., and P. H. Harvey. 1980. Why are distasteful prey not cryptic? *Nature* 286:149–150.

Godard, R. 1991. Long-term memory of individual neighbors in a migratory songbird. *Nature* 350:228–229.

Gomulkiewicz, R., and M. Kirkpatrick. 1992. Quantitative genetics and the evolution of reaction norms. *Evolution* 46:390–411.

Gould, J. P. 1974. Risk, stochastic preference, and the value of information. *Journal of Economic Theory* 8:64–84.

Greenberg, L. 1979. Genetic component of bee odor in kin recognition. *Science* 206:1059–1097.

Greenberg, R. 1983. The role of neophobia in determining the degree of foraging specialization in some migrant warblers. *American Naturalist* 122:444–453.

———. 1990. Feeding neophobia and ecological plasticity: A test of the hypothesis with captive sparrows. *Animal Behaviour* 39:375–379.

———. 1992. Differences in neophobia between naive song and swamp sparrows. *Ethology* 91:17–24.

Greene, E. 1989. A diet-induced developmental polymorphism in a caterpillar. *Science* 243:643–646.

Guilford, T., and M. S. Dawkins. 1991. Receiver psychology and the evolution of animal signals. *Animal Behaviour* 42:1–14.

Hall, J. C. 1994. The mating of a fly. *Science* 264:1702–1714.

Halpin, Z. T. 1986. Individual odors among mammals: Origins and functions. *Advances in the Study of Behavior* 16:39–70.

Harvell, C. D. 1986. The ecology and evolution of inducible defences in a marine bryozoan: Cues, costs, and consequences. *American Naturalist* 128:810–823.

Harvey, P. H., and J. R. Krebs. 1990. Comparing brains. *Science* 249:140–146.

Havel, J. E. 1985. Cyclomorphosis of *Daphnia pulex* spined morphs. *Limnology and Oceanography* 30:853–861.

Heinrich, B. 1976. The foraging specialization of individual bumblebees. *Ecological Monographs* 46:105–128.

———. 1979. Majoring and minoring by foraging bumblebees, *Bombus vagance:* An exeperimental analysis. *Ecology* 60:245–255.

———. 1984. Learning in invertebrates. In P. Marler and H. S. Terrace, eds., *The Biology of Learning,* 135–147. Berlin: Springer-Verlag.

———. 1989. *Ravens in Winter.* New York: Simon and Schuster.

Hepper, P. G., ed. 1991. *Kin Recognition.* Cambridge: Cambridge University Press.

Herrera, C. M. 1982. Defense of ripe fruit from pests: Its significance in relation to plant-dispenser interactions. *American Naturalist* 120:218–241.

Hochachka, P. W., and G. N. Somero. 1984. *Biochemical Adaptation.* Princeton, N.J.: Princeton University Press.

Holldobler, B., and C. D. Michener. 1980. Mechanisms of identification and discrimination in social hymenoptera. In H. Markl, ed., *Evolution of Social Behavior: Hypotheses and Empirical Tests,* 35–57. Berlin: Verlag.

Jerison, H. J. 1973. *Evolution of the Brain and Intelligence.* New York: Academic.

Johnson, R. A. 1991. Learning, memory, and foraging efficiency in two species of desert seed-harvester ants. *Ecology* 72:1408–1419.

Johnston, T. D. 1982. Selective costs and benefits in the evolution of learning. *Advances in the Study of Behavior* 12:65–106.

Kerby, G., and D. W. Macdonald. 1988. Cat society and the consequence of colony size. In D. C. Turner and P. Bateson, eds., *The Domestic Cat,* 67–81. Cambridge: Cambridge University Press.

Kingsolver, J. G., and D. C. Wiernasz. 1991. Seasonal polyphenism in wing melanin pattern and thermoregulatory adaptation in *Pieris* butterflies. *American Naturalist* 137:816–830.

Kirkpatrick, M. 1988. The evolution of growth patterns and size. In B. Ebenman and L. Person, eds., *Size Structured Populations,* 13–28. Berlin: Springer-Verlag.

Komi, P. V., ed. 1992. *Strength and Power in Sport.* Oxford: Blackwell Scientific.

Koshland, D. 1980. *Bacterial Chemotaxis as a Model Behavioral System.* New York: Raven.

Krebs, J. R., and A. J. Inman. 1994. Learning and foraging: Individuals, groups and populations. In L. A. Real, ed., *Behavioral Mechanisms in Evolutionary Ecology,* 46–65. Chicago: University of Chicago Press.

Kroodsma, D. E. 1983. The ecology of avian vocal learning. *BioScience* 33:165–171.

———. 1987. Contrasting styles of song development and their consequences among passerine birds. In R. C. Bolles and M. D. Beecher, eds., *Evolution and Learning,* 157–184. Hillsdale, N.J.: Erlbaum.

Kroodsma, D. E., and B. E. Byers. 1991. The function(s) of bird song. *American Zoologist* 31:318–328.

Lark, J., J. McDermont, D. P. Simon, and H. A. Simon. 1980. Expert and novice performance in solving physics problems. *Science* 208:1335–1342.

Laverty, T. M. 1980. Bumblebee foraging: Floral complexity and learning. *Canadian Journal of Zoology* 58:1324–1335.

Laverty, T. M., and R. C. Plowright. 1988. Flower handling by bumblebees: A comparison of specialists and generalists. *Animal Behaviour* 36:733–740.

Lefebvre, L., P. Whittle, E. Lascaris, and A. Finkelstein. 1997. Feeding innovations and forebrain size in birds. *Animal Behaviour* 53:549–560.

Lenat, D. B. 1995. Artificial Intelligence. *Scientific American* 273(3):80–82.

Levins, R. 1968. *Evolution in Changing Environments.* Princeton, N.J.: Princeton University Press.

Leyhausen, P. 1965. The communal organization of solitary mammals. *Symposia of the Zoological Society of London* 14:249–263.

Lindquist, N., and M. E. Hay. 1995. Can small rare prey be chemically defended? The case for marine larvae. *Ecology* 76:1347–1358.

Lively, C. 1986a. Canalization versus developmental conversion in a spatially variable environment. *American Naturalist* 128:561–572.

———. 1986b. Predator-induced shell dimorphism in the acorn barnacle, *Chthamalus anisopoma. Evolution* 40:232–242.

Lorenz, K. 1965. *Evolution and Modification of Behavior.* Chicago: University of Chicago Press.

Lynch, M., and S. J. Arnold. 1988. The measurement of selection on size and growth. In B. Ebenman and L. Person, eds., *Size Structured Populations,* 47–59. Berlin: Springer-Verlag.

Macdonald, D. W. 1983. The ecology of carnivore social behaviour. *Nature* 301:379–384.

Mace, G. M., P. H. Harvey, and T. H. Clatton-Brock. 1981. Brain size and ecology in small mammals. *Journal of Zoology* 193:333–354.

Mallet, J. 1986. Gregarious roosting and home range in *Helliconius* butterflies. *National Geographic Research* 2:198–215.

Mallet, J., J. T. Longino, D. Murawski, A. Muravski, and A. Simpson de Gamboa. 1987. Handling effects on *Helliconius:* Where do all the butterflies go? *Journal of Animal Ecology* 56:377–386.

Maloney, R. F., and I. G. McLean. 1995. Historical and experimental learned predator recognition in free-living New Zealand robins. *Animal Behaviour* 50:1193–1201.

Martin, P., and P. Bateson. 1988. Behavioural development in the cat. In D. C. Turner and P. Bateson, eds., *The Domestic Cat*, 9–22. Cambridge: Cambridge University Press.

Mayr, E. 1974. Behavior programs and evolutionary strategies. *American Scientist* 62: 650–659.

Menzel, R. 1985. Learning in honeybees in an ecological and behavioral context. In B. Holldobler and M. Lindauer, eds., *Experimental Behavioral Ecology*, 55–74. Stuttgart: Verlag.

Menzel, R., J. Erber, and T. Masuhr. 1974. Learning and memory in the honeybee. In L. Barton-Browne, ed., *Experimental Analysis of Insect Behavior*, 195–217. Berlin: Springer-Verlag.

Michener, C. D. 1974. *The Social Behavior of the Bees.* Cambridge, Mass.: Harvard University Press.

Michener, C. D., and B. H. Smith. 1987. Kin recognition in primitively eusocial insects. In D. J. C. Fletcher and C. D. Michener, eds., *Kin Recognition in Animals*, 209–242. New York: Wiley.

Minckley, R. L., W. T. Wcislo, D. Yanega, and S. L. Buchmann. 1994. Behavior and phenology of a specialist bee (*Dieunomia*) and sunflower (*Helianthus*) pollen availability. *Ecology* 75:1406–1419.

Moran, N. A. 1992. The evolutionary maintenance of alternative phenotypes. *American Naturalist* 139:971–989.

Murie, J. O., and G. R. Michener, eds. 1984. *The Biology of Ground-Dwelling Squirrels* Lincoln: University of Nebraska Press.

Nager, R. G., and A. J. van Noordwijk. 1995. Proximate and ultimate aspects of phenotypic plasticity in timing of great tit breeding in a heterogeneous environment. *American Naturalist* 146:454–474.

Newell, K. M. 1991. Motor skill acquisition. *Annual Review of Psychology* 42:213–237.

Newton, I. 1988. Age and reproduction in the sparrowhawk. In T. H. Clutton-Brock, ed., *Reproductive Success*, 201–219. Chicago: University of Chicago Press.

Nol, E., and J. N. M. Smith. 1987. Effects of age and breeding experience on seasonal reproductive success in the song sparrow. *Journal of Animal Ecology* 56:301–313.

Nottebohm, F. 1972. The origins of vocal learning. *American Naturalist* 106:116–140.

Nur, N. 1984. Increased reproductive success with age in the California gull: Due to increased effort of improvement of skill? *Oikos* 43:407–408.

Ode, P. J., M. F. Antolin, and M. R. Strand. 1995. Brood-mate avoidance in the parasitic wasp *Bracon hebetor* Say. *Animal Behaviour* 49:1239–1248.

Padilla, D. K., and S. C. Adolph. 1996. Plastic inducible morphologies are not always adaptive: The importance of time delays in a stochastic environment. *Evolutionary Ecology* 10:105–117.

Papaj, D. R. 1986. Interpopulation differences in host preferences and the evolution of learning in the butterfly, *Battus philenor. Evolution* 40:518–530.

Papaj, D. R., and A. C. Lewis, eds. 1993. *Insect Learning.* New York: Chapman and Hall.

Papaj, D. R., and R. J. Prokopy. 1989. Ecological and evolutionary aspects of learning in phytophagous insects. *Annual Review of Entomology* 34:315–350.

Parmesan, C., M. C. Singer, and I. Harris. 1995. Absence of adaptive learning from the oviposition foraging behaviour of a checkerspot butterfly. *Animal Behaviour* 50: 161–175.

Potting, R. P. J., H. Otten, and L. E. M. Vet. 1997. Absence of odour learning in the stemborer parasitoid *Cotesia flavipes. Animal Behaviour* 53:549–560.

Prokopy, R. J., A. L. Averill, S. S. Cooley, and C. A. Roitberg. 1982. Associative learning in egg-laying site selection by apple maggot flies. *Science* 218:76–77.

Pugesek, B. H. 1981. Increased reproductive effort with age in the California gull (*Larus californicus*). *Science* 212:822–823.

Pugesek, B. H., and K. L. Diem. 1990. The relationship between reproduction and survival in known-aged California gulls. *Ecology* 71:811–817.

Pyle, P., L. B. Spear, W. J. Sydeman, and D. G. Ainley. 1991. The effects of experience and age on the breeding performance of western gulls. *Auk* 108:25–33.

Rathcke, B. 1988. Interactions for pollination among coflowering shrubs. *Ecology* 69: 446–457.

Rausher, M. D. 1978. Search image for leaf shape in a butterfly. *Science* 200:1071–1073.

Real, L., and B. J. Rathcke. 1988. Patterns of individual variability in floral resources. *Ecology* 69:728–735.

Reid, W. V. 1988. Age-specific patterns of reproduction in the glaucous-winged gull: Increased effort with age? *Ecology* 69:1454–1465.

Roff, D. S. 1986. The evolution of wing dimorphism in insects. *Evolution* 40:1009–1020.

Rothstein, S. I. 1986. A test of optimality: Egg recognition in the western phoebe. *Animal Behaviour* 34:1109–1119.

———. 1990. A model system for co-evolution: Avian brood parasitism. *Annual Review of Ecology and Systematics* 21:481–508.

Rozin, P., and J. Kalat. 1971. Specific hungers and poison avoidance as adaptive specializations of learning. *Psychological Review* 78:459–486.

Rubenstein, D. I., and R. W. Wrangham, eds. 1986. *Ecological Aspects of Social Evolution.* Princeton, N.J.: Princeton University Press.

Sale, D. G. 1992. Neural adaptation to strength training. In P. V. Komi, ed., *Strength and Power in Sport,* 249–265. London: Blackwell Scientific.

Scheiner, S. M. 1993. Genetics and evolution of phenotypic plasticity. *Annual Review of Ecology and Systematics* 24:35–68.

Schmalhausen, I. I. 1949. *Factors of Evolution.* Philadelphia: Blakiston.

Schwagmeyer, P. L. 1995. Searching today for tomorrow mates. *Animal Behaviour* 50:759–767.

Seeley, T. D. 1985. *Honeybee Ecology.* Princeton, N.J.: Princeton University Press.

Seligman, M. E. P., and J. L. Hager, eds. 1972. *Biological Boundaries of Learning.* New York: Appleton.

Sergent, J., E. Zuck, S. Terriah, and B. MacDonald. 1992. Distributed neural network underlying musical sight-reading and keyboard performance. *Science* 257:106–109.

Shapiro, A. M. 1976. Seasonal polyphenism. *Evolutionary Biology* 9:259–333.

Sherry, D. F., and B. G. Galef. 1984. Cultural transmission without imitation: Milk bottle opening by birds. *Animal Behaviour* 32:937–938.

Sherry, D. F., and D. L. Schacter. 1987. The evolution of multiple memory systems. *Psychological Review* 94:439–454.

Shettleworth, S. J. 1984. Learning and behavioural ecology. In J. R. Krebs and N. B. Davies, eds., *Behavioural Ecology,* 170–194. Oxford: Blackwell Scientific.

Simmers, J., P. Meyrand, and M. Moulins. 1995. Dynamic networks of neurons. *American Scientist* 83:262–268.

Singer, M. C. 1994. Behavioral constraints on the evolutionary expansion of insect diet: A case history from checkerspot butterflies. In L. A. Real, ed., *Behavioral Mechanisms in Evolutionary Ecology,* 279–296. Chicago: University of Chicago Press.

Singley, K., and J. R. Anderson. 1989. *The Transfer of Cognitive Skill.* Cambridge, Mass.: Harvard University Press.

Smiley, J. 1978. Plant chemistry and the evolution of host specificity: New evidence from *Heliconius passiflora. Science* 201:745–747.

Smith, B. H. 1983. Recognition of female kin by male bees through olfactory signals. *Proceedings of the National Academy of Sciences of the United States of America.* 80:4551–4553.

Staddon, J. E. R. 1983. *Adaptive Behavior and Learning.* Cambridge, Mass.: Cambridge University Press.

Stamps, J. 1995. Motor learning and the value of familiar space. *American Naturalist* 146:41–58.

Stearns, S. C. 1989. The evolutionary significance of phenotypic plasticity. *BioScience* 39:436–445.

Stemberger, R. S., and J. J. Gilbert. 1987. Multiple species induction of morphological defense in the rotifer, *Keratella testudo. Ecology* 68:370–378.

Stephens, D. W. 1989. Variance and the value of information. *American Naturalist* 134:128–140.

———. 1991. Change, regularity, and value in the evolution of animal learning. *Behavioral Ecology* 2:77–89.

Stephens, D. W., and J. Krebs. 1986. *Foraging Theory.* Princeton N.J.: Princeton University Press.

Sullivan, K. A. 1988a. Age-specific profitability and prey choice. *Animal Behaviour* 36:613–615.

———. 1988b. Ontogeny of time budgets in yellow-eyed juncos: Adaptation to ecological constraints. *Ecology* 69:118–124.

————. 1989. Predation and starvation: Age-specific mortality in juvenile juncos (*Junco phaenotus*). *Journal of Animal Ecology* 58:275–286.

Thomson, J. D. 1982. Patterns of visitation by animal pollinators. *Oikos* 39:241–250.

Tinbergen, N. 1951. *The Study of Instinct.* Oxford: Oxford University Press.

Traynier, R. M. M. 1984. Associative learning in the ovipositional behaviour of the cabbage butterfly, *Pieris rapae*. *Physiological Entomology* 9:465–472.

Turlings, T. C. J., J. H. Tumlinson, and W. J. Lewis. 1990. Exploitation of herbivore-induced plant odors by host-seeking wasps. *Science* 250:1251–1253.

Turlings, T. C. J., F. L. Wackers, L. E. M. Vet, W. J. Lewis, and J. H. Tumlinson. 1993. Learning of host-finding cues by hymenopterous parasitoids. In D. R. Papaj and A. C. Lewis, eds., *Insect Learning,* 51–78. London: Chapman and Hall.

Turner, D. C., and P. Bateson, eds. 1988. *The Domestic Cat.* Cambridge: Cambridge University Press.

Vet, L. E. M., and M. Dicke. 1992. Ecology of infochemical use by natural enemies in a tritrophic context. *Annual Review of Entomology* 37:141–172.

Via, S. 1987. Genetic constraints on the evolution of phenotypic plasticity. In V. Loeschcke, ed., *Genetic Constraints on Adaptive Evolution,* 47–71. Berlin: Springer-Verlag.

Via, S., and R. Lande. 1985. Genotype-environment interaction and the evolution of phenotypic plasticity. *Evolution* 39:505–522.

Vinson, S. B. 1976. Host selection by insect parasitoids. *Annual Review of Entomology* 21:109–134.

Visalberghi, E., and D. M. Fragaszy. 1990. Do monkeys ape? In S. T. Parker and K. R. Gibson, eds., *"Language" and Intelligence in Monkeys and Apes,* 247–273. New York: Cambridge University Press.

Visscher, P. K. 1983. The honeybee way of death: Necrophobic behavior in *Apis mellifera* colonies. *Animal Behaviour* 31:1070–1076.

Visscher, P. K., and T. D. Seeley. 1982. Foraging strategy of honeybee colonies in a temperate deciduous forest. *Ecology* 63:1790–1801.

West-Eberhard, M. J. 1989. Phenotypic plasticity and the origins of diversity. *Annual Review of Ecology and Systematics* 20:249–278.

Wilson, E. O. 1971. *The Insect Societies.* Cambridge, Mass.: Harvard University Press.

Winston, M. L., and K. N. Slessor. 1992. The essence of royalty: Honeybee queen pheromone. *American Scientist* 80:374–385.

Wooller, R. D., J. S. Bradley, I. J. Skira, and D. L. Serventy. 1990. Reproductive success of short-tailed shearwater *Puffinus tenuirostris* in relation to their age and breeding experience. *Journal of Animal Ecology* 59:161–170.

CHAPTER FIVE

The Cognitive Ecology of Song Communication and Song Learning in the Song Sparrow

MICHAEL D. BEECHER,
S. ELIZABETH CAMPBELL,
AND J. CULLY NORDBY

5.1 INTRODUCTION

Song is the first line of communication among territorial songbirds. These birds post their territories with loud song that can be heard several territories away. Song undoubtedly serves additional functions in neighbor interactions beyond this purely defensive one; however, to date we have strong evidence only for the posting function (review by Catchpole and Slater 1995).

Among animals that use song, songbirds (oscines) are unusual; the form of their song is influenced critically by early learning. The general functions of song learning are debated, but considerable evidence suggests that learning permits adaptation of the song repertoire to the species' local ecology, especially the social ecology (e.g. Kroodsma 1983, 1988; Payne 1983; Slater 1989). Specifically, in many songbird species, birds sing songs distinctive of their local area. For example, in our study population of song sparrows (*Melospiza melodia*), birds sing songs distinctive of their neighborhood, which consists of six to twelve birds. In this population, birds disperse from their birthplace when about a month old and remain in a new area for the rest of life. Song learning in these birds begins at 1 month of age, after they leave the birthplace, and ends at the beginning of the first breeding season approximately 9 months later (Beecher et al. 1994b; Nordby et al. submitted). Thus, the timing of song learning—beginning after the bird enters his new home area and before he becomes fully territorial—suggests that the bird needs to learn songs that are specific to his chosen neighborhood. A variety of dialect patterns have been described for different songbirds, but the general idea that birds learn songs that are appropriate to their new neighborhood is broadly consistent with the available data (Kroodsma 1983, 1988; Baptista and Morton 1988; Nelson and Marler 1994). We will refer to this idea as the *social-ecology* hypothesis of song learning.

The first of two major themes in this chapter is this: The social-ecology

175

hypothesis has strong implications for the study of song learning and song communication. The social-ecology hypothesis suggests a particular view of song learning; namely, the rules of song learning reflect an evolved strategy that is designed to equip the bird with songs that are useful in his new neighborhood. Our goal, then, is to uncover the features of this hypothesized strategy of song learning in our study species. Similarly, the social-ecology hypothesis suggests that in interactions with their neighbors, birds will use their songs in a way that capitalizes on these being local songs. In this case also, our goal is to uncover the rules of song communication and their ecological importance.

The second major theme in this chapter is that rules of song communication and song learning may have a strong cognitive component. Until recently, a distinctly non-cognitive view has prevailed in the study of bird song. The typical models usually have not been explicit; however, on inspection, these models can be seen to be derived from the classic mechanistic models of ethology. Song is viewed typically as similar to a simple sign stimulus or releaser, and song learning is thought of as a type of imprinting. These perspectives and a desire for experimental control provided the justification for the classic experimental paradigm for the study of song learning— the "tape tutor" experiment (e.g. Marler 1970). In this experimental situation, all aspects of species- and population-specific song learning are eliminated except the song itself. Specifically, the young bird is not exposed to other birds, except for the songs he hears over the loudspeaker in his isolation chamber.

Similar perspectives have been common in the study of song communication, although these studies have been carried out routinely in the field. Field playback experiments have provided the major analysis of bird-song communication (McGregor 1992), yet these studies have often taken a quasi-laboratory approach. For example, the songs used as playback stimuli most often are from birds unknown to the subject (i.e. strangers); the assumption is that the stranger songs don't have any specific associations that neighbor songs might have. These playback experiments show their relationship to the classic ethological model experiments from which they are derived. The focus is on the magnitude of the bird's response to the manipulated parameters of the stimulus song. The implicit assumption is that the bird's response will be determined entirely by the properties of the stimulus; the context can be ignored, as it might be if the experiment were being carried out in the laboratory.

Although research influenced by these ethological models has contributed greatly to our understanding of song communication and song learning, further progress is constrained by the limitations of these models. In this chapter, we will argue that our models of song learning and song communication and our corresponding research approaches must be more realistic in two respects. First, as suggested already, research should focus on social and ecological factors instead of excluding them. Second, models should credit an animal with some cognitive capacity instead of assuming the animal to be a mere ethological automaton. We will try to show how more cognitive models, in conjunction with a social-ecology perspective, have helped us understand the details of song communication and song learning in song sparrows. Although we will develop our arguments with respect to the particular species we studied, we believe these study results can be fruitfully applied to other songbird species.

Because our adopted cognitive perspective developed out of our research program and has no strong a priori theoretical basis, we will present our perspective in the context of the our research results. The social-ecology hypothesis, on the other hand, has a long history; this hypothesis has dictated the general approach of research, and we will present its background in the following sections. Throughout this chapter, we will suggest that the social-ecology and cognitive perspectives are synergistic; we will use the phrase "cognitive ecology" of song learning and song communication to refer to this synergism.

This chapter is essentially a detailed case study in cognitive ecology. General issues of communication, learning, and interindividual interactions are discussed elsewhere in this volume, in chapters 2, 4, and 10, respectively.

5.2 BACKGROUND

Song sparrows are typical of songbird species in two key respects. First, only males sing (this is typical at least of temperate zone species). Second, a male does not sing just one species-specific song; he has a repertoire of several such song types. (About three-quarters of songbird species have song repertoires [Kroodsma 1988].) A male song sparrow typically has six to ten distinct song types (mode = eight) in his song repertoire. The song types in his repertoire are as distinct from one another as are the song types of different song sparrows. Typically, a song sparrow sings his songs with "eventual" rather than "immediate" variety; that is, he repeats a particular song type a number of times before switching to a new song type. Under conditions of free singing (i.e. singing at high rates, as most males do when unpaired or when the female is incubating), a bird usually does not return to a song type until after he has sung all (or most) of his other song types. In addition to the very large differences between song types, song sparrows also make small changes in successive renditions of a song type. Variation in a song type, however, is small compared with variation among songs, and song types are clearly defined by the eventual variety of singing (Stoddard et al. 1988; Podos et al. 1992; Nowicki et al. 1994).

All studies we describe in this chapter were carried out in a sedentary (non-migratory) population of song sparrows in an undeveloped 200-hectare park that borders Puget Sound in Seattle, Washington. We attempted to color band all (or most) of the birds in this population. Birds banded in the nest and subsequently recaptured after leaving the nest (typically a mile or so from the nest) sang song types of their new area instead of those of their birthplace. Because most birds left our study area, however, we banded most subjects when they were juveniles, after they dispersed into the population.

Song sparrows in our sedentary population usually maintained the same territorial area for life; sometimes they made small moves, but rarely did they move more than one or two territories away. The birds lived in small neighborhoods in territorial pairs; females and males both defended these territories. These neighborhoods were mixtures of long-time neighbors and first-year birds; the young birds generally established their territories sometime between their first year (following dispersal from the birth area) and the following spring.

Two birds in a neighborhood generally shared some songs (i.e., they had very similar song types). Although the number of songs shared between two neighbors varied from zero to all, the average percentage of shared songs was about 30% in our study population. An example of song sharing is shown in figure 5.1. This pattern of sharing arose, as will be described, because young birds learned the songs of the neighborhood they entered when about a month old and in which they remained for the rest of life (Beecher et al. 1994; Nordby et al. submitted). Song sparrows did not modify their song repertoire after their first breeding season; however, young birds moving into the area learned the songs of the older birds there and thus maintained song sharing in the neighborhood.

Male song sparrows used their songs in a variety of intra- and intersexual contexts. We will focus on one context of song communication—countersinging by territorial neighbors during the breeding season. Song sparrows used song in intersexual contexts as well (e.g. O'Loghlen and Beecher 1997), but we will not consider this context in this chapter.

5.3 SONG LEARNING

5.3.1 Song Learning Studies: General Procedures

The social-ecology approach dictates that we study song learning in the field; only after the key social and ecological variables have been identified does it make sense to move the study into the laboratory (for a more detailed discussion of the issues, see Beecher 1996). We attempted to record all the songs of all the birds in our study population, and we found that young birds' "song tutors" could be identified, just as in the laboratory, on the basis of the similar song types of student and tutor. In the laboratory, the songs the bird has heard

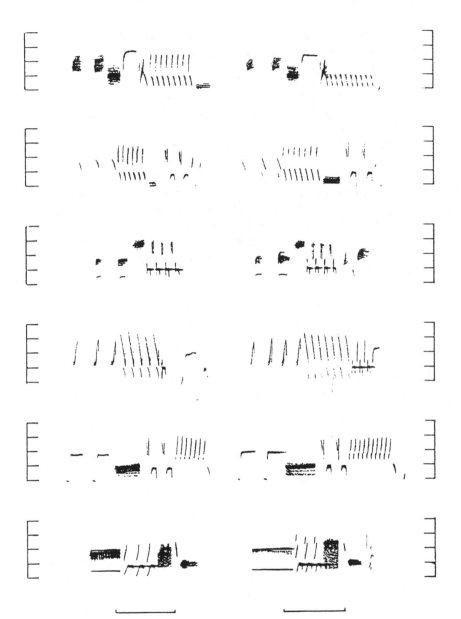

Figure 5.1. Sonagrams of six song types shared by two neighbors (the remainder of their song repertoires, not shown, were unshared song types). Frequency scale = 2–10 kHz in 2-kHz steps, time marker = 1 sec, bandwidth = 117 Hz.

can be identified with greater certainty (especially when tape tutors are used), but tutor identification, in some respects, is actually easier in the field. In our song sparrow studies, for example, song copying was more faithful and precise in the field than in the laboratory (see the following).

We considered all birds in the study population who were in the territory during the subject's hatching year as possible song tutors. We identified the bird with the most similar rendition of the song type (complete with idiosyncratic features not heard in other renditions of the song type) as the young bird's probable tutor for that song type. This judgment was rarely difficult because song-sparrow songs are complex and similar songs are conspicuous against the background of the nearly infinite variety of possible song types. In addition to the tutor–student classification, we also classified songs that were shared by tutors (i.e. two or more tutors had very similar songs) or unshared (i.e. the song type was unique to one singer in the reference group). Procedurally, this classification was fairly easy and very similar to the tutor–student classification because highly similar songs were conspicuous against the background of extraordinary song diversity.

We will summarize the results of these field studies of song learning (Beecher et al. 1994b; Nordby et al. submitted). We will contrast these results with those of conventional song-learning experiments in the laboratory.

5.3.2 Song Learning in the Field: Inferences Regarding Song-Learning Rules

Song sparrows in our sedentary population learned songs in a two-part process, which appears to fit the model developed by Marler and Nelson from laboratory and field studies (Marler and Nelson 1992; Nelson and Marler 1994). A young song sparrow left his area of birth at around a month of age. He began to learn songs during the second and third months of life in the neighborhood where he would remain and ultimately attempt to set up a territory. During this early stage, he memorized the song types of the adult birds in his new neighborhood. He did not finalize his repertoire of eight or so songs, however, until the following March. Although song memorization seemed to be completed during the sensitive period of early summer, the bird continued to be influenced by adult tutor-neighbors until the next spring; the bird generally settled next to the tutor-neighbor from whom he learned the most songs. Consequently the bird's final repertoire was more heavily influenced by tutors of the previous summer who survived the winter than those who did not. There was also usually greater influence from nearer tutor-neighbors than more distant tutor-neighbors. In this second phase of learning, the bird did not appear to learn new songs de novo, but was greatly influenced as to which of the many songs he had memorized during the early sensitive period he kept for his final repertoire of eight or so songs. Social interactions appeared to influence most of the songs that were retained; the bird tended to keep those song types that were shared by neighbors with whom the bird interacted most.

The pattern of song learning we observed in the field can be summarized in terms of the following rules of song learning.

5.3.2.1 The Young Bird Learns Songs from the Birds Who Will Be His Future Neighbors

As already indicated, the young song sparrow learned the songs of the adult males in his new neighborhood, and he preferentially retained the songs of birds that survived into his first breeding season. A representative example is shown in figure 5.2. When one examines the details of song learning, a number of interesting additional rules emerge.

5.3.2.2 The Young Bird Copies Whole Songs Precisely

In the field, young song sparrows usually learned to sing nearly perfect copies of the songs of their older neighbors (fig. 5.3). The similarities were striking; the differences between songs of tutor and student were often no greater than what is normally heard among repetitions of the same song sung by one bird. In the example in figure 5.3, which shows five songs of the young bird and four of his tutors, the biggest difference between songs of student and tutor is in the third song. The young bird appears to have simplified this song by dropping the high-frequency section near the end. The student's rendition of the fifth song is a blend of two tutor songs (this song is discussed later in this chapter).

These field results differ substantially from those of laboratory tape-tutor studies of song sparrows (Marler and Peters 1987, 1988). In the laboratory, young song sparrows copied elements of particular tutors, but they commonly combined elements of the different songs of different tutors to form hybrid songs, or songs made up of parts of different song types.

5.3.2.3 The Young Bird Preserves Type and Tutor in His Songs

In our field population, two exceptions to the perfect-copy rule actually clarified the rule and suggested an additional principle. The first exception occurred when the young bird blended two tutors' somewhat different versions of the same song type (instead of copying one or the other song type). These songs were not true hybrids because song elements—although selected from two different tutors—were selected from the same or very similar song type. The second exception, which was rare, occurred when the young bird combined elements from two dissimilar song types of the *same* tutor (fig. 5.4). The prin-

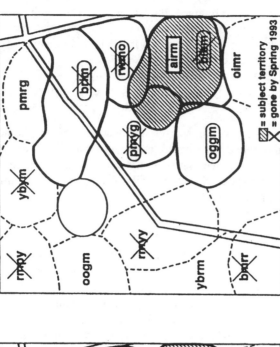

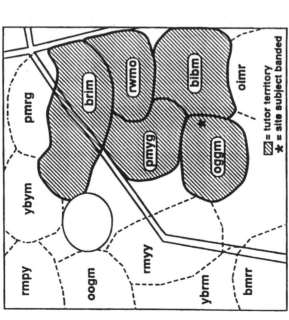

Figure 5.2. A young bird's tutors are neighbors in the area where he settles after leaving his birthplace in his first year; however, not all of these tutors survive into the young bird's first breeding season the next spring. *Left:* Map of territories of thirteen adult birds in the area where the young male AIRM settled in his hatch year (1992). Star indicates where AIRM was banded. Territories of the five birds who were subsequently identified as AIRM's tutors on the basis of his final repertoire in 1993 are outlined and hatch-marked. *Right:* Same configuration overlaid with AIRM's 1993 territory (*hatched*), dead birds are crossed out (eight of the 13 birds, including four of the five tutors). Although the actual 1993 territories of birds other than AIRM are not shown, birds OGGM (the sole-surviving tutor) and OIMR remained in approximately the same place. A pond (*circle*) and two intersecting paths are indicated; open areas were unoccupied (e.g. steep hills and meadows) (from Beecher 1996).

ciple suggested by these two exceptions is the following: Song elements of different songs are combined only if they are from (1) different tutors' versions of the same type; or (2) different song types sung by the same tutor. We summarize this principle as the student "preserving type and/or tutor" in his songs. We have yet to find a clear example of a bird hybridizing a song type of one singer with a distinctly dissimilar song type from a different singer; yet, in the laboratory, song sparrows do this commonly.

5.3.2.4 Young Birds Preferentially Learn Shared Songs

As we discussed, neighbors in our song sparrow population usually shared a portion of their song repertoires, on average about four of their eight to nine song types. We discovered that the young bird preferentially learned (or retained) song types shared by his tutor-neighbors. Typically, the bird's version of the song type most resembled that of one or another of his tutors, but sometimes the bird blended features of the versions of two or more tutors. An example of this learning preference for shared song types is shown in figure 5.5. The young bird shown retained seven types that were shared by two or more of his tutors and only two that were unique to one of these tutors, despite the existence of 11 shared and 13 unshared types in the tutor group. In the full sample analyzed by Beecher et al. (1994b), birds learned (retained) 84% of tutor-shared types and only 21% of tutor-unique types, although only 37% of the tutors' song types were shared.

5.3.3 Song-Learning Rules Imply Cognitive Principles

The song-learning rules we derived from our field observations suggest that a young song sparrow classifies songs by type and singer identity; this two-way classification appears to be central to the construction of the bird's final repertoire. We recapitulate these rules: (1) Song types shared by two or more tutors are preferentially learned. (2) Different tutors' versions of the same song type are often blended, but different tutors' versions of different song types are not. (3) The rare cases in which a bird combines elements of different song types to form a hybrid-song type occur only when the song types are sung by the same singer. The strategy of song learning defined by these rules is possible only if the young bird classifies tutor songs by type and singer identity.

It is instructive to contrast these rules in the field with those in the laboratory. In the Marler and Peters (1987) tape-tutor study, there was no evidence of this two-way classification. With tape tutors, however, the necessary conditions for defining song type and singer identity are eliminated. With tape recordings, individual identity can be preserved only to the extent that the different song types of a song sparrow have common "signature" or "voice"

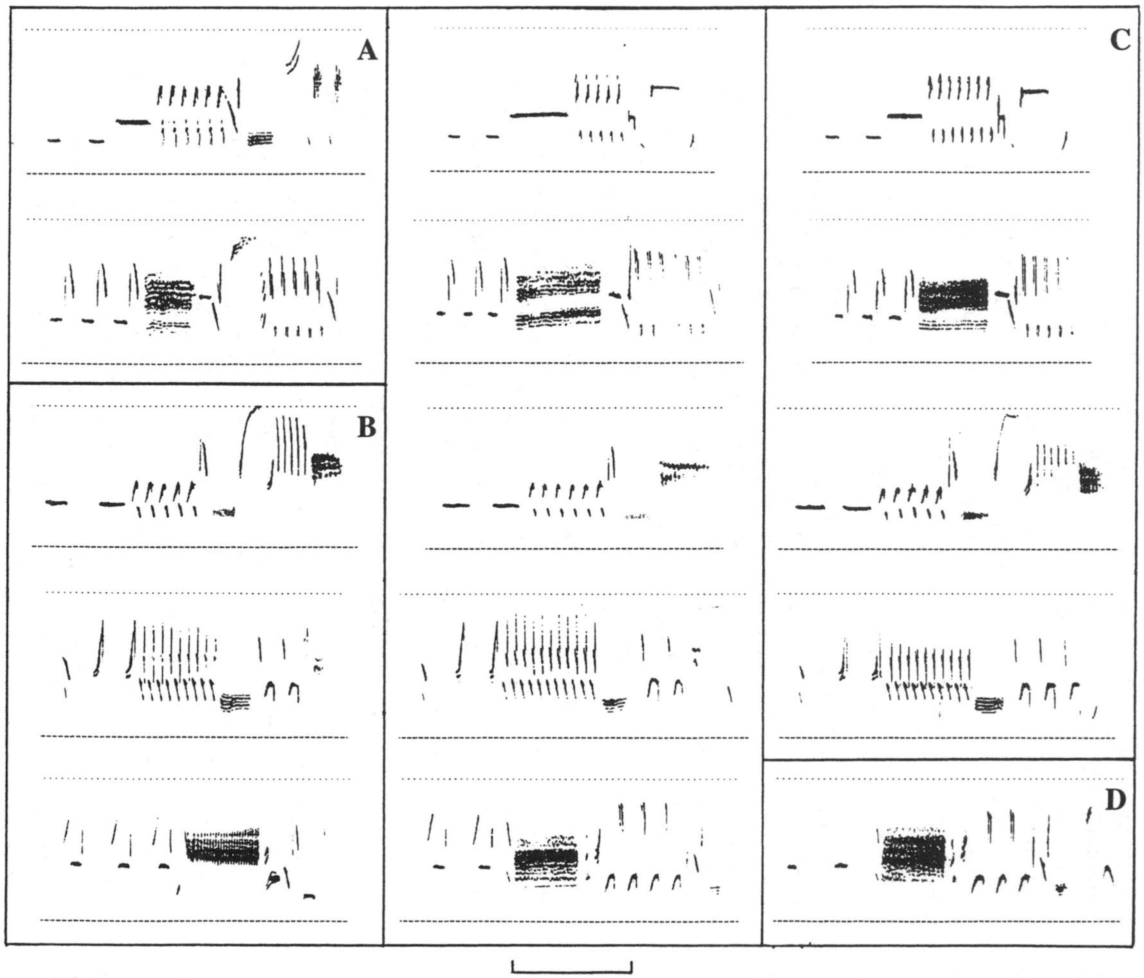
A
B
C
D

Figure 5.4. A young bird's hybrid song from two songs of one tutor. The young bird's hybrid song (*right*) incorporates the first four elements of one song type (*top left*) of the tutor and all but the first two elements of another song type (*bottom left*) of the tutor. Frequency scale is 2–10 kHz in 2-kHz steps. Time marker = 1 second. Bandwidth = 117 Hz.

characteristics; we showed in perceptual experiments that the song types do not have these characteristics (Beecher et al. 1994a). Moreover, tape-tutor studies have not used different singers' versions of the same type. The results of these tape-tutor studies contrast with those of our field studies. The most obvious discrepancy is that cross-tutor hybrid songs are common in tape-tutor studies but essentially are not sung in the field. We suggest that a bird preserves a whole song type not simply because he has frequently heard those particular elements sung together. He preserves a whole song type because (1) he has heard that song type contrasted with other distinct song types within the repertoire of a particular singer, and (2) he has heard that song type sung by different birds. In other words, a song type is defined by the song being sung by both multiple birds and a particular bird that sings multiple types.

Figure 5.3. Sonagrams of five of the nine song types of one young bird are shown (*middle column*). Left and right columns show the matching song types of his four tutors; each tutor is denoted with a box (*left,* two song types of tutor A and three of tutor B; *right,* four songs of tutor C and one of tutor D). Song types not shown include those unique to a particular tutor and those shared by three or more tutors. In some cases, the young bird's version of the song type was closer to that of one tutor or the other. For example, his renditions of the top two song types were more like that of the tutor's songs on the right. Song classification was based on several versions of each song type from each bird (not all shown). Song sparrows vary their song types from one occasion to another, and song types, therefore, are better described as song classes. Song endings are most variable and least diagnostic of song type. Frequency markers at the bottom and top of each sonagram are 0 and 10 kHz, respectively. Time marker denotes 1 second. Bandwidth = 117 Hz.

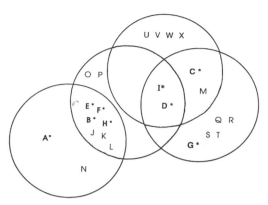

Figure 5.5. Young males preferentially learn songs shared by two or more tutors. Venn diagram of four tutors (*circles*) of one young bird. Each song type is indicated with a letter, and shared songs of two or more birds are indicated with the same letter. Shared song types are in the overlapping sections of the circles (e.g. types I and D were shared by three tutors). The nine song types learned by the young bird are indicated (bold type and *asterisk*). The young bird learned seven of the eleven tutor-shared types but only two of the thirteen tutor-unique types (from Beecher 1996).

Thus, if conditions in the laboratory eliminate or muffle these features (as a tape-tutor experiment inevitably does), the bird will not copy whole songs but will recombine the learned song elements into hybrid songs.

Simply replacing tape tutors with live tutors in the laboratory is not sufficient to simulate adequately the natural song-learning environment and, therefore, replicate field results. In our first attempt at such a simulation, we exposed young song sparrows in the laboratory to the songs of four live tutors (Beecher et al. unpublished). Otherwise, our paradigm had no more ecological validity than the typical tape-tutor experiment, and our results were closer to those of the comparable tape-tutor study (Marler and Peters 1987) than those of the field study. In particular, our young birds developed repertoires of nearly all hybrid songs.

We offer the following as one likely reason for the number of hybrid songs observed in our laboratory study. In this study, we used four tutors each of whom had come from a different area; these tutors, therefore, had no song types in common. Thus, we presented the young birds with a set of tutors that were very abnormal in our bird population. As discussed, song types were poorly defined in this situation; there were no song types common to more than a single tutor.

Moreover, if a young song sparrow has a preference for learning shared song types, what does he do if he is presented with tutors that share no song types? Suppose the song-learning bird has a predisposition to pay attention to song types that are sung by two or more tutors. The bird must, therefore, make a judgment about the similarity of or differences among songs of different tutors. If the songs are similar enough, he categorizes them as the same type; and if the different tutors' versions of the type are different enough, the bird must blend the versions. The more different the blended songs are, the more the resulting song will sound like a hybrid song to the experimenter. Yet, from the bird's perspective, he has produced a version of a tutor-shared song type. A set of tutor songs that are dissimilar may cause the bird to stretch his similarity criterion. We think this is the most likely (and certainly the most interesting) explanation for the hybrid songs produced by song sparrows in our laboratory experiment (and perhaps for those produced in the tape-tutor study of Marler and Peters 1987). It is a testable proposition: Hybrid songs should be uncommon when there are many shared songs in the set of tutor songs, but hybrid songs should be common when there are few shared songs among tutors.

In our study population in the field hybrid songs were exceptionally rare, probably because tutor-neighbors invariably shared many clearly similar songs. Occasionally, we observed a tricky case, in which the bird appeared to regard two songs of his tutor-neighbors as 'shared,' but we did not classify the songs as shared. For example, figure 5.6 shows two examples of a young bird's blending of similar songs from two tutors. In each case, we can infer that the young bird regarded the pair of tutor songs as the same type (i.e. the songs were similar enough to blend). Cases like these were relatively rare in our field population because probably the bird was usually exposed to more similar songs from which to choose.

Finally, some possible support for the interpretation that song learning is partially guided by the bird's concept of song type comes from a recent tape-tutor experiment by Nowicki et al. (in preparation). They found that song types that were presented with variations were preferentially learned over songs that were presented without variations. A possible interpretation of this result is that the bird's concept of song type may be stimulated by hearing variations of the type. This effect may be analogous to the bird's tendency in the field to preferentially learn song types sung by more than one tutor (i.e. each tutor sings the song type somewhat differently). We do not know, of course, if the learning preference for tutor-shared types is because of the greater number of variations of a type or the greater number of repetitions of that type; however, these propositions are testable.

In summary, we are suggesting that the song-learning bird be viewed as a cognitive decision maker; he decides how to lump or split the many tutor songs that he hears. These decisions are guided in part by the bird's concepts of singer individuality and song type, and the bird will not blur these categories; he will preserve tutor and type distinctions unless the conditions of song learning are ambiguous. Further research, in which these variables are experi-

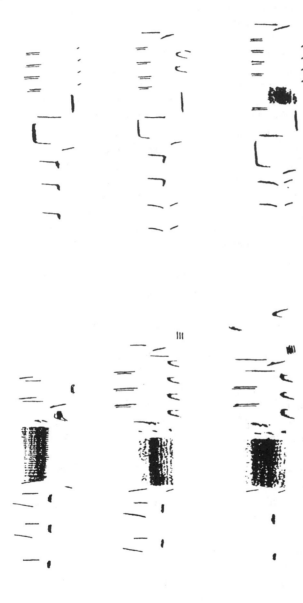

Figure 5.6. Two examples of a young bird blending the songs of two tutors: the two song types were similar but not easily classified as shared. The young bird's sonagram (*middle*) and the two tutors' sonagrams (*above* and *below*) are shown. In the example on the left (also shown in fig. 5.3), the young bird takes the first part of his song from one tutor song (*above*) and the rest is from the other tutor song (*below*). The similarity between the two tutor songs is not in the microstructure of the elements, but the songs have the same sorts of elements in the same order. Specifically, both begin with two or three notes at frequencies of about 4,000 Hz (birds often vary the number of notes within a song type) spaced about 40 msec apart. The notes of one tutor song but not the other are embellished with higher grace notes. The introductory notes are followed by a buzz about 0.5 seconds in duration. In both tutor songs, the buzz is preceded by a down sweep and followed by a two-voiced upsweep. The end of the two songs differ, although not re-markably so (from Beecher 1996). In the example on the right, the young bird takes the following introductory notes from both tutors: the shared four-note complex (the buzz of one tutor [*below*] is omitted); the shared trill (with modification); and the final note complex of one tutor (*below*). The two tutor songs are difficult to classify as shared, despite strong similarities in microstructure, because they differ in terms of the introductory notes and the presence of a buzz in one tutor song only. Note, however, that if you remove the buzz from the one tutor's song (*below*), it is virtually identical to the other tutor's song (*above*) except for the introductory notes (the notes at very end of the song are not diagnostic because the song is invariably changed from one performance to the next) (from Beecher 1996).

mentally manipulated, will be required to test the validity of this cognitive interpretation.

5.3.4 Simulating Key Ecological Song-Learning Variables in the Laboratory

We argued that laboratory studies of song learning can be misleading because key social variables and context are removed from the laboratory situation. It should follow that adding these variables back into the laboratory will provide results that are more similar to those of field studies. Our first attempt (described) failed in this regard because our simulation was unrealistic (apart from substituting live birds for tape tutors). We recently completed a second simulation; this time, we brought into the laboratory more features thought to be critical to the pattern of song learning in the field (Nordby et al., unpublished).

In this experiment, we put four adult song sparrows in aviaries at four corners of the roof of a building; we attempted to simulate four neighboring territories. The adults were neighbors in our population, and consequently, they shared some song types. Young birds were taken from the nest in another area at approximately 4 days of age and raised in the laboratory. At fledging (about 1 month), they were placed in cages next to the tutors. The young bird could see only the tutor he was stationed next to, but he could hear the other tutors at a distance. During their second and third months, the young birds were rotated among the four tutors on a weekly basis. That fall and following spring, half the subjects were again rotated among the four tutors, and each of the other subjects was stationed next to only one tutor. Final song repertoires of the young birds were recorded in late spring. The pattern of song learning was very similar to that we observed in the field. First, most songs the birds developed were good matches to song types of tutors. Only rarely did birds produce hybrid songs (i.e. elements of different tutor songs rearranged into a new song type). Second, although our analysis is not yet complete, it appeared that birds preferentially learned or retained song types shared by tutors. Third, birds learned songs from multiple tutors. Fourth, a bird stationed next to a particular tutor learned, or retained, more songs of that tutor than of the other tutors.

5.4 COMMUNICATION BY SONG IN MALE–MALE INTERACTIONS

5.4.1 Design of the Playback Experiments

We examined the nature of singing interactions primarily with playback experiments in which we experimentally manipulated one part of the singing interaction. In these experiments, we played neighbor or stranger songs near the boundary between the subject's and neighbor's territories and, in some cases, at an "inappropriate" territory boundary (i.e. that of another neighbor) or within the subject's territory. When song was played from a neighbor's territory, the neighbor was drawn off during the trial. Playback trials were 3 minutes long and songs were repeated at the typical rate of the species, approximately one song every 10 seconds. We measured five components of the subject's response to the playback: (1) closest approach to the playback speaker; (2) number of flights; (3) number of songs; (4) song types sung (usually only one); and (5) displays such as wing waves and "quiet song." Closest approach and number of flights are clear correlates of aggressive responses. In song sparrows, the number of loud songs sung during the trial is not a simple correlate of aggressive response; although birds sing very little when uninterested in the playback, typically they are equally silent when in "attack mode," searching for the singer in the bushes near the playback speaker. Which particular song types the birds sing may be related to aggressive response and will be discussed further. Wing waves and quiet song (which is qualitatively different from normally loud song in respects besides amplitude) are very intense responses, but they were seen mainly when the playback occurred within the subject's territory. Loud song was rarely sung in these cases; these cases indicate that loud song is truly a long-distance signal.

5.4.2 Results

5.4.2.1 Neighbor Song from the Neighbor's Territory Evokes a Weak Response

If the playback song was presented from the neighbor's territory near the boundary, the subject usually responded less intensely if the playback song was that of the neighbor versus that of a different neighbor (i.e. a neighbor in the wrong place), a stranger, or the subject (the latter perhaps is perceived as a stranger song) (fig. 5.7) (Stoddard et al. 1990, 1991; Beecher et al. 1996). It makes sense, of course, that the subject responded less strongly to a neighbor's song from the neighbor's territory than to other songs; singing of these other songs may suggest territorial intrusion to the subject.

5.4.2.2 Otherwise, Responses to Neighbor and Stranger Songs Are Equally Strong

The subject responded equally strongly to neighbor and stranger songs when the songs were played back from a boundary of an inappropriate territory or within the subject's territory (fig. 5.7, Stoddard et al. 1991). A ceiling effect is a possible interpretation of the response to the playbacks within the territory

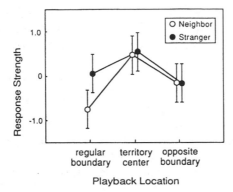

Figure 5.7. Mean response to playback of fourteen male song sparrows to songs of neighbors and strangers played in three locations: (1) the regular boundary (where the neighbor's song is ordinarily heard); (2) the center of the subject's territory; and (3) the opposite boundary (i.e. opposite from where the neighbor's song is ordinarily heard). Bars are ± 2 standard errors of the mean and represent 95% confidence intervals for the mean estimates (from Stoddard et al. 1991).

but not the response to the playback from the inappropriate boundary. The response at the boundary was reduced compared with that in the territory; the most conservative interpretation of the equality of the response at the boundary is neighbors and strangers were viewed by the subject as equally threatening. This interpretation is consistent with field studies of another sedentary population of song sparrows; these studies showed that a bird's territory may be usurped by either a new bird (floater) or one of his current neighbors (Arcese 1987, 1989).

5.4.2.3 Birds Reply to Neighbor Songs with Songs Specific to Those Neighbors

The prevailing view on species-specific repertoires is that different song types in a repertoire are interchangeable and exist primarily to provide diversity (e.g. Kroodsma 1988). If this view were correct, we might expect the bird to reply to a stimulus song with a song type that is chosen randomly from his repertoire. We found, however, that a song sparrow did not reply with a randomly chosen song, at least under most circumstances. Instead, he typically replied to a neighbor's song (from the neighbor's territory) with one of the song types he shared with that particular neighbor (remember, neighbors in our population shared an average of about 30% of their eight to nine song types). If the particular stimulus song was in his repertoire he usually would not reply with that song but with another one of the song types he shared with the neighbor (assuming he shared more than the one). We call this pattern of song selection "repertoire matching" because usually the bird matched some song in the stimulus bird's repertoire (Beecher et al. 1996). Repertoire matching implies knowledge of the singer's repertoire; the bird must know which songs he shares with the singer.

A bird responded to the song of a stranger very differently. If he had a similar song type in his repertoire, he usually responded with that type (type matching) (Stoddard et al. 1992). We cannot describe how he selected his reply song if he did not have a similar type in his repertoire (the usual case); however, we made the following observation: If the stranger song was played from a neighbor's territory, the subject avoided replying with a song type he shared with the neighbor. The bird appeared to reserve the shared songs for that neighbor. This interesting response suggests a concept of "not neighbor." In one study, we played back the same stranger song to the subjects on two different days; the birds tended to respond with different songs on the two days. This suggests that the subject birds may not have been selecting a particular reply song (Stoddard et al. 1992).

To summarize these findings of type matching, song sparrows generally matched type in response to stranger songs (that were similar enough) and their own playback songs, but they rarely matched type in response to neighbor songs. The same pattern has been observed in western meadowlarks (Falls 1985) and a similar pattern has been observed in great tits (Falls et al. 1982).

Under one specific set of conditions, however, birds matched type in response to neighbors songs: This occurred when neighbors were new that year and the breeding season was still early (Beecher et al. unpublished). Longtime neighbors, on the other hand, rarely matched type; birds replied with another song they shared with that neighbor.

5.4.3 Another Instructive Contrast

Just as we unwittingly performed a laboratory study with minimal ecological validity that happened to provide an instructive contrast to its field counterpart, so have we performed ecologically dubious playback studies, and they have proved to be equally instructive. The concept behind these playback experiments was closer to that of the classic ethological model described earlier than to that of the cognitive-ecology model we now favor. Both studies were designed to reveal which types of songs are more threatening (i.e. better releasers) (Krebs et al. 1981). In the first experiment, we compared unmatchable songs (stranger songs which did not resemble any of the subject's songs) with songs that could be matched (the subject's own songs, or self songs). (We used self songs because previous studies showed these songs were equivalent to very similar stranger songs; however, similar stranger songs were much harder to find because usually, in our population, only neighbors sung very similar songs.) We used a two-speaker design in which two speakers were placed at equivalent locations, about 50 m apart, within the bird's territory; on different days, we switched which speaker played which song. In the second experiment, we compared matching vs. nonmatching song. In this case, we used a single speaker; one stimulus was used on one test day, and another was used on another test day. The song was presented from a neighbor-free area (an open field) outside the bird's territory. We used self songs as playback stimuli; the same song the subject was singing was played back on one test day, but a different song not in his repertoire was played on another test day. With these two experiments, we attempted to answer similar questions: (1) Are matchable songs more or less threatening than unmatchable songs? (2) Is having your song matched more or less threatening than having your song replied to with an equally familiar, but nonmatching song?

In both studies, we found that the birds responded equally strongly, on average, to both stimuli. Although ceiling effects potentially cloud interpretation of both experiments, the birds simply could have found these songs equally threatening. Whether songs were matchable or not, or matched or not, birds evidently perceived the songs as those of strangers or neighbors in the wrong place (more on this point below); to the birds, these situations are equally dangerous.

These two studies have not been submitted for publication because, on reflection, the results are difficult to interpret. When viewed from a cognitive-ecology perspective, the results don't appear to make much sense. Usually, the only songs a bird hears are those sung by his neighbors. Thus, the bird's best hypothesis concerning a playback song—whether in truth it is a neighbor's song, a stranger's song, a dead neighbor's song, a computer-manipulated song, or one of the bird's own songs—is that it is a neighbor's song. If the hypothesis fits, i.e., if a nearby neighbor has a similar song in his repertoire, the bird should respond appropriately: the response should be weak if the song comes from the hypothesized neighbor's territory and strong if the song does not. Because these two experiments were performed within the subject's territory or at the boundary of a neighbor-free territory, the subject should have responded strongly. If the hypothesis does not fit, i.e., if no neighbor has a song like the playback song, then the bird should assume that the song is that of a stranger and respond appropriately: strongly. Thus, if we take the social-ecological context into account and suppose that the bird takes a cognitive perspective (e.g. "Who is that bird, and does he belong there?"), we see that the logical outcome of these two experiments should be precisely what we found: On average, the bird should have responded equally strongly to both classes of songs because an intruder is an intruder. If this view is correct, these results are trivial; we would be better off working systematically with neighbor songs or designing reasonable simulations of the kinds of stranger intrusions that typically occur in this population of birds.

5.4.4 Conclusions

Although our understanding of the rules of song communication in neighbor interactions of song sparrows is still rudimentary, one aspect of the system seems very clear. The song sparrows in our study population live in relatively stable neighborhoods, and our playback experiments suggest the birds know one another on the basis of the songs. The birds' knowledge is much more sophisticated than we expected. We were not surprised to find that a bird gave a milder response to a neighbor's song from the neighbor's territory than to a stranger's song from the same place, nor were we surprised to find that the bird responded equally to these songs from other locations (Stoddard et al. 1991, 1992). These results simply imply that the bird recognizes this neighbor. We were surprised, however, to find that the bird replied in a qualitatively different fashion to a neighbor's song from the neighbor's territory than to a stranger's song from the same place: to the neighbor's song with another one of the songs he shares with the neighbor (repertoire matching), to the stranger song with one of the song types he does *not* share with the neighbor. Repertoire matching indicates a more detailed knowledge of the song repertoires of neighbors than we had expected. Repertoire matching suggests that birds classify songs by type (the bird knows which songs he shares or does not share with each of his neighbors) and by individual identity (the bird knows which songs each of his neighbors has in his repertoire).

5.5 Discussion

In this chapter, we developed two points. First, the key variables that determine how a song sparrow constructs his song repertoire (during the period he learns songs) and how he uses the songs in his repertoire are variables that pertain to the bird's social ecology. Second, song communication and song learning in song sparrows are largely organized into two main cognitive categories—individual singer and song type. This two-way classification helps explain which songs birds learn and retain in their first year and the way in which birds use these songs as adults in countersinging interactions.

The major unanswered questions are (1) Why do song sparrows use their songs as they do? and (2) What is the advantage of a learning strategy that provides the bird with songs he shares with his neighbors? We showed that song sparrows preferentially communicate with neighbors with shared songs, but the function of this repertoire matching is unknown. The posting function remains the only function of song repertoires in male–male interactions that we are certain of, but how repertoire matching or having shared songs relates to this function (if at all) is not obvious.

A major impediment to developing hypotheses about the function of song in neighbor communication is our limited understanding of the territorial neighbor relationship. The field has advanced beyond theories like the "Beau Geste" hypothesis (Krebs 1977), which suggests that a bird moving into an area and prospecting for territory misperceives a single bird with several song types as several different birds and consequently avoids that high-density neighborhood. We can now be reasonably sure that these birds know not only how many birds are in the neighborhood, but who they are, and probably the details of their song repertoires as well. However, the working model of the territorial neighbor relationship remains a simple competition model, and most theories of song function in this context do not go beyond that assumption. For example, consider the various hypotheses that have been advanced to explain the function of song repertoires in this context. All of these hypotheses propose that the bird uses his song repertoire as a weapon against his neighbor in one way or another (the hypothesis usually is based on whether the bird's songs match or do not match those of his neighbors). For example, according to the ranging hypothesis (Morton 1986), the bird uses his repertoire to disturb his neighbor; according to the threat hypothesis (Krebs et al. 1981), repertoire is used to threaten the neighbor. According to the antihabituation hypothesis (Kroodsma 1988), repertoire is used to keep the posting signal fresh and hold the neighbor's attention; according to the Beau Geste hypothesis, repertoire is used to confuse and discourage prospective neighbors (Krebs 1977).

It is possible, however, that song serves functions in neighbor interactions beyond the purely defensive or offensive ones suggested with the existing theories. In particular, the relationships of territorial neighbors may have cooperative and competitive aspects (e.g. Getty 1987; Beletsky and Orians 1989). The kinds of interactions implied (e.g. cooperative defense against intrusions by new birds) may have different influences on singing and song learning. At the very least, a long-time neighbor represents a different sort of competitor than a new bird; yet, existing theories make no such distinction. We suggest that future developments in our understanding of the cognitive ecology of song communication and song learning will hinge on the development of a general theory of the social ecology of territorial songbirds (see Dugatkin and Sih, this vol. chap. 10).

In this chapter, we focused on our particular study species, the song sparrow; however, we expect that many of the general conclusions made apply more broadly, at the very least to many other songbird species. For example, considerable evidence supports the hypothesis that the function of song learning in many species is to equip the bird with songs he shares with his neighbors. In addition to studies that show birds learn their new neighbors' songs after leaving their birthplace (Bewick's wrens, Kroodsma 1974; saddlebacks, Jenkins 1978), a number of studies have indicated that birds may later modify their song repertoires to increase song sharing with neighbors in the first or subsequent breeding seasons (indigo buntings, Payne 1982, 1983; white-crowned sparrows, Baptista and Morton 1988; great tits, McGregor and Krebs 1989; field sparrows, Nelson 1992; American redstarts, Lemon et al. 1994; cowbirds, O'Loghlen and Rothstein 1995; European starlings, Mountjoy and Lemon 1995).

In this chapter, we concentrated on the aspect of the song sparrow's social ecology that pertains to social interactions among neighboring males. Male–female interactions undoubtedly are also of great importance in shaping song learning and song use in this species and other songbirds (e.g. Searcy 1984; Catchpole 1986, 1987; West and King 1988; King and West 1989; Searcy and Yasukawa 1990). We have recently begun to look at female influences on song in our study species, and we have completed several studies which suggest that song sharing may be very important with respect to female sexual preferences (O'Loghlen and Beecher, 1997; in preparation). It will be interesting to see if the study of intra- and intersexual selection provides reinforcing or opposing arguments for the cognitive ecology of song learning and song communication in this species.

5.6 SUMMARY

In song sparrows, the strategy of song learning in young birds and communication with song between neighboring territorial males are shaped by two major sets of variables: (1) cognitive factors at the proximate level; and (2) variables in the species' social ecology at the ultimate level. At the proximate level, song learning and song use appear to be shaped and guided by the bird's concepts of song type and singer identity. At the ultimate level, the function of song learning appears to be the acquisition of songs that will be shared with territorial neighbors. Thus, a young male song sparrow develops a song repertoire of eight or so song types that are taken from the repertoires of neighboring males. Songs of the nearest neighbors and those of birds that survive the winter between the young bird's hatching summer and his first breeding season are more likely to be adopted. In singing interactions with his neighbors, the bird not only recognizes his individual neighbors but selectively uses just those songs he shares with each particular neighbor. We argue that a cognitive-ecology perspective—which focuses on cognitive factors and social-ecological variables—captures the key features of song learning and song communication in this species and will probably do so as well in other species of songbirds.

ACKNOWLEDGMENTS

Our colleagues in the studies described in this chapter were Philip Stoddard, John Burt, Adrian O'Loghlen, Christopher Hill, Cindy Horning, Patricia Loesche, Michelle Elekonich, and Mary Willis. We thank Discovery Park for hosting our field work and National Science Foundation for supporting this research. Finally, we thank Reuven Dukas for patience beyond his years.

LITERATURE CITED

Arcese, P. 1987. Age, intrusion pressure, and defence against floater by territorial male song sparrows. *Animal Behaviour* 35:773–784.

———. 1989. Intrasexual competition, mating system, and natal dispersal in song sparrows. *Animal Behaviour* 37:45–55.

Baptista, L. F., and M. L. Morton. 1988. Song learning in montane white-crowned sparrows: From whom and when. *Animal Behaviour* 36:1753–1764.

Beecher, M. D. 1996. Bird song learning in the laboratory and the field. In D. E. Kroodsma and E. L. Miller, eds., *Ecology and Evolution of Acoustic Communication in Birds,* 61–78. Ithaca, N.Y.: Cornell University Press.

Beecher, M. D., S. E. Campbell, and J. M. Burt. 1994a. Song perception in the song sparrow: Birds classify by song type but not by singer. *Animal Behaviour* 47:1343–1351.

Beecher, M. D., S. E. Campbell, and P. K. Stoddard. 1994b. Correlation of song learning and territory establishment strategies in the song sparrow. *Proceedings of the National Academy of Sciences of the United States of America* 91:1450–1454.

Beecher, M. D., P. K. Stoddard, S. E. Campbell, and C. L. Horning. 1996. Repertoire matching between neighbouring songbirds. *Animal Behaviour* 51:917–923.

Beletsky, L. D., and G. H. Orians. 1989. Familiar neighbors enhance breeding success in birds. *Proceedings of the National Academy of Sciences of the United States of America* 86:7933–7936.

Catchpole, C. K. 1986. Song repertoires and reproductive success in the great reed warbler *Acrocephalus arundinaceus. Behavioral Ecology and Sociobiology* 19:439–445.

———. 1987. Bird song, sexual selection, and female choice. *Trends in Ecology and Evolution* 2:94–97.

Catchpole, C. K., and P. J. B. Slater. 1995. *Bird Song: Biological Themes and Variations.* Cambridge: Cambridge University Press.

Falls, J. B. 1985. Song matching in western meadowlarks. *Canadian Journal of Zoology* 63:2520–2524.

Falls, J. B., J. R. Krebs, and P. K. McGregor. 1982. Song matching in the great tit (*Parus major*): The effect of similarity and familiarity. *Animal Behaviour* 30:997–1009.

Getty, T. 1987. Dear enemies and the prisoner's dilemma: Why should territorial neighbors form defensive coalitions? *American Zoologist* 27:327–336.

Jenkins, P. F. 1978. Cultural transmission of song patterns and dialect development in a free-living bird population. *Animal Behaviour* 26:50–78.

King, A. P., and M. W. West. 1989. Presence of female cowbirds (*Molothrus ater ater*) affects vocal imitation and improvisation in males. *Journal of Comparative Psychology* 103:39–44.

Krebs, J. R. 1977. The significance of song repertoires: The Beau Geste hypothesis. *Animal Behaviour* 25:475–478.

Krebs, J. R., R. Ashcroft, and K. van Orsdol. 1981. Song matching in the great tit (*Parus major* L.). *Animal Behaviour* 29:918–923.

Kroodsma, D. E. 1974. Song learning, dialects, and dispersal in the Bewick's wren. *Zietschrift der Tierpsychologie* 35:352–380.

———. 1983. The ecology of avian vocal learning. *Bioscience* 33:165–171.

———. 1988. Contrasting styles of song development and their consequences. In R. B. Bolles and M. D. Beecher, eds., *Evolution and Learning,* 157–184. Hillsdale, N.J.: Erlbaum.

Lemon, R. E., S. Perrault, and D. M. Weary. 1994. Dual strategies of song development in American redstarts, *Setophaga ruticilla. Animal Behaviour* 47:317–329.

Marler, P. 1970. A comparative approach to vocal learning: Song development in the white-crowned sparrow. *Journal of Comparative Physiology and Psychology Monograph* 71:1–25.

Marler, P., and D. A. Nelson. 1992. Neuroselection and song learning in birds: Species universals in culturally transmitted behavior. *Seminars in Neuroscience* 4:415–423.

Marler, P., and S. Peters. 1987. A sensitive period for song acquisition in the song sparrow, *Melospiza melodia:* A case of age-limited learning. *Ethology* 76:89–100.

———. 1988. The role of song phonology and syntax in vocal learning preferences in the song sparrow, *Melospiza melodia. Ethology* 77:125–149.

McGregor, P. K. 1992. *Playback and Studies of Animal Communication.* New York: Plenum.

McGregor, P. K., and J. R. Krebs. 1989. Song learning in adult great tits (*Parus major*): Effects of neighbours. *Behaviour* 108:139–159.

Morton, E. S. 1986. Predictions from the ranging hypothesis for the evolution of long distance signals in birds. *Behaviour* 99:65–86.

Mountjoy, D. J., and R. E. Lemon. 1995. Extended song learning in wild European starlings. *Animal Behaviour* 49:357–366.

Nelson, D. A. 1992. Song overproduction and selective attrition lead to song sharing in the field sparrow (*Spizella pusilla*). *Behavioral Ecology and Sociobiology* 30:415–424.

Nelson, D. A., and P. Marler. 1994. Selection-based learning in bird song development. *Proceedings of the National Academy of Sciences of the United States of America* 91:10498–10501.

Nowicki, S., J. Podos, and F. Valdes. 1994. Temporal patterning of within-song type and between-song type variation in song repertoires. *Behavioral Ecology and Sociobiology* 34:329–335.

O'Loghlen, A. L., and M. D. Beecher. 1997. Sexual preferences for mate song types in female song types. *Animal Behaviour* 53:835–841.

O'Loghlen, A. L., and S. I. Rothstein. 1993. An extreme example of delayed vocal development: Song learning in a population of wild brown-headed cowbirds. *Animal Behaviour* 46:293–304.

Payne, R. B. 1982. Ecological consequences of song matching: Breeding success and intraspecific song mimicry in indigo buntings. *Ecology* 63:401–411.

Payne, R. B. 1983. The social context of song mimicry: Song matching dialects in indigo buntings (*Passerina cyanea*). *Animal Behaviour* 31:788–805.

Podos, J., S. Peters, T. Rudnicky, P. Marler, and S. Nowicki. 1992. The organization of song repertoires of song sparrows: Themes and variations. *Ethology* 90:89–106.

Searcy, W. A. 1984. Song repertoire size and female preferences in song sparrows. *Behavioral Ecology and Sociobiology* 14:281–286.

Searcy, W. A., and K. Yasukawa. 1990. Use of the song repertoire in intersexual and intrasexual contexts by red-winged blackbirds. *Behavioral Ecology and Sociobiology* 27:123–128.

Slater, P. J. B. 1989. Bird song learning: Causes and consequences. *Ethology Ecology Evolution* 1:19–46.

Stoddard, P. K., M. D. Beecher, S. E. Campbell, and C. Horning. 1992. Song-type matching in the song sparrow. *Canadian Journal of Zoology* 70:1440–1444.

Stoddard, P. K., M. D. Beecher, C. L. Horning, and S. E. Campbell. 1991. Recognition of individual neighbors by song in the song sparrow, a species with song repertoires. *Behavioral Ecology and Sociobiology* 29:211–215.

Stoddard, P. K., M. D. Beecher, C. H. Horning, and M. S. Willis. 1990. Strong neighbor-stranger discrimination in song sparrows. *Condor* 97:1051–1056.

Stoddard, P. K., M. D. Beecher, and M. S. Willis. 1988. Response of territorial male song sparrows to song types and variations. *Behavioral Ecology and Sociobiology* 22:125–130.

West, M. J., and A. P. King. 1988. Female visual displays affect the development of male song in the cowbird. *Nature* 334:244–246.

Cognitive Ecology of Navigation

FRED C. DYER

6.1 INTRODUCTION

On the most general level, the adaptive significance of spatial orientation is obvious: It is easy to imagine why natural selection has equipped animals with mechanisms that enable them to (1) acquire information about their position and orientation relative to fitness-enhancing resources, such as food or mates; and (2) guide movements in search of better conditions. Accordingly, most biologists interested in orientation have focused on the underlying sensory and neural mechanisms and have treated questions about the adaptive importance of these mechanisms only superficially, if at all.

It is clear, however, that the flexibility and accuracy used to solve similar navigational problems vary enormously among species. It is also clear that the navigational tasks faced by animals vary with species-specific ecological differences. Such patterns of variation demand an evolutionary explanation that focuses not only on the variation in behavioral performance but also on the mechanisms that underlie the behavior. By analogy, consider foraging behavior, another class of biological traits whose general adaptive importance of foraging behavior is obvious, and yet which varies widely among species. For 30 years behavioral ecologists have attempted to understand the foraging decisions of animals in relation to (1) the selection pressures associated with specific foraging tasks, and (2) the constraints imposed on the design of mechanisms that underlie these decisions (reviews by Stephens and Krebs 1986; Krebs and Kacelnik 1992; Bateson and Kacelnik, this vol. chap. 8; Ydenberg, this vol. chap. 9). To the degree that the study of foraging focuses on the adaptive design of mechanisms used to process information about the profitability, distribution, riskiness, and palatability of food resources, it is already a relatively mature branch of cognitive ecology. Conceivably, it could provide a model for developing a cognitive ecology of the mechanisms used by animals to process spatial information while navigating.

This chapter is one of two in this volume that examines the prospects for a cognitive ecology of spatial orientation. Sherry (this vol. chap. 7) stresses well-developed vertebrate model systems that have been used to explore orientation and spatial memory on a small spatial scale, such as during the final approach to a familiar food source. With comparative and experimental approaches, this work has provided important insights into the evolutionary modification of spatial learning and memory. I adopt a somewhat broader approach to the evolution of orientation mechanisms. The chapter is organized into three main sections. In the first section, I adopt a broad problem-oriented approach by describing the diverse manifestations of the general problem of finding the way to a goal and considering how the diversity of navigational tasks helps account for the diversity of navigational abilities. In the second section, I outline the components of a trait-oriented approach to the study of navigation by arguing for the importance of hypothesizing specific fitness benefits and costs associated with different navigational strategies and exploring various constraints that may influence the course and outcome of selection on such strategies. These first two sections provide a conceptual framework for making sense of the diversity and adaptive design of navigational mechanisms. In the final section, I analyze several specific studies that illustrate the potential for an evolutionary-ecological approach to the cognitive mechanisms that underlie navigation.

Before proceeding, I would like to explain briefly my emphasis on the cognitive aspects of navigation. The hallmark of a cognitive approach—and indeed its main advantage over the behaviorist tradition with which it is often contrasted—is a concern with the internal processes that mediate behavior and not merely the observable correlations between an animal's experiences and its responses. There is considerable variation in opinion among researchers, however, about what mechanisms are encompassed with a cognitive approach.

In the most restrictive view, cognitive processes are those involved in humanlike mental qualities such as self-awareness. This seems to be the view of many so-called "cognitive ethologists" (Ristau 1991), who are largely concerned with investigating the role of humanlike mental processes in animals. Like other authors in this volume, I use the term "cognition" in a broader sense that refers to information processing in general by nervous systems and that encompasses simple perceptual and learning processes as well as more complex mechanisms used to acquire internal representations of the world (Dukas, this vol. sec. 1.1). Studies of animal orientation provide scant reason to invoke humanlike mental processes such as self-awareness. These studies do, however, provide abundant evidence for sophisticated internal processing of information acquired with the senses and reveal the inadequacy of simple stimulus-response accounts of the behavior. Indeed in navigation, as in foraging behavior, the differences in behavioral performance among species can often be understood only by considering how these species detect and organize information about features of the outside world.

An important caveat to bear in mind is the danger of committing to a particular cognitive hypothesis that is suggested by the behavioral data. Often one overestimate the sophistication of the mechanisms underlying the flexible, adaptive behavior of animals (for a detailed discussion see Dyer 1994). I will describe some impressive feats of navigation that are accomplished with surprisingly simple responses to environmental cues. Any attempt to understand the adaptive design of the information-processing mechanisms that underlie navigation must start with an accurate depiction of what those mechanisms are.

6.2 DIVERSITY IN THE PROBLEM OF GOAL ORIENTATION

6.2.1 General Considerations

In this section, I outline a classification system for describing the specific navigational tasks faced by different animals. This system involves a set of distinctions that define the parameters of the general problem of finding the way to the goal. This classification system is analogous to the categorization of foraging tasks into prey models or patch models or variants thereof (Stephens and Krebs 1986). Such categorization clarifies the general and specific features of the problem that natural selection has presumably designed the animal to solve. Control systems used in orientation are often clearly tuned to the parameters of the navigational task, and so a classification based on these parameters provides an initial way to account for the variation among species in navigational ability.

Attempts to classify the abilities of animals to orient themselves in space date back at least to the time of Loeb's (1918) hypothesis that all behavior can be described as manifestations of simple orienting responses to environmental stimuli (tropisms). Kühn (1919) and Fraenkel and Gunn (1940) developed more elaborate systems, which nevertheless invoked only very simple control mechanisms (taxes and kineses). These taxis and tropism theories inadequately accounted for the impressive discoveries in avian and insect orientation that began in the 1940s. In the 1970s and 1980s, improved classification systems were advanced (e.g. Jander 1975; Baker 1978; Schöne 1984). The system I present is most similar to that of Jander (1975) in relating the design features of different orientation systems to their function in the life history of the species.

In the classification of orientational mechanisms, Jander (1975) and Schöne (1984) agreed on a major distinction between (1) rotational adjustments to stabilize the body axes relative to external stimuli, and (2) goal orientation, or movement toward a different place where the animal can find better conditions (e.g. more food or less stress). This distinction encompasses most of the controlled, nonrandom movements that animals make.

I will focus entirely on goal orientation, which I define very broadly to encompass such widely different navigational tasks as (1) heading for a nearby object in plain view, (2) heading for an unseen feeding or nesting site within a familiar home range, and (3) heading for home from a distant, unfamiliar location. Following Griffin (1952), many researchers—especially those who study birds—restrict the term "navigation" to mean the last of these tasks, but I will use the term "navigation" synonymously with goal orientation. By doing so, I explicitly recognize the fundamental similarities among different types of goal orientation, and I believe this recognition makes it easier to understand the differences.

All types of goal orientation necessitate that the animal, at its starting point, obtain two general sorts of information about the environment: First, the animal must be able to discriminate among different directions (body orientations) relative to some external reference (e.g. a celestial body, a landmark, or a chemical cue). Second, the animal must be able to determine its position in space relative to its goal. The animal may need to measure its angular position and distance relative to some directional reference, or as we shall see, the animal may need to use a simpler measure of location. The crucial point is that information about position indicates to the animal which direction— among those that it can discriminate—is the correct direction that will lead to the goal.

Bear in mind that the dichotomy between positional and directional information is sometimes indistinct. For example, if an animal heads toward a clearly visible conspecific or prey item, the mechanisms involved in recognizing the item's position (i.e. the direction that will lead to the goal) may be hard to separate from the mechanisms involved in choosing the correct direction. On the other hand, for many navigational challenges, such as homing from a long distance, the tasks of obtaining positional and directional information may entail very different problems, perhaps even different sensory modalities.

To characterize the specific problem faced by an animal that is orienting to a particular goal, we can identify three parameters that affect in some way the procurement of positional and directional information and therefore the setting of a course to the goal: (1) the geometry of the space containing the goal; (2) the distance of the goal from the starting point; and (3) the variability (either spatial or temporal) of the environmental cues that provide navigational information. Classifying goal orientation with these parameters allows us to see more clearly how navigational challenges among animals are similar or different and thus provides a foundation for examining the similarities and differences in the mechanisms used by animals to meet these challenges.

6.2.2 Geometry of the Goal

Jander (1975) and others recognized that the nature of the navigational task is strongly influenced by the geometry of the region of the environment in

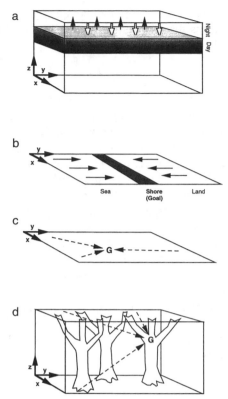

Figure 6.1. Geometries of goals. (*a*) Goal is a stratum in a volume. Shaded area is preferred stratum into which zooplankton (for example) move to escape predation during the day and out of which they move to feed at night. This orientation is commonly referred to as "diel vertical migrations" (e.g. Ringelberg and Flik 1994); here it is called z-axis orientation. (*b*) Goal is a linear zone on a surface, for example a preferred strip of beach where sandhoppers find salubrious conditions (y-axis orientation) (see Able 1980). (*c*) Goal is a point on a surface, for example the nest of an insect or a bird relative to the surface of the ground (x,y-orientation). (*d*) Goal is a point in a volume, for example the nest of an insect or a bird in a particular portion of the forest canopy (x,y,z-orientation).

which the navigating animal will find improved conditions. The geometry of the goal profoundly influences the problem of obtaining positional and directional information and thus determines how sophisticated the mechanisms used to obtain this information must be. We can distinguish four main goal geometries (fig. 6.1, where the x + y axes define the horizontal plane and the z-axis defines the vertical). I describe these relative to a three-dimensional Cartesian coordinate system, but I make no assumptions about whether an animal actually measures its orientation relative to such a coordinate system.

The point is to show that, in general, more complex goal geometries require navigational mechanisms that allow for more degrees of freedom of spatial position relative to the goal.

6.2.2.1 Goal Is a Plane or Stratum in a Volume

This situation is best illustrated by the "diel vertical migrations" of marine and freshwater zooplankton (reviews by Baker 1978; Dingle 1980, 1996). In the three-dimensional space of their environment, these animals must move along the z-axis from one stratum to another. For example, to escape predators during the day, they move deeper; to feed at night, they move toward the surface (see fig. 6.1*a*). The animals are presumably indifferent about their positions along the x- or y-axis when they are in the appropriate stratum because conditions within this stratum are essentially uniform. I refer to this orientation as "z-axis orientation."

The positional and directional cues that allow animals to perform z-axis orientation are potentially easy to ascertain (review by Schöne 1984). To determine position, the animal must decide if it is above or below the preferred stratum. Conceivably, this can be done by measuring light intensities (which is likely the most important cue for most species; Ringelberg and Flik 1994), water temperature, water pressure, or chemical concentrations in the water. To discriminate vertical directions, the animal can swim along the vertically oriented gradients of these stimuli or orient to gravitational forces.

6.2.2.2 Goal Is a Line or a Zone across a Surface

This situation applies to an organism with a preferred region in the environment that is a narrow strip of land containing food refugia, or optimal levels of temperature or humidity (see fig. 6.1*b*). A common example is the search of many littoral organisms for a particular zone along a shore. In this context, littoral amphipod crustaceans (sandhoppers) and amphibians have been particularly well studied (reviews by Able 1980; Baker 1978; Pardi and Ercolini 1986). When displaced from their preferred zone, such animals orient in the direction that is perpendicular to the shore and appear indifferent about where along the shore they end up. This type of orientation is often called y-axis orientation; the x-axis is parallel to the shore and the y-axis is perpendicular.

In y-axis orientation, the animal obtains positional information relatively simply by detecting which side of the preferred zone it is on. Direct cues may be available (e.g. visual features of the shore). Also, there is abundant experimental evidence that many animals use indirect cues. For example, if it is wet and cold, a sandhopper behaves as if it is in the water and heads toward drier land; if it is dry and hot, the sandhopper behaves as if it is inland and heads in the opposite direction (reviewed by Able 1980; Pardi and Ercolini

1986). These behavioral responses are exhibited even in experimental arenas far from the shore. Conceivably, a more sophisticated measure of position includes the distance from the goal as well as the direction. I know of no evidence, however, that animals use information about distance for y-axis orientation; nor is it likely that a navigational strategy that employs distance information is superior to one that uses only directional information.

To discriminate the two relevant directions in y-axis orientation (e.g. in littoral animals, toward water versus toward land), animals may again use direct cues—features of the shore itself—or indirect cues (Pardi and Ercolini 1986). Direct cues are gradients (e.g. slope or temperature) or visual cues (e.g. landmarks) that are aligned with the y-axis (Craig 1973; Hartwick 1976). Conceivably, the mechanisms that underlie a response to a local gradient are no more sophisticated than those used in z-axis orientation; in fact, mechanisms that underlie z-axis orientation presumably would result in y-axis orientation if the animal could only move in a plane perpendicular to the preferred stratum.

Indirect information about the y-axis direction is provided by compass references such as the sun or the earth's magnetic field; both references can be used by amphipod crustaceans (Ugolini and Pardi 1992) and newts (Phillips 1986). With use of such a compass, the major challenge for the animal is to determine which compass direction corresponds to the y-axis of the local shoreline. In some species, naive individuals behave as if they are innately informed about the y-axis direction that corresponds to their natal beach, and animals from different populations (from differently oriented beaches) exhibit hereditary differences in the innate compass response (e.g. Pardi and Scapini 1985). These results imply that gene flow among populations is relatively slight; this implication is surprising given that individuals are probably often carried off by ocean currents to other shores. The compass direction of the local y-axis is learned in at least some species of sandhoppers (Ugolini and Scapini, 1988). Also, a learned response can override the innate response (Ugolini et al. 1988); learning would allow for appropriate orientation if an individual drifts to a differently oriented beach. Learning the direction of the local y-axis has also been documented in amphibians (e.g. newts) that live along the shores of ponds (Phillips 1986).

6.2.2.3 Goal Is a Point on a Surface

This situation applies to those organisms that need to find the location of a specific nesting or feeding site on a two-dimensional surface (see fig. 6.1c). This task may be called x,y-orientation, although some animals may locate goals in terms of polar coordinates (not Cartesian coordinates) relative to a central location such as a nest. Orienting to a point on a surface is performed

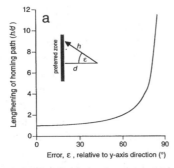
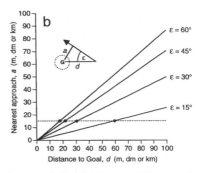

Figure 6.2. Implications of error in setting the homeward direction for two different goal geometries. (a) In the y-axis orientation, the error (ε) in setting the course toward shore results in an increase in the length of the homing path, h. The animal should eventually reach the shore if ε < 90°. The length of the homing path is not doubled until ε = 60°. (b) When the goal is a point (G in inset) on a surface, even a fairly small error in setting the course imposes a risk of missing the goal altogether. For a given error, ε, the distance, a, of the nearest approach to the goal increases with the distance, d, between the starting point and the goal. Suppose the animal must come within a certain distance to detect the goal (shown with the dotted circle around G in the inset and the horizontal line set at 15 distance units on the graph). The homing accuracy required to ensure that the animal comes this close to the goal must increase as the distance between the starting point and the goal increases (compare symbols on horizontal dotted line).

by most terrestrial animals and many flying animals. Although flying animals can move vertically and may need to orient in three dimensions (see section 6.2.2.4), the critical task for many is to find locations relative to the earth's surface (the earth's curvature can generally be neglected). Well-studied examples include homing and migrating birds (reviews in Berthold 1991) and hymenopteran insects returning to their nests after finding food (reviews by Dyer 1994, 1996; Wehner et al. 1996). Also, flying animals typically maintain a relatively stable altitude (e.g. nesting insects, a few meters; migrating birds, a few hundred meters), and so they perform the basic navigational tasks in two dimensions.

If the goal is a point on a surface instead of a linear zone across a surface, the setting of a course to the goal is greatly complicated. To determine position in y-axis orientation, the animal merely has to decide which of two opposite directions is correct. Errors of up to just less than 90° do not prevent the animal from reaching the preferred zone. Such errors slow the approach to the preferred zone, but this cost is not severe if small errors are made (fig. 6.2a). If the goal is a point, however, the goal can be in any direction, and a much more precise ability to determine position is required. Small errors in positional or directional measurements can result in more drastic consequences because the animal can miss the goal altogether (fig. 6.2b). Furthermore, unlike an animal that targets a linear zone, an animal that sets a course for a point may benefit considerably from information about its distance as well as its direction from the point. With distance information, the animal would be able to engage in a search strategy after traveling the expected distance and not reaching its goal (Wehner and Srinivasan 1981).

When determining their position relative to a point, many animals do determine both the direction and the distance of their starting point relative to their goal. As I will discuss in the next section, however, the specific strategies used for determining position are strongly related to the distance over which the x,y-navigation takes place.

Information about direction can be obtained from cues directly associated with the goal (e.g. visual or odor cues), but only if the animal is already close to the goal. For setting and maintaining a direction of travel over greater distances, the animal must use either a compass (celestial or magnetic) or landmarks. Landmarks are stable features of the environment; they are usually detected visually, but in principle animals obtain information about landmarks with other sensory modalities (Bennett 1996). As discussed in section 6.2.3, the directional references available to the animal may depend on its distance from the goal. Both celestial compasses and landmarks require learning, as I will discuss later in the chapter.

6.2.2.4 Goal Is a Point in a Volume

Consider (1) a sea otter that seeks to return to a familiar patch of sea urchins; or (2) a forest-dwelling bird or bee that needs to find its nest or a patch of food at the correct stratum in the forest canopy or at the correct (x,y) position on the forest floor; or (3) an ant that forages in dense vegetation and must locate its nest or familiar food sources by negotiating a three-dimensional maze of branches and leaves. In each of these cases, the organism needs to find a particular point within a three-dimensional space (fig. 6.1 d); this situation may be referred to as x,y,z-orientation. The navigational strategies animals use to solve such a problem have scarcely been studied at all, and I mention this situation mainly for the sake of completeness.

Orienting to a point in three dimensions need not be much more complicated than orienting to a point in two dimensions. For example, the sea otter may first locate the correct x,y position on the surface of the water (with use of landmarks on the shore) and then dive. The ant may follow odor trails and use guidelines (topographic features) that are essentially the same as those used in a two-dimensional environment (Jander 1990). On the other hand, some animals may use more complex strategies for fixing their position relative to their goal in three dimensions and then set a direct course. In general, it seems likely that comparisons between closely related taxa that live in two-

and three-dimensional environments may reveal interesting differences in the designs of the mechanisms that underlie goal orientation.

6.2.3 Distance of the Goal

The importance of distance is threefold. First, distance affects the geometry of the task. For an animal (such as a sandhopper) that is heading for a beach from just offshore, the goal is a line on a plane, and the task is y-axis orientation. For an animal (such as a sea turtle) that is heading for the same beach from hundreds of kilometers away, the goal is effectively a point on a surface, and the task is x,y-orientation. Similarly, a bird that is heading for its nest in a forest canopy must find the correct height and the x,y position of the nest on the earth's surface (x,y,z-orientation). If the forest is tens or hundreds of kilometers away, only the x,y position of the goal is important.

Second, for point locations, the distance of the goal affects the accuracy required for successful homing. Figure 6.2b shows how, as homing distance increases, the error in the setting of the course to the goal must decrease so the animal can come within a certain distance of the goal.

Third, distance affects what strategies are available to an animal for determining its position relative to its goal. The effects of distance on the information available for navigation are particularly critical (and best understood) if the goal is a point on a surface. The following is a summary of the most important strategies that an animal uses to determine its position relative to a point at different starting distances (fig. 6.3). Note that distance in this case is a relative dimension that depends on such factors as the sensory modality used by the animal, the animal's travel speed, and the structure of the environment. What is a long distance to a walking animal with poor vision may be a short distance to a flying animal with good vision.

6.2.3.1 Short Distance: Final Approach to a Goal

From a short distance (fig. 6.3, zone 1) an animal may orient to stimuli emanating from the goal. A nearby goal may serve as a visual beacon, or it may emit sounds (e.g. a calling male frog luring a female) or odors (e.g. the pheromone plume originating from a female moth). In these cases, the task of determining position relative to the goal is a simple matter of recognizing and turning toward the relevant stimulus.

If the animal is guiding its final approach to a hidden goal, the animal must use indirect information about its position and direction of travel relative to the goal. Many animals can use surrounding landmarks to pinpoint the location of a hidden goal. This use of landmarks requires an ability to learn the spatial relationship between the landmark array and the goal. Such abilities have been studied intensively in invertebrates and vertebrates. Indeed, all of the major

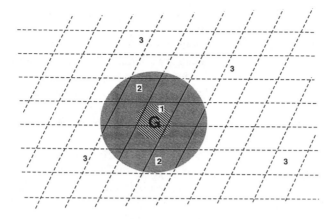

Figure 6.3. Determining positions relative to a familiar goal (G) on different spatial scales. In zone 1, the animal can detect G, or landmarks around it, directly. In zone 2, the animal is within its home range and relies on familiar landmarks that indicate a path to G; alternatively, if the animal traveled to its current location from G, it can use path integration to compute its displacement from G. In zone 3, the animal is outside its home range and establishes its position relative to G only by extrapolating previously experienced, geophysical stimulus gradients within its home range. Lines represent isoclines of two roughly orthogonal stimulus gradients. Solid sections are known values; hatched sections are values not previously experienced.

paradigms used for studying spatial learning in vertebrates deal with short-distance navigational tasks: these paradigms include the radial-arm maze, the Morris water pool, and other cue-controlled arenas used in behavioral and neurophysiological studies of rodent spatial memory (reviewed by Leonard and McNaughton 1990; McNaughton et al. 1996), and the cache-recovery tasks used to study spatial memory capacities of certain songbirds (Sherry, this vol. sec. 7.3.1.1). Studies of short-range spatial navigation in insects date to Tinbergen's famous studies on the abilities of digger wasps to learn the configuration of landmarks around their nests; more recent intensive studies include those of landmark learning in bees and wasps (for reviews, see Wehner 1981; Gallistel 1990; Dyer 1994; Collett 1996).

6.2.3.2 Intermediate Distance: Goal Orientation within Familiar Home Range

If an animal ranges widely, it may face the problem of heading toward a familiar part of the environment it cannot detect directly (fig. 6.3). To determine its initial position relative to the unseen goal, the animal must refer to environmental features at the starting point and along the way to the goal.

Animals have evolved two main mechanisms for determining their position within a home range. Use of both mechanisms necessitates learning. First, many animals can, during their travels through the environment, keep track of their position relative to a particular goal such as the nest. This process is known as path integration or "dead reckoning." The animal measures the angles (relative to an external compass or an internal directional reference provided with vestibular or tactile stimuli) and distances traveled over successive path segments; the animal then computes its net direction and distance from the starting point (reviews by Wehner and Wehner 1990; Gallistel 1990; Dyer 1994). I will discuss this process in a following section.

Second, many animals can determine their position by recognizing familiar landmarks that they have previously used to move to their goal. Obviously, this strategy is available only if the animal is familiar with the landmarks that are visible from the starting point. The ability to use landmarks to move to a distant, unseen goal is interesting because the initial landmarks that the animal uses may no longer be available once the animal has traveled some distance; therefore, different landmarks must be used to maintain the homeward course.

Although many animals use both path integration and landmarks for determining position within a familiar home range, the two strategies are potentially independent of one another. Path integration can be used to determine position even if the animal's path enters unfamiliar terrain. Landmarks can be used to determine position even if the animal has not traveled the path to its current location. For example, the animal may be displaced by wind or the experimenter's hand, and the animal, therefore, has no opportunity to determine position via path integration (reviewed by Wehner 1981; Dyer 1994). Despite their potential independence, the two strategies do reinforce one another under normal conditions.

6.2.3.3 Long Distance: Goal Orientation outside a Familiar Home Range

This is the task faced by homing and migratory birds and long-distance migrants of various non-avian taxa, such as sea turtles (Lohmann and Lohmann 1996) and monarch butterflies (Brower 1996). The animal's goal is tens or hundreds of kilometers away, and the animal may never have been at the starting point (as in the case of experimentally or wind-displaced homing birds) or at the goal (as in the case of migratory birds making their first trip to their winter habitat). Thus, path integration is not an option, and the animals do not have any opportunity to learn the landmarks along the route from the starting point to the goal. In the scheme put forth originally by Griffin (1952), this is the task classified as "true navigation" or "complete navigation."

How does the animal determine its position relative to its goal in this case? Two general strategies have been identified. The first, which is relatively simple, is exemplified by migratory birds that must determine if they are in the north and must head south or in the south and must head north. This decision is influenced strongly by hormone levels that vary seasonally as a result of the photoperiod. Thus, the animal's internal state effectively indicates the animal's approximate geographical position by affecting the animal's response to directional references (e.g. flying north or south) (reviews by Berthold 1991; Gwinner 1996). Experiments in which birds in the northern hemisphere are exposed to artificial photoperiods demonstrate that this navigational process is not dependent on actual measurements of geographical position. For example, birds in the lab will head north in the fall if they are exposed to increasing day lengths that are typical of spring. Birds that winter in the lab will also head north in the spring although the birds are already within their summer range.

The second, and much more complicated, strategy is employed in situations in which the animal is displaced a great distance to an unfamiliar starting point in an unknown direction, and the animal must determine its actual geographical position relative to its goal (fig. 6.3). This is what pigeons do when setting a homeward course after being displaced to a novel location far from their home loft (Keeton 1974). It is also what experienced starlings do when correcting for a lateral displacement from their normal migratory route that leads from northern Europe to southern England (Perdeck 1958). Just how animals accomplish such feats remains deeply puzzling and controversial. The general consensus, if there is one, is that animals exploit large-scale (global) stimulus gradients; animals generalize from gradient patterns that are experienced in a familiar area (e.g. around their home nest) to determine their position in the gradient when they are in an unfamiliar area. Furthermore, in this case, there is widespread agreement that animals would require at least two large-scale stimulus gradients that are not parallel to each other to fix their position unambiguously (review by Wallraff 1991). By this view, animals are assumed to locate their positions relative to an extrapolated bicoordinate map in a manner similar to how human explorers find their position in new lands relative to the system of latitude and longitude that is linked to geographical coordinates of home.

The controversy arises over the sensory basis of the extrapolated map used by animals. Among those studying bird navigation, debate has centered around two main hypotheses: use of olfactory cues versus use of geomagnetic cues. I will not review this controversy in this chapter (for recent reviews, see Able 1996; Walraff 1996; Wiltschko 1996). The crucial point is that in both the olfactory and geomagnetic hypotheses, it is assumed birds find their position on a large-scale map extrapolated from environmental patterns experienced on a much smaller scale. The sensory basis of the extrapolated map is a little less controversial in sea turtles, which also need to find their way over enormous distances. Lohmann and Lohmann (1996a, 1996b) recently provided evidence that young loggerhead sea turtles use a bicoordinate geomagnetic map (with use of the inclination and intensity of the geomagnetic field) to determine their position in the North Atlantic Ocean. Olfactory cues do not likely explain the long-distance navigation of sea turtles.

6.2.4 Variability of Navigational Cues

The environmental features that animals use to obtain positional or directional information may vary spatially or temporally. Temporal variation means that, in a given location, a fixed strategy for use of a particular environmental feature for navigation does not work equally well at different times of day or in different seasons. Spatial variation means that a fixed strategy does not work equally well in all locations. From the perspective of most animals, spatial variation can effectively be viewed as temporal variation because navigational cues vary at different times as a consequence of movement between different locations. Variation of such cues may be an important component of the navigational problem that an animal must solve. In particular, an animal confronted with a potentially variable navigational reference must be able to learn an appropriate response given the current state of the reference. Moreover, differences in the time scales over which animals experience environmental variability may affect the design of different learning mechanisms that deal with this variability.

Learning is commonly thought to be beneficial because it allows animals to adjust behavior in response to changes in particular environmental features. As Stephens (1993) and others have recognized, however, learning is not necessarily advantageous if environmental conditions fluctuate very rapidly relative to the time scale in which the animal makes successive responses to the environment. Thus, there must be some persistence of environmental conditions for learned responses to be of any value. There would be no point in learning features of a food that changed every time the animal encountered the food. Learning, however, may also be disadvantageous if environmental features fluctuate very slowly such that the appropriate responses to stimuli are the same for each generation. In this case, animals that adopt fixed responses may benefit over those that incur the costs of learning (e.g. the time and energy needed to sample or the cost of neural machinery; see Dukas, this vol. section 4.4). As I will discuss, the time scale of environmental fluctuation that favors learning depends heavily on species-specific life-history variables (e.g. lifespan).

In the context of navigation, animals tend to exhibit experience-independent responses to environmental features that change slowly (and hence are the same for each generation) and to depend on learning for features that vary in a shorter time scale. For example, naive amphipod crustaceans respond innately to a slope by heading down, which is the direction that allows escape to the sea for all sandhoppers everywhere (Scapini and Mezzetti 1993). Also, hatchling sea turtles, in their rush to the sea, exhibit innate responses to a slope (heading down), to landmarks (heading away), and to waves after entering the water (heading into waves); all of these responses predictably lead the hatchlings offshore (reviewed by Lohmann and Lohmann 1996a).

Other examples show how learning allows animals to adapt their behavior to unpredictable environmental features that, once encountered and learned, persist long enough to serve as reliable navigational references. The use of landmarks by animals that are heading to a familiar goal at a point location (x,y-orientation) illustrates this idea well. Although landmarks reliably indicate the y-axis direction of a shoreline, the spatial relationships between landmarks and a particular point are often completely unpredictable, and so a fixed innate response to the landmarks is of no value. Because landmarks do not move, however, a learned response will enable repeated visits to a location.

A second example arises in connection with use of the sun as a compass. As I will discuss in this chapter, animals that use the sun must compensate for the movement of its azimuth (the sun's compass angle) during the day. The exact pattern of change of the azimuth (the solar ephemeris function), which varies with latitude and season, is unpredictable; in species that are active in a wide range of latitudes and seasons, individual animals cannot predict exactly which ephemeris function they will need to use. Once a given ephemeris function is learned it will be useful for several days, because the ephemeris function changes only gradually from day to day. Both insects and birds learn the current, local ephemeris function early in life. Because the ephemeris function changes relatively slowly, short-lived animals such as insects may never need to relearn it (von Frisch 1967). Long-lived or wide-ranging animals such as birds, however, may need to recalibrate their memory of this pattern, and indeed there is evidence of such relearning in pigeons (Schmidt-Koenig et al. 1991).

A final example is the learning of the compass direction of the local y-axis by hatchling sea turtles that are heading seaward (reviewed by Lohmann and Lohmann 1996a). As mentioned, naive hatchlings can head in the y-axis direction with use of a succession of highly reliable cues, but hatchlings are uninformed about the compass orientation of the y-axis (a presumably unpredictable variable given the possibility that mothers lay eggs on beaches in various orientations). The baby turtles rapidly learn their compass heading, however, using their outbound orientation relative to the waves as a reference. The feature used to measure the compass heading (the local geomagnetic field) is stable; their learned response to this feature guarantees consistent movement away from the shore.

Animals that initially exhibit an innate response to a particular cue may

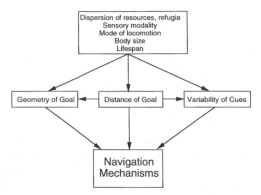

Figure 6.4. Hypothesized causal relationships among various factors that could have shaped the evolution of navigational mechanisms.

subsequently modify their behavior through learning. This is illustrated in the y-axis orientation of amphipod crustaceans that exhibit an innate initial response to the sun or the geomagnetic field and adopt the compass heading that is aligned with the y-axis of their natal beach. The innate response implies that populations are sedentary enough so that, for each new generation, a given direction relative to the sun compass or the magnetic compass predictably corresponds to the local y-axis. If, however, sandhoppers are displaced to a differently oriented beach, they can learn a new y-axis orientation relative to a compass reference. Apparently sandhoppers do this by using the direction of the slope—a highly reliable indicator of the y-axis direction—to calibrate their response to celestial or magnetic cues (Scapini and Mezzetti 1993; Ugolini and Pezzani 1995).

6.2.5 Biological Factors That Set the Parameters of the Navigational Task

This classification of the diversity of navigational tasks has so far ignored the question of why animals face different navigational tasks. Here I list several general biological and ecological traits that are likely to be relevant for defining the navigational problems an animal faces (fig. 6.4). To the extent that these traits are phylogenetically conserved, one could argue that they have played a causal role in the evolution of particular navigational abilities. Of course, the direction of causation could be reversed if improvement in an animal's navigational mechanisms creates opportunities or imposes demands that were not experienced by ancestors, and hence leads to adaptive modifications in these traits.

6.2.5.1 Diet

The food resource typically exploited by an animal is dispersed in a characteristic way, which requires a particular set of behavioral strategies (including navigational strategies) to find and exploit it. For example, a resource distributed in widely spaced patches necessitates a different navigational strategy than one distributed more uniformly. The rates of depletion and renewal of resources in patches influence the need to learn the locations of new patches.

6.2.5.2 Mode of Locomotion

Flying animals can cover more area in a given period of time than walking animals, and thus, flying animals may have to navigate toward goals over greater distances. Because distance affects the geometry of navigational tasks faced by animals, the mode of locomotion may indirectly influence the complexity of the navigational strategies needed. Also, flying (or swimming) animals potentially face navigational problems in three dimensions, whereas walking (or benthic) creatures face navigational problems in essentially two dimensions. Furthermore, animals whose mode of locomotion allows them to range widely are also more likely to be exposed to variable environmental conditions, and hence, these animals more likely need to learn new responses to variable navigational cues.

6.2.5.3 Sensory Modality

Because the positional and directional information that is necessary for goal orientation must enter an animal's brain by way of its senses, the sensory modalities used to detect environmental features may influence profoundly the navigational problems the animal faces and its capacity to solve a particular problem. For an obvious example, landmarks (fixed features of the environment) provide spatial information through any sensory modality, but landmarks are most easily used by highly visual animals. Even animals that rely on the same sensory modality may differ in their navigational abilities because their sensory organs are phylogenetically different. The compound eye of the insect, compared with the vertebrate eye, probably restricts greatly the visual information available to the insect's brain (Land 1981).

6.2.5.4 Body Size

Body size profoundly affects the animal's resource needs and hence the size of its home range, which in turn affects the distance over which the animal typically has to navigate to feeding sites. Body size also affects travel speed, which by definition affects the distances that can be covered by an animal daily. As mentioned, the scale of an animal's movements possibly influences the geometry of the navigational task and the variability of the navigational cues encountered by the animal. At the same time, a faster animal can sample the environment more rapidly and thus may accumulate a richer knowledge of relevant spatial relationships.

6.2.5.5 Lifespan

The most important consequence of lifespan (which generally correlates positively with body size; Calder 1984) is the extent of environmental variation an animal experiences. An animal that lives longer more likely needs to recalibrate its responses to environmental features (when important spatial relationships change) during its life (see Dukas, this vol. section 4.5.6).

6.2.6 Potentialities and Limitations of This System

In this section, I attempted to organize the diversity of navigational problems faced by animals. I hope this classification system not only describes known patterns but also stimulates new research by providing a clearer framework for comparative studies. This system establishes criteria for identifying species that face similar or different navigational problems and may help us frame questions about the similarities and differences in the underlying mechanisms. Furthermore, given the wide variation of navigational problems faced by animals, this system may help us understand the adaptive design of particular navigational mechanisms.

The various mechanisms that underlie animals' navigational abilities are not elucidated with this system; it only allows determination of similar or different navigational problems that are encountered. Indeed, some species that may be classified as having similar navigational problems are likely to solve these problems in different ways. For example, like many songbirds that spend the summer in the northern hemisphere and head south for the winter, monarch butterflies migrate thousands of kilometers on a path that effectively leads to a point location in the subtropics (reviewed by Brower 1996). The geometry of the goal and the scale of movement are essentially the same in the cases of songbirds and butterflies, but the underlying mechanisms of navigation are, in all likelihood, very different. Of course, discovering that animals solve comparable problems in different ways raises interesting biological questions. This classification system clarifies the similarities of the problems that two organisms face, and thus may better guide the search for an explanation of the different navigational solutions that are used.

6.3 CURRENCIES AND CONSTRAINTS

In this section, I present the components of a trait-oriented approach to the adaptive design of navigational traits. The goal of a trait-oriented study is to understand why selection has produced specific phenotypes in an organism. Ideally, we would like to know how and why phenotypes have been selected. Often, however, the history of a phenotypic trait cannot be constructed, and so we can only ask how the phenotypic end points observed may be maintained by natural selection.

Behavioral ecologists have attempted to answer this question by asking two related questions (Stephens and Krebs 1986). One question concerns the fitness benefits and costs of variants of the trait: Does the trait's design reflect the selection to maximize certain benefits, minimize certain costs, or both? The goal is to identify the benefits and costs as specifically as possible and understand the causal link between phenotype and fitness. The notion of fitness "currencies" is often used to summarize the benefits and costs that correlate with different phenotypic variants. A given currency expresses (ideally in precise quantitative terms but sometimes in qualitative terms) the relationship between a particular phenotypic variable and particular benefits and costs. For example, a currency commonly hypothesized for the adaptive design of foraging behavior is the "net rate of energy gained" (NREG) by the animal while it forages: energy gained per unit time (the benefit) minus energy expended per unit time (the cost). Such a currency is a measure of performance that is assumed to correlate with fitness, such that fitness is highest when the currency is maximized (for a currency such as NREG that emphasizes benefits relative to costs) or minimized (for currencies that emphasize costs relative to benefits (see Ydenberg, this vol. section 9.2, for further details).

The second question concerns the biological and physical constraints that limit the course and outcome of selection. Understanding the constraints on phenotypic design is difficult. Several different constraints have been proposed, and there has been considerable debate about whether some constraints are more important than others (Arnold 1994; Dukas, this vol. section 3.2). To simplify the discussion, I focus on two relatively well-defined classes of constraints. The first I have already introduced (6.2.5): phylogenetically conserved or "ground-plan" traits that determine the general nature of the challenges of survival and reproduction. Such traits are constraints in the sense that they determine the selection pressures that operate on another trait being studied. For example, the traits that make a zooplankton a zooplankton—its dietary and habitat requirements (which determine how sources of food and shelter are distributed in space), its size, its mode of locomotion—jointly determine that the zooplankton's most difficult navigational problems are

(1) whether to go up or down, and (2) which way is up. The traits that make a bee a bee determine that the bee must learn to find, in a two-dimensional terrain, the widely separated point locations of its nest and nectar and pollen sources.

The other class of constraints comprises those features of the organism or its environment that slow or prevent a trait from reaching a particular phenotypic optimum. Invoking constraints in this context assumes not only that selection is inherently an optimizing process (an assumption some dispute) but also that we can identify the optimal phenotype—the peak in a fitness landscape—that selection is "trying" to reach (an even more controversial assumption). The failure of the organism to exhibit the optimal phenotype may mean that there is some biological or physical reason for why the optimum can never be reached (e.g. lack of genetic variation in the organism) or why the optimum has not been reached yet (e.g. genetic correlations with other selected traits that slow the course of evolution toward the optimum or limit an organism's ability to acquire information that is needed for optimal performance). In the case of navigational mechanisms, commonly invoked constraints of this sort are the animal's sensory modality (which hypothetically limits the ability to acquire information that enables better performance) and brain size (which hypothetically limits the ability to process and store the information that is obtained).

Ideally, one can incorporate hypotheses about benefits, costs, and constraints into a formal quantitative model. This is standard practice in studies of foraging. (For examples in this volume, see Dukas, sections 3.3.2.2, 3.4.2; Bateson and Kacelnik, sections 8.5, 8.6; Ydenberg, sections 9.2.7, 9.2.8.) The few explicit hypotheses about the adaptive design of navigational mechanisms, by contrast, are presented in qualitative, verbal terms. One of these is Baker's (1978) "principle of least navigation." Baker uses "navigation" to refer to the process(es) that must occur in the animal's brain for the setting and maintenance of a correct course, and not to the behavioral outcome of this process(es). The idea is that animals have evolved to use those sensory cues and information-processing strategies that will allow them to meet some critical level of performance while minimizing investment in processing: Thus, natural selection does not favor use of a supercomputer when a pocket calculator will do. This hypothesis assumes that (1) the currency for the adaptive design of navigational mechanisms is dominated by the costs associated with "more navigation," and (2) the selection favors mechanisms that minimize this cost while also producing the benefits associated with reaching a given goal. The costs could include, for example, the time required for relevant calculations or the energy needed for neural processing. This hypothesis also assumes there are constraints on the capacity of animals to handle navigational information such that more complex solutions require more time or energy.

There is much evidence that animals tend to use approximations and short-cuts that guarantee adequate performance, instead of processes that may produce (with a greater investment of time and energy in computation) more accurate navigation (Dyer 1994; Wehner 1987, 1991; Wehner et al. 1996). At the same time, however, this hypothesis of minimized investment provides little explanation about why the navigational abilities of a given species are limited in specific ways or why some species exhibit more sophisticated navigational abilities than others. Perhaps more important, this hypothesis considers only a narrow subset of the possible currencies that may affect the adaptive design of navigational mechanisms and offers a very superficial account of why animals are constrained in the ways they apparently are. I shall conclude this section by outlining ways in which the study of navigational currencies and constraints may be broadened.

There are a number of possible currencies for the natural selection of navigational mechanisms: homing speed (maximize), homing accuracy (maximize), probability of disorientation (minimize), energy expenditure (minimize), risk of predation en route (minimize; i.e., follow the safest path), or probability of contacting possible resources (maximize). One can imagine how each of these may be favored in particular circumstances—depending on the ecological and life-history characteristics of the species—and produce very different behavioral consequences.

Furthermore, by drawing an analogy to foraging currencies, one can also propose more complex navigational currencies that reflect the fitness trade-offs between competing needs. A widely studied trade-off in the foraging literature is that between foraging and predation: For some species, increases in the foraging rate may occur at the expense of a higher predation rate, and animals behave in ways that balance the need to feed well and stay safe (Ydenberg, this vol. section 9.4). By analogy, navigational mechanisms in some species may reflect the competing needs of maximizing homing speed and minimizing exposure to predators. Confronted with the general task of finding the way from point A to point B, a species that is relatively immune to predation (for example, because it is distasteful) may be selected to find the shortest path; however, a more vulnerable species may be selected to find the shortest *safe* path (as exemplified with the thigmotactic [e.g. wall-hugging] behavior of mice or cockroaches).

With regard to constraints, I have already outlined a broad framework for identifying the ways in which biological or physical factors external to the trait that is under selection may place limits on the course of selection. I want to focus my attention on a specific constraint that is addressed by Baker's least-navigation model and many other explanations about navigational limitations in animals. This constraint is the inherent limitation in the ability to obtain and use information from sensory stimuli; this constraint in turn limits the speed or accuracy of navigational performance. No doubt all animals are limited in their capacity to handle information about the environment (Dukas, this vol. chap. 3). To verify this idea, however, we must determine what has prevented natural selection from producing a more sophisticated animal than what we observe.

The difficulty of answering this question becomes apparent when we attempt to understand the role of brain size as a constraint. Brain size is widely invoked as a constraint on the cognitive capacities of animals, both as a general constraint on basic learning ability and as an explanation for the failure to adopt certain strategies. Brain size is invoked as a constraint any time it is assumed that an animal can learn more or faster or remember longer if it had more neural tissue. This assumption is made, for example, when we contrast the sophistication of big-brained animals with the limitations of small-brained animals (Wehner 1991). It is also made in any study that correlates the sizes of brain regions with the apparent complexity of a cognitive task (reviews by Fahrbach and Robinson 1995; Sherry, this vol. section 7.4.4).

The assumption that brain size constrains cognitive performance is reasonable. A computational device with more computing elements (be they transistors or synapses) must be able to process larger amounts of information at a time, store more information in memory, and possibly handle more tasks simultaneously. The problem is to translate this eminently reasonable assumption into an account of the specific limitations on cognitive performance (see Dukas, this vol. section 4.3). Correlations between brain size and performance abound (see Sherry, this vol. section 7.4.4). Because we do not know how the brain does most of what it does, however, there is no theory that allows us to predict how much cognitive sophistication can be manifested with a given amount of neural tissue or explain why an animal's cognitive performance and its consequent behavior are limited in the ways they are.

Furthermore, overemphasis on the constraints imposed by brain size may lead one to overlook other explanations for an apparent failure to employ a presumed optimal strategy. Perhaps the cognitive strategy that we regard as superior—because it appears more sophisticated, more accurate, or more reliable—would actually be inferior to the one used by the animals. For example, the more sophisticated strategy may necessitate more processing steps or greater sampling of the environment and hence may cost the animal more time or greater exposure to predation. If so, then the animal, by adopting its strategy, may actually be balancing optimally the competing demands of computational accuracy and computational speed.

Most hypotheses proposed about the adaptive significance of navigational abilities apply to particular situations only, plausibly identifying the likely functions of particular mechanisms. This is not necessarily bad—we have to start somewhere. I believe, however, the rigor of hypotheses about the adaptive design of particular navigational mechanisms can be improved considerably.

6.4 CASE STUDIES

As I stressed, most investigators have focused on how animals carry out their amazing feats of navigation. In the following, I will consider a number of examples of navigational processes that either have been or could be analyzed from an adaptative standpoint. My goal is to illustrate the potential for new insights into the biology of specific navigational mechanisms and the evolution of general cognitive mechanisms. I will consider only cases in which there is good behavioral evidence of the underlying mechanisms. There is little point in speculating about the adaptive design of mechanisms if their basic properties are a matter of debate (e.g. the navigational maps of birds).

6.4.1 Path Integration by Nesting Insects

As mentioned, path integration is a mechanism by which an animal can determine its position on the basis of information acquired along the path. A particularly well-studied example is the ability of desert ants to determine the direction and distance to home from a feeding place after moving on a circuitous, outward searching path (reviews by Wehner and Wehner 1990; Gallistel 1990; Dyer 1994). Similar abilities have also been studied in honeybees (von Frisch 1967), funnel-web spiders (which need to retreat after running onto the web to find prey) (Görner and Claas 1985), hamsters (Etienne et al. 1996), rats (McNaughton et al. 1996), and humans (Loomis et al. 1993). Such navigational abilities provide clear evidence of internal representations that are derived partly with computation of sensory data. The homing ant behaves as if she has encoded a vector that corresponds to her distance and direction from home. Usually, however, this vector does not correspond to any segment of the outward path or any path previously traveled; therefore, we can exclude the hypothesis that the animal merely encodes a neural replica of sensory information that is acquired during the movement on the outward path. Instead, we must conclude that the animal computes the homing vector with use of distance and directional information acquired during the twists and turns of the outward path (fig. 6.5).

Recent studies have provided new insights into the nature of the computations that ants perform and have raised important questions about the adaptive design of the mechanisms used to perform these computations. Before the recent work on ants, investigators of path integration traditionally assumed that animals compute the homing vector with use of the equivalent of trigonometric vector summation (e.g. Mittelstaedt 1985). This computation would require

224   Fred C. Dyer

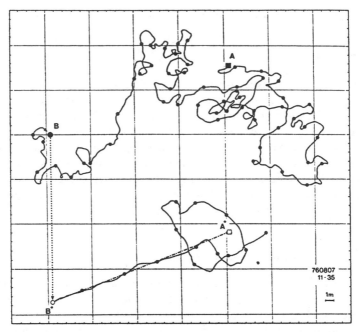

Figure 6.5. Track of a desert ant (*Cataglyphis* sp.) during its search for food. Track originates at nest (A) and terminates at B (dots give position at 10-second intervals). After experimental displacement from B to B*, the ant searches for home (A). The ant travels in the direction from B* that would have led it to A from point B. The ant also travels roughly the same distance from B* that would have led to A from B; the ant then wanders in search of A. The ant's ability to set a homeward course is not dependent on any cues that emanate from its nest or previous experience of the homing route; instead, the ant integrates the directions and distances traveled over the outward path and maintains a continuously updated internal representation of its net displacement from home (from Wehner 1982).

neural mechanisms to decompose the vector that corresponds to each segment of the outward path into its sine and cosine components and to compute from the summed components the net direction and distance of displacement. The implicit assumption is that natural selection equipped animals to perform path integration in this way because other possible methods introduce systematic biases in the calculation of the direction home.

Wehner and Müller (1988) found, however, that ants actually do exhibit systematic errors of up to 30° in their estimate of the homeward direction. The pattern of these errors was evident when ants were constrained experimentally to travel to food along a two-segment outward path. These ants chose a homeward direction that deviated consistently to one side of the actual homeward direction. Ants that made a right turn from the first segment to the second segment set a homeward course slightly to the right of the actual homeward

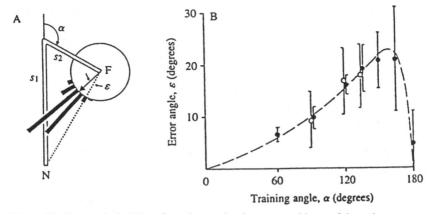

Figure 6.6. Systematic deviations from the true homing course with use of the path-integration system in desert ants (*Cataglyphis fortis*). (*a*) Homing directions selected by ants after travel on a two-segment outward path in an experimental channel that leads from the nest (*N*) to food (*F*). From *F*, the ants were displaced to a featureless plain and their homing orientation was measured. The angle ε is the deviation of the ants' mean path from the compass direction that would have led to *N* from *F*. (*b*) As the angle (α) between s_1 and s_2 increased, the ants' angle of deviation increased (ε) (although ε is low when α = 180°) (from Wehner et al. 1996).

direction. Ants that made a left turn erred to the left. The degree of error varied systematically with the angle of the turn (fig. 6.6).

These systematic errors suggest that these ants do not carry out the equivalent of trigonometric vector summation. Furthermore, the bias in the homeward direction cannot be accounted for by random errors in a trigonometric calculation because random errors would produce only scatter around the correct homing vector. By analyzing the errors, Müller and Wehner (1988) deduced an alternative computational algorithm that better explains the ants' behavior. The alternative algorithm is not a trigonometric solution. In the alternative algorithm, the animal maintains an arithmetic running average of the directions traveled along the outward path; the contribution of each path segment to the average direction is weighted by that segment's length. Reanalysis of published path-integration data revealed that other invertebrate species, including honeybees, exhibit systematic errors that can be explained with the same algorithm (Wehner 1991).

Wehner (1991) suggested that use of a nontrigonometric computation for path integration represents a "small-brained strategy." Wehner implied that trigonometric vector summation is inherently more difficult to perform than the computations in the algorithm of Müller and Wehner. We encounter a problem I alluded to and one stressed recently by Wehner et al. (1996): There is no theoretical or empirical justification for the assumption that a brain the size of an ant's cannot perform the equivalent of trigonometric computations.

Conceivably, all that is required is a neural unit with an output signal that scales as a trigonometric function of the input signal (Gallistel 1990). Wehner et al. (1996) pointed out that mammals (including humans) performing path integration make systematic errors inconsistent with the predictions of trigonometric vector summation. Should we therefore assume that the processing capabilities of the mammalian brain are too limited to support an exact solution to the problem of path integration?

An alternative possibility is that the systematic errors in path integration observed in so many species reflect not an inherent limitation of computing power but an adaptive strategy that is designed to solve a specific problem in path integration. Any path-integration mechanism is subject to random error (which results from imprecise measurements of directions and distances during the outward trip). Furthermore, these errors accumulate over successive segments of a long path. The systematic errors exhibited by ants and other animals could be viewed as strategies used to mitigate the effects of accumulated random errors (Hartmann and Wehner 1995). The direction in which homing ants deviate from the actual homeward direction tends to lead them across the initial segment of the outward path; this deviation would bring them into visual contact with familiar landmarks that could be used for reorientation. Even if there were more scatter around this (systematically biased) homing direction, most ants would come within view of familiar landmarks. Hence, an ant's probability of missing the nest to the other side and heading into unknown terrain would be reduced. This explanation may account for the observation that there is no systematic error in the ants' homing direction if the outward path doubles back toward home (Müller and Wehner 1988).

The hypothesis that the systematic deviations in the ant's path-integration system are adaptive raises new questions that could be examined by means of an optimality analysis. Presumably, the systematic deviations from the homeward direction are advantageous only if they are not too great. Very large deviations may waste energy or time or even increase the probability of getting lost. What determines the optimal degree of systematic deviation? One important factor might be the degree of random error that occurs during travel on a typical foraging path (fig. 6.7). Greater random error would produce greater scatter of the homing vectors calculated by ants, and would favor a greater deviation in the homeward direction to reduce the probability of missing the nest to the wrong side (fig. 6.6). Another factor might be the size of the familiar area where familiar landmarks could direct the ant home. The size of the familiar area would be determined by such factors as the visual structure of the environment, the visual acuity of the animal, and the animal's capacity to learn spatial relationships between landmarks and the nest. If the familiar area were larger, then larger random errors of path integration would be tolerated, and there would be less need for compensation with systematic biases.

It is premature to say if such an optimality analysis would provide new

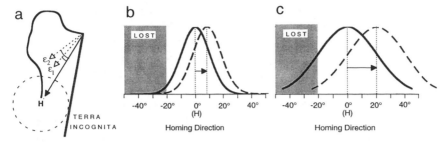

Figure 6.7. Optimal tuning of systematic deviations in the path-integration system of desert ants. (*a*) Systematic deviations possibly compensate for random error in the computation of the homing direction by ensuring that the homeward path leads across the first segment of the outbound path instead of into unfamiliar territory (*terra incognita*) (Wehner et al. 1996). If so, the magnitude of systematic deviation sufficient to eliminate the possibility of getting lost should be smaller when random error produces a relatively small scatter in the computation of the homing direction (*b*) than when random error produces a relatively large scatter (*c*).

insights into the design of path-integration mechanisms. My aim is to illustrate that study of this fascinating system can be moved beyond a consideration of mechanisms toward the development of testable ideas about why mechanisms have the properties they do.

6.4.2 Sun-Compass Orientation by Insects

In both y-axis orientation (navigation perpendicular to a shoreline) and x,y-orientation (navigation to a goal at a point location), animals must hold a straight-line course once they have determined their position relative to their goal. A common solution is to use an external feature of the environment as a compass. The most important sources of compass information for animals and humans are celestial cues (e.g. the sun, polarized light patterns, and stars) and the earth's magnetic field. In this section, I will consider sun-compass orientation in insects. Honeybees have also been found to have a magnetic compass (Schmitt and Esch 1993; Collett and Baron 1994). I will discuss the honeybees' magnetic compass later in the context of the function it serves in landmark learning.

Insects use their sun compass to hold a straight course during path integration, and honeybees, specifically, use the sun as a reference to communicate the direction of food with dance (von Frisch 1967). The sun is ideal as a directional reference because it is reliable and too far away for parallax error to occur. To maintain a straight course, the animal simply holds a constant body angle relative to the sun's azimuth. The problem, however, is that the sun moves. An animal that must maintain a straight course over an extended period of time or travel along a given route at different times of day must compensate for changes in the sun's direction relative to the direction of travel.

Many animals can compensate for changes in the solar azimuth (even during periods when they cannot see the sun directly); this compensation means that animals must be informed about the pattern of solar movement during the day. One major question regarding cognition concerns the nature of the mechanisms by which animals estimate changes in the sun's position over the course of a day.

The task of compensating for the sun's movement is complicated further because the rate of the sun's movement is not constant (fig. 6.8), which precludes the possibility of compensating at a fixed rate. The rate of change of the azimuth is relatively slow in the morning and afternoon and relatively rapid at midday. In addition, this daily pattern, or the ephemeris function, varies with the season and latitude. Clearly, an animal that uses the sun as a compass would do well to use the current, local ephemeris function. We have known since the 1950s (Lindauer 1957, 1959) that insects learn the solar ephemeris function early in life. Recently, investigators have focused on understanding how this learning takes place, and, in the process, they have provided clues about the adaptive design of the underlying learning mechanisms.

To begin, we should consider what learning the sun's course entails. At the minimum, the animal must associate different positions of the solar azimuth, as measured relative to some earthbound coordinate system (e.g. landmarks or the earth's magnetic field), with the times of day at which they are measured, so that time can cue the retrieval of the correct azimuth. Conceivably, the entire course of the sun can be learned by simply encoding time-linked measurements of the azimuth into a neural "look-up table." We have long known, however, that animals do something a bit more impressive. As Lindauer (1957, 1959) first showed in honeybees, some animals can estimate the sun's position at times of day when they have never seen the sun. It is as if the animals can somehow compute unknown segments of the sun's course. Lindauer reared bees in an incubator and allowed them to see only the afternoon segment of the sun's course. During their afternoon flight time, he trained these bees to find a feeding station south of their hive. When he tested the bees in the morning (in a different terrain so that familiar landmarks could not be used), he found that bees could use their sun compass to find the food. Other studies have shown that honeybees (Lindauer 1957; Dyer 1985) and desert ants (Wehner 1982) can estimate the sun's course at night. Bees refer to the nocturnal position of the sun when performing waggle dances to communicate the location of food at night (Dyer 1985); in ants, compensation for the sun's nocturnal course has been demonstrated experimentally with use of an artificial sun (Wehner 1982); however, why ants have this ability is not clear.

Thus, in insects, learning the sun's course entails not only the recording of known time-linked positions of the sun's azimuth but also filling in unknown

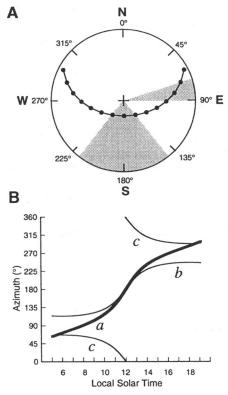

Figure 6.8. Solar ephemeris function. (*a*) The sun moves at constant rate 15°/h along its arc (symbols show sun's position at hourly intervals starting at 0500 hours on the June solstice in East Lansing, Michigan, ≈43° N); however, the azimuth, or the compass direction of the sun, changes at a variable rate during the day. Shaded sectors show the change in the azimuth for two 2-h time intervals—one early in the morning when the azimuth changes relatively slowly (≈7°/h) and one spanning noon when the azimuth changes quickly (≈40°/h). (*b*) Alternative method for representing the sun's pattern of movement, illustrating seasonal and latitudinal variation in ephemeris function. Thick line (*a*) shows the ephemeris function for East Lansing, Michigan, on the June solstice. Other lines show functions for the equator on two dates: (*b*) June solstice, when the azimuth shifts right-to-left (counterclockwise relative to cardinal compass coordinates) along the northern horizon; and (*c*) December solstice, when the azimuth shifts left-to-right (clockwise) along the southern horizon (from Dyer and Dickinson 1996). Function *c* appears in two segments; the break occurs at noon when the azimuth is in the north.

positions of the azimuth with some sort of computation. The computation can be modeled with the following general equation: $A_{\tau+t} = A_\tau + R \cdot t$, where A_τ is a known time-linked azimuth of the sun, $A_{\tau+t}$ is the unknown azimuth after an elapsed time interval t, and R is the rate of compensation (degrees of azimuth/time) during t. Most investigators who have sought to understand how insects perform this computation have asked how insects determine the rate of compensation, R. The prevailing assumption has been that insects compute R on the basis of the position or the rate of movement of the azimuth when they have observed it. For example, much of the available data can be explained with the hypothesis that insects interpolate at a linear rate to find unknown positions of the azimuth (New and New 1962; Wehner 1982); thus, the value of R is determined for a gap between two known positions of the azimuth by dividing the difference in the azimuth by the difference in time. Other data suggest, however, that insects extrapolate the sun's position on the basis of the rate of azimuthal movement observed at other times of day (Gould 1980; Dyer 1985).

Recent studies of desert ants and honeybees have suggested that the actual learning mechanism used by insects is considerably different from the mechanism that any previous investigators had assumed. The properties of these mechanisms were revealed in experiments that used improved assays to infer how experience-restricted insects estimate unknown segments of the sun's course. For example, Dyer and Dickinson (1994) studied the dances of bees; with dance, the bees indicated a known feeding site to which they had flown under a cloudy sky. The bees used the dance to indicate the direction the dancer had flown to reach food, relative to the sun. Human observers can use dances to infer where the bees have determined the sun to be relative to the line of flight. By testing bees on an overcast day, one can ensure that bees base their dances on an internally generated estimate of the sun's position instead of a direct measurement taken during flight (Dyer 1987).

Examination of dances during an extended period of cloudy weather revealed that experience-restricted bees did not use any of the proposed linear computations to estimate the sun's course at times of day when bees had not seen the sun. Instead, bees behaved as if they used an internal ephemeris that approximated the actual nonlinear pattern of movement of the azimuth (fig. 6.9a). The internal ephemeris was an exaggeration of the actual ephemeris at the time of the experiment: In the internal and actual ephemerides the sun rises opposite from where it sets, the azimuth changes little throughout the morning and afternoon, and the azimuth moves rapidly from the eastern to the western half of the sky at midday. The data, and hence the mechanism that produced the data, can be described as a step function in which the azimuth changes by 180° at midday (fig. 6.9b). More importantly, with additional experience throughout the day, this approximation of the sun's course is trans-

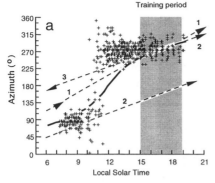
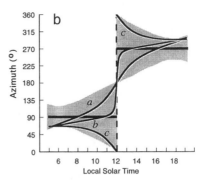

Figure 6.9. Partially experienced bees estimate the sun's course at times of day when they have never seen it. (*a*) Solar positions inferred from dances that indicate the same feeding place by incubator-reared bees that had seen the sun only from 1500 until sunset (training period) on days before the test. Data from 554 dances performed over the day by 44 different bees are shown. Bees were tested on a cloudy day so they could not base their dances on a direct perception of the sun during the flight to the food; bees had to estimate the sun's position on the basis of their experience on previous afternoons. The thick line is the actual solar ephemeris function at the time of the experiment. Lines 1–3 show the predictions of previous computational models proposed to explain the ability of insects to fill gaps regarding the sun's course: 1 = linear interpolation of the sun's position at times between sunset and the beginning of the daily training period, 2 = forward extrapolation (through the night and into the next day) of the rate of solar movement measured during the training time, 3 = backward extrapolation of the rate observed during the daily training period to earlier times during the day. All three predications account poorly for the bees' behavior. The data can be described by a step function in which the azimuth used in the morning is 180° from the azimuth the bees experienced on previous afternoons. Data from two bees that used the afternoon angle in the morning and the morning angle in the afternoon are excluded (see Dyer and Dickinson 1994 for these data). (*b*) Comparison of 180° step function with actual ephemeris functions. Curving lines show the ephemerides for three latitudes on the June solstice: *a*, 43° N (East Lansing, Michigan); *b*, 25° N; and *c*, 15° N. Shaded region covers all possible solar-ephemeris functions that could be observed on the earth's surface (i.e. all latitudes and all days of the year). The mean of all these functions is a 180° step function that resembles one used by the bees.

formed so that it more accurately reflects the actual movement of the azimuth (see Dyer and Dickinson 1994).

With evidence of a similar phenomenon in desert ants (Wehner and Müller 1993), the results suggest that the development of the insects' sun compass involves an interplay of innate and individually acquired information about the dynamics of solar movement. I suggest that the learning process starts when bees, during their first flights outside the nest, record time-linked positions of the sun relative to landmarks and thereby associate their approximate, innate ephemeris with the local terrain. This association allows them to use remembered positions of the sun at times of day when they have seen it and estimate, with reasonable accuracy, the azimuth at other times of day. With additional experience, a bee refines its internal ephemeris to resemble the actual ephemeris.

The innate ephemeris uncovered by these experiments has some intriguing properties that could represent adaptive design features facilitating the bees' learning of the sun's course. Note that the solar ephemeris observed at latitudes near the equator (where honeybees evolved) closely resembles the step function developed by the experience-restricted bees (fig. 6.9b). This means that a bee that has sampled only a small segment of the sun's course is able to estimate the sun's position at other times of day to within about 30°. Even at higher latitudes, a bee's errors would exceed 45° only during a couple of hours near noon. Although such errors degrade the accuracy of compass orientation and path integration based on the sun, the approximate ephemeris may nevertheless allow foraging bees to begin to use the sun, perhaps in conjunction with familiar landmarks, before they have extensively sampled the sun's course throughout the day (Dyer and Dickinson, 1996).

An even more intriguing feature of the innate ephemeris function revealed by the experiments is that the shape of the ephemeris—a 180° step function—exactly matches the average of all possible solar ephemeris functions (considering all latitudes and all days of the year). The learning task faced by a bee is to develop an internal ephemeris function that resembles the ephemeris function at the time and place at which the bee lives. The function to be learned cannot be predicted in advance because a developing bee cannot anticipate the latitude and season in which she will be active; however, consider the advantages of beginning with a default ephemeris that resembles the average ephemeris function instead of, for example, substituting random angles for unknown positions of the azimuth. As shown in figure 6.10, the estimates produced with a 180° step function deviate far less on average from any actual ephemeris function than do random estimates. Thus, the underlying neural network that encodes the sun's course presumably has to change less to conform to the actual ephemeris and thus approximates more quickly an accurate representation of the ephemeris. The faster learning speed may be a considerable advantage for an animal that, like a honeybee, spends at most 2 weeks gathering food for her colony before she dies (reviewed by Winston 1987).

In summary, the behavioral evidence suggests that the insect brain was designed for rapid learning of the sun's changing position relative to the terrain. The development of the sun compass exploits regularities in the sun's general pattern of movement. The innate representation essentially encodes this general pattern; this encoding allows bees to use the sun for orientation relatively accurately (even at times of day when they have never seen the sun)

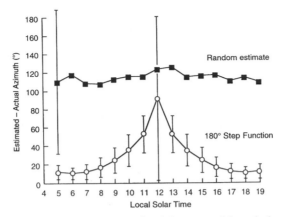

Figure 6.10. Deviation from the actual solar azimuth in two possible methods used for estimating the azimuth at unknown times of day. At each time of day, symbols show the mean difference ($n = 500$ comparisons; error bars, ± 1 standard deviation) between the estimated azimuth and the actual azimuth as defined by the solar ephemeris function from a randomly selected latitude and date. The values for the random estimate were based on a randomly selected angle between $0°$ and $360°$. The error bars for only one time of day are shown; the others are comparable in magnitude. The values for the $180°$ step function were based on angles of $90°$ before noon, $270°$ after noon, and either $0°$ or $180°$ (each with probability 0.5) at noon. This simulation suggests that a $180°$ step function, like that used by partially experienced bees and ants, would, on average, estimate unknown positions of the sun with far greater accuracy at most times of day than a process that generates a random estimate (Dickinson and Dyer in preparation).

and possibly facilitates subsequent development of a more accurate representation.

6.4.3 Compass Orientation in Birds

Compass orientation has been intensively studied in birds since Gustav Kramer (1950) presented evidence of a time-compensated sun compass in starlings. His discovery occurred the same year von Frisch (1950) reported the existence of sun-compass orientation in honeybees, and so the studies of compass orientation in these two species have developed simultaneously. The research has revealed certain parallels between these two species in addition to the role played by celestial cues in both groups. Most important, compass orientation in both insects and birds depends on an interaction of innate and individually acquired information, leading to the development of an internal record of relevant features of the compass reference. Furthermore, in birds as in insects, the plasticity of this developmental process allows the animal to develop a reliable compass reference despite geographic and seasonal variation in the environmental features that provide the directional information.

These parallels notwithstanding, compass orientation in birds appears to be far more complicated than that in insects; in birds, multiple interacting compass systems and complicated developmental trajectories are involved. This complexity, combined with interspecific variation in how compass senses develop and are used, has led to a rich literature that defies easy summary. I will attempt to provide an overview of the most important design features of avian compass systems and relate these design features to the navigational problems that the animals must solve. For more extensive reviews, see Able and Able (1996), Wilschko and Wilschko (1991, 1996), and Schmidt-Koenig et al. (1991). For work on the navigational compasses of other vertebrates, see Lohmann and Lohmann (1996a, 1996b) and Phillips and Borland (1994).

6.4.3.1 Sun Compass

The avian sun compass has been best studied in pigeons, which use the sun compass for homing, and in various songbirds (e.g. starlings), which use the sun compass to learn directions of food in a way that is analogous to that of insects. Like insects, birds need to contend with variations in the rate of movement of the azimuth. Studies by Wiltschko and Wiltschko (1981) and others have shown that the pigeon sun compass is learned, but many puzzles remain unsolved regarding the nature of this learning process. It is still unclear just what geographic references can be used for measuring successive time-linked positions of the sun. Various evidence supports a role for the earth's magnetic field (Wiltschko et al. 1976). Other nonexclusive possibilities include the "pole point" (the center of rotation of the sky vault; see section 6.4.3.3) and of course landmarks. The location of the pole point can be determined with the patterns of skylight polarization that are detected at sunrise and sunset (Phillips and Waldvogel 1988).

Another issue is whether partially experienced birds can solve the problem of filling in unknown segments of the sun's course. I have already discussed that insects can do this (Wehner and Müller 1993; Dyer and Dickinson 1994). In one study, pigeons with experience limited to the afternoon seemed unable to use the sun as a compass when their homing ability was tested in the morning; instead, they used their magnetic compass (Wiltschko et al. 1981). On the other hand, Schmidt-Koenig (1961) showed that pigeons used an artificial sun for finding the compass direction of food at night and behaved as if they compensated for the solar movement after sunset. Also, direction-trained starlings (Hoffmann 1959) and pigeons (Schmidt-Koenig 1963) correctly used the midnight sun in the arctic sky for orientation when they only had been exposed to the diurnal course of the sun at middle northern latitudes. Such abilities, which closely resemble those of insects, strongly suggest that birds can fill in gaps in their experience of the sun's course. Pigeons that did not exhibit such an ability in a homing task (Wiltschko et al. 1981) may simply have been using the magnetic compass instead of the sun compass in this context.

Birds, unlike insects, appear to be able to recalibrate their internal compass as the solar ephemeris changes seasonally or as the bird moves to a different latitude. An ability to recalibrate the sun compass is sensible for a long-lived, wide-ranging species. One study examined pigeons living near the equator (reviewed by Schmidt-Koenig et al. 1991). There the sun's movement varies seasonally in a particularly dramatic way; the sun passes north of the zenith (with a counterclockwise shift of the azimuth) during part of the year and south of the zenith (with a clockwise shift of the azimuth) during the other part of the year (fig. 6.8b). The homing orientation of pigeons was tested at noon, and they behaved as if their internal ephemeris was out-of-date by about 7 weeks. The pigeons sometimes sought home in the wrong direction, as if (for example) they had interpreted the noontime sun to be in the south when it was really in the north.

This result may underestimate the flexibility with which birds recalibrate their sun compass when confronted with changes in the sun's movement. At tropical latitudes, seasonal changes in the local ephemeris function are very subtle. Each day, the azimuth stays in the east for most of the morning and in the west for most of the afternoon (fig. 6.9b); only during a few hours (or even minutes) near noon does the ephemeris differ from that on other days. Thus, the pigeons may not have had adequate information to keep their internal ephemerides up-to-date. One might expect transequatorial migrants, which would be faced with far more obvious changes in the ephemeris function, to be able to recalibrate their memory of the sun's course more rapidly.

I have stressed the ability of birds to develop an internal representation that matches the pattern of movement of the sun, and the problems that arise for sun-compass orientation as a result of changes in season or latitude. Alerstam (1996) recently discussed interesting issues that arise with regard to movements in longitude. Specifically, he suggested that the linkage of the sun compass to an internal clock allows birds to solve the problem of setting an energetically economical route over long distances.

As airline travelers know, the shortest (and, ceteris paribus, the least costly) route between two points on the globe follows a great circle, the line on the earth's surface formed by an imaginary plane that intersects those two points and the center of the earth. Following a great circle presents challenges to a navigator, however, because one may need to continuously change one's geographical heading (i.e. the angle of travel relative to latitudinal and longitudinal coordinates or relative to a compass that is linked to geographical coordinates). An alternative strategy is to follow a line of constant geographical heading, which is referred to as a "rhumb line." Although the rhumb-line and great-circle routes between two points can be identical (when both points lie on the same line of longitude or the equator), the rhumb-line route is always longer if the routes differ. The difference between the rhumb-line and great-circle routes between two points is greatest near the poles. For an extreme example, imagine setting a course from 1 km on one side of the north pole to 1 km on the other side of the north pole. The rhumb-line course may be set by walking 3.14 km east (or west) along the 89.99th parallel, maintaining a constant orientation relative to the pole. Or, one could walk the 2-km great circle across the north pole; one would have to head first north and then south and, hence, change orientation relative to the pole (and other geographic coordinates) midcourse. If the goal were at the same latitude but not at exactly 180° of longitude from the starting point, the great-circle route would not intersect with the north pole, and continuous changes in compass course would be required.

It is apparent that an animal changing longitude may save considerable time and energy by taking the great-circle instead of the rhumb-line route, if the animal can solve the compass-course adjustments required for great-circle navigation. Alerstam (1996) provided evidence that some arctic birds face this very problem and use the sun compass to solve it. For example, populations of waders (phalaropes and sandpipers) travel from their breeding grounds in northern Siberia to winter quarters in South America. The initial segment of this journey apparently takes the birds roughly 2,000 km across the Arctic Ocean to Alaska—a trip that spans more than 60° of longitude at latitudes of 65°–75°. Radar tracks of these migrating flocks suggest that a great circle is followed instead of a rhumb line (fig. 6.11). By following a great circle, the birds save at least 200 km in travel distance over this initial segment.

Alerstam proposed a surprisingly simple mechanism to account for this ability. If a bird sets its initial course relative to its sun compass and does not reset its internal clock as it crosses successive time zones, its geographical compass course changes during the trip to closely resemble a great-circle route. This happens because, at the same subjective time, the bearing of the sun is different in different time zones. For example, for trekking across the north pole (assuming travel occurs close to the June solstice), one could head northward by starting at local noon and keeping the sun at one's back. As the north pole is crossed, the longitude changes by 12 time zones so that it is now local midnight. By ignoring the time change (at least temporarily), one can switch to a southbound course by continuing to walk with the sun at one's back. At lower latitudes, the same process produces a continuous veer in the compass course.

A potential problem with this strategy is that flight paths would veer in the same way as that of birds as they move through time zones near the equator, where this effect would not be so useful. The rhumb-line and great-circle routes are essentially the same at low latitudes, and they are identical along

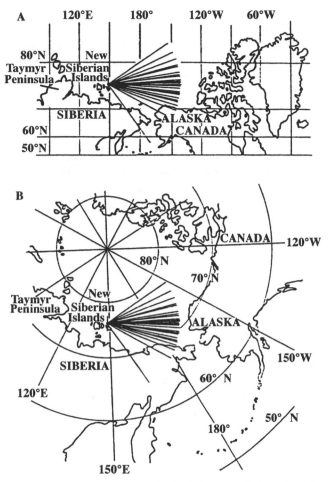

Figure 6.11. Do birds that migrate at high latitudes follow great-circle or rhumb-line routes? Maps show radar tracks from arctic waders interpreted as rhumb-line routes on a Mercator projection (*a*) and great-circle routes on a polar projection (*b*). The situation in *b* is a more likely interpretation because the routes carry the birds over the Alaskan peninsula, where populations of migrating waders are known to pass (from Alerstam 1996).

the equator. The veer produced with a sun compass linked to a different time zone would cause the bird to deviate from both of these equally efficient routes. For a given distance of travel, however, the effect would be much less pronounced near the equator than near the poles because the same linear distance spans fewer time zones at the equator than near the poles.

This example provides a useful lesson for demonstrating how an apparently difficult problem—one which human navigators can solve only with considerable computational effort—may be solved by animals without special cognitive abilities. Certainly, the bird must obtain positional information that allows it to set off on the correct course, and obtaining this information may entail a complicated assessment of its spatial location. However, following an efficient great-circle route over a great distance may be no more difficult than using a sun-compass orientation over short distances.

6.4.3.2 Magnetic Compasses

As human navigators have long known, the earth's magnetic field provides an extremely reliable directional reference. Therefore, it is scarcely surprising that animals have evolved the ability to use this reference. The evidence for magnetic-compass orientation is overwhelming in a wide variety of avian species. Most work has focused on migratory songbirds, which can be easily studied in controlled conditions because a behavior called "migratory restlessness" is exhibited by captive birds (reviewed by Berthold 1991; Gwinner 1996). During the migration season, caged birds increase their activity levels at the time of day (or night) they would be flying along their migratory route. When active, the birds tend to cluster toward the side of the cage that corresponds to the migratory direction, as if the birds are trying to take off in that direction. This clustering makes it very easy to manipulate stimuli experienced by the birds and thus explore the sensory basis of their ability to discriminate directions. In addition to the studies on migrant songbirds, magnetic-compass orientation has been explored in pigeons that are heading homeward after being experimentally displaced from their loft (reviewed by Wiltschko and Wiltschko 1996).

To understand the problems that must be solved in the design of the avian magnetic compass, it is important to understand that the avian magnetic compass responds to a different feature of the earth's magnetic field than magnetic compasses used by people. The earth's magnetic field lines are not parallel to the ground over most of the earth; instead, the field lines penetrate into the surface of the earth at an angle (fig. 6.12). The inclination, or dip, of the field lines varies systematically over the globe: the field lines are horizontal (parallel to the ground) at the magnetic equator and progressively steeper near the magnetic poles, where they are nearly vertical. With conventional compasses, human navigators detect the polarity of the field lines by means of the alignment of a magnetized needle with the horizontal component of the magnetic field; these conventional compasses are essentially insensitive to the dip. The avian magnetic compass, by contrast, detects inclination. The compass can detect if the animal is heading toward a magnetic pole (the direction in which the field lines form a small angle relative to the direction of gravity) or toward the equator (the direction in which the field lines form a large angle relative to the direction of gravity). The compass is insensitive to the polarity of the

North

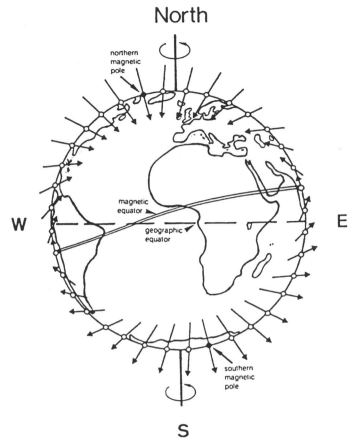

Figure 6.12. The earth's magnetic field with inclination of field lines relative to the earth's surface (from Wiltschko and Wiltschko 1991).

field per se; experiments in which the horizontal and vertical components of the field were manipulated independently revealed this insensitivity (reviewed by Wiltschko and Wiltschko 1996).

The use of an inclination compass that is insensitive to magnetic polarity has interesting implications when the performance of such a compass at different latitudes is considered. The same compass will indicate north as the poleward direction in the northern hemisphere and south as the poleward direction in the southern hemisphere. Thus, without any evolutionary modification, the compass works in both northern and southern hemispheres to lead birds toward the pole in the spring and back toward the equator in the fall (fig. 6.13).

Such a compass presents problems, however, for birds with migratory routes that span the magnetic equator. The compass cannot resolve directions at the

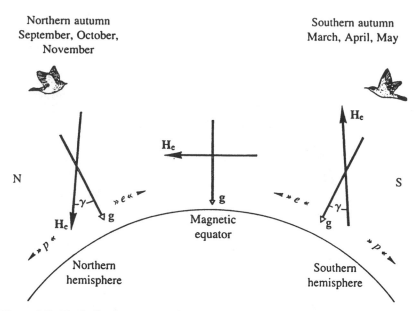

Figure 6.13. The inclination compass of birds. In this system, the direction toward the pole ($\bar{p}$) is the direction in which the magnetic field lines (H_e) make the smallest angle with gravity (g). The direction toward the equator ($\bar{e}$) is opposite that toward the pole. The field lines are horizontal at the magnetic equator. Birds from both northern- and southern-hemisphere populations can use an identical compass to fly toward the equator during their respective autumns. Birds with migratory paths that cross the magnetic equator, however, must change their response to the magnetic compass to stay in the same geographical heading. (From Wiltschko and Wiltschko 1996.)

magnetic equator because the field lines are horizontal. Migrant songbirds do indeed become disoriented when placed in a horizontal field. Transequatorial migrants evidently have some way of compensating for the ambiguity in their magnetic-compass information during their passage across the magnetic equator. One possibility is that these birds may temporarily rely on a celestial compass (Wiltschko and Wiltschko 1996).

Another problem is that a bird with a migratory course that passes the magnetic equator cannot employ a constant response to the dip of the field lines throughout the journey. The response that leads the bird southward in the northern hemisphere to the equator causes the bird to turn northward (to the equator) again after the bird crosses the magnetic equator (see fig. 6.13). To keep flying south, the bird must adopt a poleward response relative to the magnetic compass. Wiltschko and Wiltschko (1992) obtained strong evidence for such a switch in the compass response of garden warblers by exposing caged birds to a simulated crossing of the magnetic equator. Southbound birds heading toward the equator relative to an inclined magnetic field in the laboratory were placed in a horizontal magnetic field (the condition at the magnetic equator) for two days. When returned to the inclined field, the birds adopted a poleward response, which was opposite to their previous heading and opposite to the orientation of birds that had not been exposed to the horizontal field. This remarkable result suggests, among other things, that birds do not learn the new response to the magnetic field by referring to an independent compass reference that indicates the direction of true south. Instead birds appear to experience a relatively simple programming change when exposed to the horizontal field. Thus, an elegant bit of engineering solves a problem that may not have been solved any more reliably with a learning process.

Why do birds rely on a magnetic-inclination compass instead of a magnetic-polarity compass for migration? This question is especially perplexing given that there is evidence for a polarity compass in newts; therefore, a polarity compass is not a neurobiological impossibility for vertebrates. Indeed, newts appear to use both types of magnetic compasses; an inclination compass is used for y-axis orientation toward their local shoreline, and a polarity compass is used for long-distance x,y-orientation toward their natal pond (Phillips 1986). Baker (1984) suggested that development of biological polarity compasses would have been heavily disfavored during the periodic changes in the polarity of the earth's magnetic field. The newt's polarity compass argues against Baker's hypothesis. The hypothesis is also unsatisfying because it is unlikely that such an infrequent event (the last polarity reversal occurred over 700,000 years ago) has played an important role in the adaptive design of the avian compass. An alternative possibility is that magnetic inclination provides more reliable or accurate information than that provided with magnetic polarity; however, it is not obvious what the specific advantages may be. Indeed, we have seen that there are serious problems with use of an inclination compass near the magnetic equator because the field lines there have no inclination. Migrants may well deal with this problem by relying briefly on a celestial compass. What about birds that live throughout the year in that part of the world? Conceivably, such species rely less on a magnetic compass than migrants do, or perhaps some species have evolved the ability to use a polarity compass.

6.4.3.3 Stellar Compasses

Most songbirds migrate at night, and several species have been shown to use the pattern of stars in the sky as a compass reference during migration (reviews by Able 1980; Able and Able 1996; Wiltschko and Wiltschko 1991). Stellar-compass orientation, like magnetic-compass orientation, is exhibited by caged birds during migratory restlessness; therefore, stellar-compass orientation can be easily manipulated for the study of its properties. Like humans, birds can identify north by the static configuration of stars in the sky. Learning plays a crucial role in the development of this ability. North is defined by the pole point, which is the point in the sky around which the sky appears to rotate. The azimuth of the star closest to the pole point, Polaris, does not change during the night. Birds probably do not cue specifically on Polaris as we do, but birds can recognize the configuration of stars that rotate around Polaris. Naive birds can be trained to recognize another star as the pole star if they are reared in a planetarium in which the stars rotate around a new pole point. Birds can even learn the pole point in an arbitrary configuration of artificial stars (e.g. Able and Able 1990). Other than that learning plays a role, we know little about the integrative mechanisms by which birds measure celestial rotation and identify the pole point.

6.4.3.4 Interactions among Compass Systems

Some of the most exciting work on bird orientation in recent years has focused on how different compass systems interact during migratory orientation and how these interactions develop. In compass systems, we see most clearly the interplay of innate and individually acquired information in the formation of internal representations of spatial relationships. In addition, we can find examples of how the variable and predictable features of navigational cues may favor learning in the development of compass systems.

The following developmental pattern is, in many ways, typical of the processes by which a naive songbird acquires the ability to use different compass references (reviewed by Wiltschko and Wiltschko 1991 and by Able and Able 1996). When birds are completely naive (as in the case of birds that are hand raised in the lab), their orientation during migratory restlessness is controlled with an innate response to the earth's magnetic field. Southbound birds head in the direction relative to magnetic north that corresponds to the direction taken during actual migration. For many songbirds from central and northern Europe (two well-studied species are garden warblers and pied flycatchers), this innate heading is either southwest or southeast—the directions that allow birds to bypass the Alps. In eastern North America, the innate heading of savannah sparrows is southeast, which corresponds roughly to the direction actually traveled by migratory flocks bound for South America. (Although, oddly, on some nights caged savannah sparrows reverse their direction and head toward the northwest.)

The innate response of naive birds to the magnetic field is modified as birds acquire experience with celestial cues. Specifically, after birds have learned the pole point (which corresponds to true geographical north), they use this reference to calibrate their response to the magnetic reference. This calibration process may entail a drastic modification of the magnetic heading. For example, Able and Able (1990, 1995) reared savannah sparrows under an artificial sky in which the pole point deviated by 90° from magnetic north (instead of only a few degrees as in the actual sky) so that celestial north was now in the magnetic east (see Wiltschko and Wiltschko 1991 for similar experiments with pied flycatchers). Birds reared under such conditions behaved as if the shifted pole point redefined the position of magnetic north. During migratory restlessness in an artificial magnetic field, they adopted a new angle relative to the magnetic compass and behaved as if magnetic east was aligned with geographic north. Similar effects were observed when the shifted pole point was defined with patterns of polarized light observed at sunrise and sunset (Able and Able 1993).

This experiment showed that the celestial compass not only serves as a navigational reference but also affects calibration of the initial response of naive birds to the magnetic field. Why is it that the celestial compass initially calibrates the magnetic compass instead of vice versa? The answer presumably is that the magnetic field varies over the earth's surface. The variation consists of both large-scale systematic patterns and small-scale anomalies in which the local field departs from a regional pattern. This variation means that a fixed response to local magnetic cues could result in a magnetic heading that deviates markedly from the appropriate geographic heading. Given the distances that birds are traveling, errors in heading of a few degrees may result in a course that ends over open ocean instead of land. The magnetic compass is useful for the same reasons to migrating birds as it is to humans: The magnetic compass generally provides reliable directional information on a global scale, and its reliability is not affected by atmospheric conditions that obliterate celestial references. The usefulness of the magnetic compass depends, however, on its correlation with true geographical north. Because local anomalies in the earth's magnetic field may produce substantial deviations in this relationship, it needs to be calibrated through learning.

This tidy account of the adaptive importance of this calibration process gets a little messier when plasticity in the compass systems of birds after they have begun to migrate is examined. As I discussed, in very young birds the response to the earth's magnetic field is calibrated with reference to the celestial pole point. In older birds, responses to celestial compass cues can be recalibrated with reference to the magnetic field! In experiments with warblers, Wiltschko and Wiltschko (reviewed 1991) caught experienced birds during migration and exposed them to a view of the nighttime sky in an artificial magnetic field that was shifted 120° from the normal field. The birds adopted their typical magnetic heading relative to the artificial field, and hence, they appeared to ignore the stars. When these birds were later tested with a view of the sky but in the absence of magnetic information (an artificially nulled field), they adopted the new angle relative to the stars. Thus, the birds had not been ignoring the stars in the shifted magnetic field; they were learning a new response to the stars with reference to the shifted magnetic field.

How can we make sense of the evidence that (1) the celestial compass initially serves to calibrate rasponses of naive birds to the magnetic field; and (2) the magnetic field serves to recalibrate responses of experienced birds to the celestial reference? The Wiltschkos (1991) suggested the following interpretation. Although the stars that surround the pole point provide a reliable reference for learning the correct initial migratory heading relative to the magnetic field, they cannot continue to guide the bird along its journey because they gradually disappear below the northern horizon as the bird moves south. New patterns of stars emerge from the southern horizon, but the bird can use these only if it has learned the proper geographical heading relative to them. In this situation, the magnetic field provides a reliable reference. (An alternative possibility is that birds calibrate their responses to newly encountered stars by referring to familiar stars before they disappear below the horizon.)

Thus, the ontogeny of avian compass systems involves a complex interplay of innate and individually acquired responses to navigational references in multiple sensory modalities. We can make sense of this process by invoking plausible arguments about the reliability and variability of different compass references birds use in different stages of their lives. The challenge is to understand in greater detail the evolutionary forces that have shaped these compass systems.

6.4.4 Landmark Learning by Insects

The stereotypical image of a rat running in an artificial maze alludes to the long-standing interest that behavioral scientists have had in the ability of animals to use landmarks for spatial orientation. Insects too have long been studied for their abilities to learn landmarks. Because most studies of insects have been performed in the natural environment, these studies have provided a richly detailed picture of landmark-learning abilities in the context of natural navigational tasks. I will summarize what has been learned about the design of these learning mechanisms, which have been revealed mainly with studies of honeybees (*Apis mellifera*) and desert ants (*Cataglyphis* spp.).

These species, like other nesting animals, face the navigational problem of finding their nest in the terrain. In addition, like other animals that exploit rich patches of food, they often must return to a feeding site that is depleted only after many foraging trips. The scale of their movements demands the use of visual cues, which are far superior to olfactory cues for efficient long-distance orientation to a point goal. (Although many ants use odor trails deposited on the substrate, *Cataglyphis* and other desert ants live in a dry, windswept environment in which odor is less useful.) Furthermore, the animal must learn the relationships between visual cues and the locations of repeatedly visited goals.

We have already seen how the sun and sun-linked patterns of polarized light serve extremely important functions in the orientation of insects within their foraging range. The celestial compass has certain limitations, however. Clouds may obscure celestial cues. Furthermore, the celestial compass provides good directional information but, by itself, no positional information. Thus, if an animal is passively displaced in the environment (e.g. by the wind or the hand of an experimenter) and cannot perform path integration to update its change of position during the displacement, the celestial compass is of little help to the animal. Finally, as we have seen, the path-integration system used by insects is subject to errors (both systematic and random) that limit the accuracy of pinpointing the location of a goal, such as the nest. For orientation under such circumstances, insects rely on a well-developed ability to use landmarks.

In this section, I will discuss a system for classifying the cognitive sophistication of the landmark-learning abilities of animals. Then I will discuss experimental studies of spatial cognition on small and large scales and consider what these results indicate about the adaptive design of the underlying processes.

6.4.4.1 Characterizing the Contents of Spatial Memory

In recent years, much of the research on landmark orientation has focused on how animals memorize and use spatial information that landmarks provide. Generally, an animal must link the retinal images of landmarks to motor responses that will lead toward the goal. It is not immediately obvious how a brain can do this. At each point in the environment, an animal that views a set of landmarks can record only the two-dimensional retinal pattern. If the animal is moving, it can also record movement across the retina. Different images can be recorded at different points in the environment. Use of landmarks for navigation through the environment, therefore, must involve integration of these separately recorded images.

Gallistel (1990) outlined a three-part system that describes the capacity of animals to encode and use spatial relationships among landmarks. This system is useful for comparing the performance of different species with respect to the sophistication of the cognitive mechanisms that are required to encode spatial information. The first component of this system is the *geometry* of the spatial representation. Does the animal visually encode a rich metric, or Euclidean, representation of the angles and distances among elements it has seen? Alternatively, does the animal encode spatial relationships in a simpler geometric system? For example, an animal that encodes spatial patterns in an affine geometric system would confuse shapes such as squares, rectangles, and parallelograms that are readily distinguished with use of a metric geometric system.

The second part of Gallistel's system is the *coordinate system* used by the brain to encode shapes. One common distinction is that between an egocentric representation (in which the animal learns directions and distances of different landmarks relative to a coordinate system defined by its body axes) and an allocentric representation (in which the animal learns directions and distances of different landmarks relative to an external coordinate system, such as one based on an external compass). An allocentric representation is assumed to be more complicated because it must be assembled from the egocentric images recorded with the visual system.

The third part of Gallistel's system is the *spatial scale* of the representation. Many nesting insects must learn spatial relationships on a small scale (e.g. the panorama of surrounding landmarks that pinpoint the location of the nest) and a large scale (e.g. the sequences of landmark panoramas seen in different parts of the foraging range). As I will discuss, the properties of the spatial representations that animals form on different spatial scales may be very different.

To illustrate some of what is known about the capacities of insects to use landmarks, I will focus on experiments that have been performed on two spatial scales.

6.4.4.2 Use of Surrounding Landmarks to Pinpoint a Goal

Work on the small-scale task began with Tinbergen's classic studies of nest finding by digger wasps (e.g. Tinbergen and van Kruyt 1938). These studies provided the first clear evidence of landmark learning in insects and defined some of the properties of landmarks that insects learn. More recently, Cartwright and Collett (1983), who worked with honeybees, and Wehner and colleagues (Wehner et al. 1996 [review]), who worked with desert ants, began examining systematically the nature of landmark learning. The general assumption has been that insects arriving at the nest or feeding site stored an image of the surrounding landmarks in memory and then used this memory image to guide their search on return to the goal. The questions are (1) What spatial information do these images encode? (2) How are they used? and (3) How are they acquired by naive insects?

Several studies have demonstrated that insects visually encode the angles that separate different elements in a scene. For example, bees trained to find food in the middle of a square array of landmarks will subsequently search for food in a square array instead of a rectangular array of roughly similarly sized landmarks (Cartwright and Collett 1983). Furthermore, bees can encode the distances to familiar landmarks at least in the following way. If bees are trained to find food relative to an array of landmarks, some of which are close to the food and some of which are farther away, their subsequent searching behavior is more strongly influenced with the landmarks that were near the food—even if the distant landmarks were larger and presented the same size of retinal image (Cheng et al. 1987). Thus, nearby landmarks (which specify the location of the goal with greater precision), are weighted more heavily in the memory of a particular landmark array. The source of distance information is probably the motion parallax produced with near and far landmarks.

These experiments thus indicate that insects encode metric information— angles and distances—about familiar landmarks around a feeding or nesting site. This finding is not surprising; the compound eye is ideally suited to measure the relative angles and image motions that correspond to different landmarks that surround the animal.

It is more difficult to understand in what coordinate system this information is encoded. The simplest hypothesis, which accounts for much of the evidence, is that bees encode landmark positions in egocentric instead of allocentric coordinates. The memory image that corresponds to a particular array of landmarks is recorded as a neural "snapshot" of the image as it falls on the retina (Cartwright and Collett 1983). It is relatively easy to see how such a snapshot, fixed relative to retinal coordinates, indicates to the animal that it has reached a familiar goal. The animal need only recognize that its current view of the landmarks matches the snapshot stored during previous visits. The challenge is to understand how a snapshot encoded in egocentric coordinates can guide the animal when it has not yet reached the goal. Perhaps the animal encodes a sequence of snapshots, each slightly different, at successive points along its approach to the goal. However, encoding a sequence of snapshots would require considerable memory capacity, especially if the insect needed to learn how to approach a goal from a variety of directions (see Dyer 1994).

Cartwright and Collett proposed a relatively simple model that would allow a single (distance-filtered) snapshot recorded at the goal to guide the insect's approach from a variety of starting positions relative to the landmark array (fig. 6.14). In essence, the model assumes that the visual system makes multiple local comparisons of discrepancies between the current retinal image of the landmarks and the image stored (in retinal coordinates) at the goal. Each local comparison generates rotational and translational responses to reduce the discrepancy. These responses are added to steer the animal. A computer simulation showed that this model could lead an insect to its goal from a variety of starting positions relative to a simple landmark array. A major difficulty, however, was that the model necessitated that the snapshot remain oriented in the same compass direction throughout the process (fig. 6.14a, b).

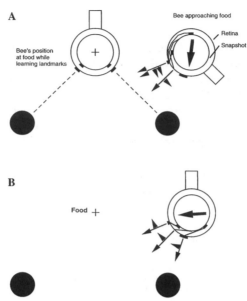

Figure 6.14. How bees may use memorized landmarks to pinpoint the location of a familiar goal such as a feeding site (+), according to the model of Cartwright and Collett (1983). A bee's visual field and body axis are shown schematically. (*a*) The bee records a neural snapshot of landmarks as seen from the food source. On subsequent return (*right*), the bee guides herself by comparing the current view of landmarks to the snapshot, which remains in the original position relative to retinal coordinates. Each element of the snapshot (the two landmarks and the space between them) is compared with the nearest element of the image on the retina. Each local comparison generates a unit translational vector and a unit rotational vector (*thin arrows* and *arrowheads,* respectively). These unit vectors are summed to determine the bee's new heading (heavy arrow) from its current position. If, as in this case, the bee is not facing in the direction in which the snapshot was originally recorded, the model may not compute a heading that points to the goal. (*b*) The model provides a better simulation if the bee is facing in the direction in which the snapshot was originally recorded. Bees apparently do face in a constant direction, when close to the food source, using an external compass reference adjust their orientation (Collett and Baron 1994).

Bees can approach a familiar goal from a variety of directions. Since their visual field is fixed relative to their body axes and thus oriented relative to their line of flight (it was thought), the model seems to require that the visual image, linked to an allocentric reference frame, be used independently of how the visual image projects to retinal coordinates.

In a follow-up study to that of Cartwright and Collett (1983), Collett and Baron (1994) found that landmark learning by bees does involve use of an external compass reference, but the representation of landmarks is probably encoded in an egocentric reference frame. By measuring the orientation of (videotaped) bees as they approached a familiar feeder, Collett and Baron found that bees maintained their bodies (and hence their visual field) in a more or less constant compass orientation by using the magnetic field as a reference. Dickinson (1994) found that bees can also use their celestial compass in this context.

In bees, the effect of maintaining a constant body orientation is a constant orientation of an egocentrically recorded snapshot of the landmarks (Cartwright and Collett 1983). Hence, the task of matching a recorded snapshot to the current retinal image of the landmarks is simplified (fig. 6.14b). Note that this is not at all the same as recording the landmarks relative to an external compass and hence in an allocentric reference frame. Instead, the bees use the compass to standardize their viewing angle and then encode (and later use) a snapshot that was recorded relative to egocentric coordinates.

These interpretations have received further support from observations of the "locality study" that a flying insect does prior to leaving a goal (such as the nest entrance) to which it will return. This behavior, which has also been called a "turn-back-and-look" response (Lehrer 1996 [review]), allows the bee to learn visual features of the goal and the surrounding landmarks. In careful video analyses of yellow jacket wasps, Collett (1996 [review]) showed that the insects' orientation during the turn-back-and-look response closely matched the orientation maintained during subsequent approaches to the food. Thus, the spatial arrangement of the landmarks can be stored and retrieved with the orientation of the animal's visual field in the same direction, which presumably greatly facilitates the recognition process. To see the advantages of maintaining the pattern to be recognized in a standardized orientation relative to the visual field, rotate this book by 90° and notice how much the rotation slows your reading speed. Although we can learn to read rotated text, the recognition processes (even in humans) are affected by the orientations of patterns relative to retinal coordinates.

These studies of landmark learning on a small spatial scale thus suggest that insects store a relatively simple representation of the landmarks that surround the goal and can employ various tricks to ensure efficient recognition of the pattern and correct orientation toward the goal. In section 6.4.4.4, I will consider the implications of these observations for understanding the adaptive design of the learning mechanisms.

6.4.4.3 Landmark Learning on the Scale of the Foraging Range

Insects face the task just described—narrowing the search for a goal relative to a surrounding array of landmarks—only when close to the goal. But the foraging ranges of some nesting insects are very large (10–20 km) (reviewed by Wehner 1981). In these cases, the forager must head for a part of the landscape it cannot directly see. An ability to use landmarks for navigation over such distances implies that the animal can in effect use landmarks at the starting point to determine its position relative to the goal and discriminate different directions. Then the animal can repeatedly perform this process as it encounters each successive visual scene along a given route (reviewed by Dyer 1994, 1996; Wehner et al. 1996).

Evidence that landmarks can be used as a directional reference was provided by experiments in which insects were given conspicuous landmarks along a familiar foraging route and then were displaced to a different location with similar landmarks. For example, von Frisch and Lindauer (1954) trained bees to fly along a line of trees to reach food. They tested the bees where a similar line of trees was aligned in a different compass direction. Thus, the bees were faced with a conflict between their celestial compass and the familiar landmark configuration. Although some bees flew to food in the direction indicated with the celestial compass, most searched for food along the landmarks. However, bees tested in environments without conspicuous and familiar landmarks searched for food with use of the celestial compass. Thus, landmarks can be powerful enough to override the celestial compass as a directional reference.

Evidence that landmarks provide insects with positional information comes from experiments in which insects were displaced passively from their nest. Passive displacement denies insects the opportunity to perform path integration. The ability to fly home after such a displacement implies that insects determined their position with use of information obtained at the release site. A variety of evidence suggests that insects, unlike birds, cannot find their way home from release sites beyond their usual foraging range (see Wehner 1981 for a review). Instead, insects need to see previously encountered landmarks that led them home from feeding places in the vicinity of the release site. Insects can reach home after displacements of hundreds or (in some bee species) thousands of meters (Wehner 1981). Such displacement experiments therefore suggest that these insects learn visual features in an enormous area of the terrain around the nest.

Most of these results can be explained with a very simple model, which was first proposed by Baerends (1941) on the basis of his studies of digger wasps. The insect follows a sequence of visual images that correspond to successively encountered stages of a given familiar route. This model is similar to the snapshot model proposed much later by Collett and Cartwright. In essence, orientation to landmarks on a large scale entails the formation of a route map that consists of a string of snapshots. The route maps need not be linked in memory to the celestial compass, although directions learned relative to landmarks and celestial cues would typically provide equivalent navigational information. With a few route maps radiating from its nest, a bee would be able to navigate efficiently in a large area around the nest (fig. 6.15*a, b*).

About 10 years ago, Gould (1986) triggered a flurry of excitement by suggesting that bees could integrate their experience on separate foraging routes into a common, geometrically accurate map of the terrain (fig. 6.15*c*). Experimental evidence suggested that bees estimate a novel shortcut from one familiar site to another. This is analogous to what we may do if we want to find the most direct route between two distant locations that we have visited separately from our home but have never traveled to along a connecting route. This navigation requires placing each location in a common geometrical frame of reference. For example, we keep track of the directions and distances traveled to each place from our home and then use the same frame of reference in which this information was obtained to compute the path connecting the sites. Bees have certain prerequisite abilities for performing the same task: They learn landmarks along multiple routes, and they have a directional reference (the sun compass) and a measure of distance that can serve as the basis for a large-scale coordinate system. Thus, Gould's proposal that bees also have the capacity to perform the necessary computations to determine the relative positions of separately visited locations was not far-fetched.

Gould's results have been very hard to replicate. Some investigators did not find evidence that bees compensate for their displacement from a route by heading along the shortcut to their goal. I found that bees could set a shortcut, but only if (1) they had a view of large-scale landmark features that could be seen during the flights from the nest to the goal, or (2) they had previously flown along the shortcut (Dyer 1991, 1994 [review]). In neither of these circumstances would bees require a large-scale map of the sort Gould envisioned. Instead, their behavior can be easily explained with the same sort of egocentrically referenced pattern-matching model that Cartwright and Collett (1983) proposed for orientation on the smaller scale. When I presented bees with a more stringent test of the cognitive-map hypothesis—by denying them large-scale landmark features or previous experience along the shortcut—they failed to head in the direction of the shortcut. In more recent experiments (reviewed by Dyer 1996), I found that honeybees lack a key ability that is needed to organize experience into a common large-scale map. This is the ability to learn the orientation of landscape features relative to a compass reference. Thus, bees do not guide their movements by means of cognitive maps as proposed by Gould (1986), nor do they have a way to form such maps.

6.4.4.4 Constraints on Landmark Learning by Insects

The landmark learning that results from both small- and large-scale orientation tasks suggest that insects store spatial information derived with use of land-

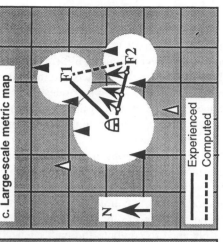

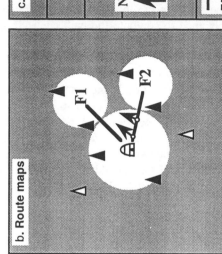

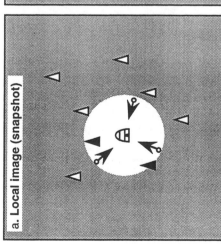

Figure 6.15. Alternative mapping schemes for encoding large-scale spatial relationships in a foraging range. Shaded area represents the unexplored part of the environment, and white areas indicate regions in which bees have learned visual features of the terrain. Triangles represent familiar (*black*) and unfamiliar (*white*) landmarks. (*a*) The local image, similar to the snapshot of Cartwright and Collett (1983), allows for efficient homing from multiple locations that are surrounded by the same panorama of landmarks. (*b*) Route maps allow for navigation to unseen portions of the terrain with use of sequences of snapshots compiled during previous flights. Flexibility in compensating for displacements from a particular route may be based on the same sorts of mechanisms that allow for flexibility with use of local images (snapshots). (*c*) Metric maps are constructed with route maps referenced to a common coordinate system (e.g. an external compass). Such a map allows an animal to compute novel routes between separately visited sites (from Dyer 1996). Gould (1986) suggested that insects form large-scale metric maps of a familiar terrain; however, most recent work suggests that insects cannot form these maps, although they clearly form route maps (Dyer 1991; Wehner et al. 1990).

marks in a relatively simple manner. Unlike humans, bees do not seem to encode landmark images in an allocentric reference frame and cannot form or use large-scale mental maps that encode the metric relationships among separately traveled routes. Insects can obviously respond to familiar landmarks with considerable flexibility (reviewed by Dyer 1994), but consider the implications of their limitations. Again, the limitations of insect behavior have been attributed to the small size of their brains (Wehner 1991, Dyer 1994). The assumption, however, that brain size constrains performance begs various questions: Why should a particular strategy, such as the ability to form a large-scale metric map of the terrain, be beyond the capacity of a brain the size of the bee's? How much more neural tissue, if any, is needed to develop and use such a strategy? We simply do not have the empirical or theoretical perspectives that allow us to answer such questions.

Furthermore, as in the case of path integration, a focus on neural constraints may blind us to the possible advantages of the insects' apparently simpler strategies for encoding and using familiar landmarks. One possible advantage of an egocentric representation is that it allows more rapid acquisition of relevant information as landmarks are learned; it also allows more rapid recognition of familiar landmarks when returning to a goal. But what constraint limits the speed of processing allocentric instead of egocentric spatial representations? It is possible that such a constraint applies equally to a larger brain. If so, then perhaps the small size of the insect brain may not limit the evolution of richer cognitive capacities; the selective disadvantages of such capacities in an animal with a short lifespan may be the limiting factors.

Clearly, we need more information about how brain size constrains behavior in invertebrates. A valuable starting point would be to identify correlations between the sizes of brain regions and behavioral performance that are analogous to the correlations between hippocampal volume and behavioral flexibility discovered in mammals and birds (Sherry, this vol. sec. 7.4.4). Recent studies of honeybees (reviewed by Fahrbach and Robinson 1995) have revealed that the mushroom bodies, a region of the insect brain that has been implicated in learning and sensory integration, increase in size relative to the rest of the brain as the bee ages. This neuroanatomical change parallels behavioral changes that take place as bees move through a succession of different jobs in the colony. Young bees work mainly inside the nest while older bees— during roughly the last third of their lives—work outside the nest collecting nectar and pollen for the colony. A reasonable interpretation for the increase in size of the mushroom bodies is that they are associated with the greater information-processing demands that foraging bees face, including the need to learn features of the environment necessary for orientation and the colors, odors, and shapes of rewarding flowers. If so, then studies of the insect mushroom bodies and other neuropils in the insect brain may contribute to the development of a more neurobiological cognitive ecology of spatial memory in insects, which is similar to that described in vertebrates (Sherry, this vol. chap 7).

6.5 CONCLUSIONS

One major lesson in this chapter is the cognitive ecology of animal navigation is still relatively immature. Given that most research has been on the behavioral level, our understanding of the information-processing *mechanisms* that underlie the navigational abilities of animals is very sophisticated. In some species, we can specify in considerable detail the capacities and limitations of such mechanisms and relate performance to sensory and neural function. Concerning the *evolutionary forces* that have shaped such mechanisms, however, our understanding, for the most part, remains conjectural. Certainly our understanding of the cognitive ecology of navigation lacks sophistication compared with our understanding of other adaptive behavioral traits, such as those involved in foraging or life-history strategies. In this chapter, I have provided an outline of the issues that must be considered in the development of a more rigorous study of the adaptive design of cognitive mechanisms. In particular, I reiterate the importance of clearly identifying navigational problems that animals face and developing rigorous hypotheses about the fitness benefits, fitness costs, and constraints that may have guided the evolution of navigational mechanisms.

ACKNOWLEDGMENTS

This chapter was written during a sabbatical at the University of California, San Diego, and I thank Jack Bradbury and Sandy Vehrencamp for their gracious hospitality. I thank Reuven Dukas, Lee Gass, and Don Wilkie for comments on the manuscript and the National Science Foundation for financial support.

LITERATURE CITED

Able, K. P. 1980. Mechanisms of orientation, navigation, and homing. In S. A. Gauthreaux, Jr., ed., *Animal Migration, Orientation, and Navigation,* 283–373. New York: Academic Press.
———. 1996. The debate over olfactory navigation by homing pigeons. *Journal of Experimental Biology* 199:121–124.
Able, K. P., and M. A. Able. 1990. Calibration of the magnetic compass of a migratory bird by celestial rotation. *Nature* 347:378–380.

————. 1993. Daytime calibration of magnetic orientation in a migratory bird requires a view of skylight polarization. *Nature* 364:523–525.

————. 1996. The flexible migratory orientation system of the savannah sparrow (*Passerculus sanwichensis*). *Journal of Experimental Biology* 199:3–8.

Alerstam, T. 1996. The geographical scale factor in orientation of migrating birds. *Journal of Experimental Biology* 199:9–19.

Arnold, S. P. 1994. Constraints on phenotypic evolution. In L. A. Real, ed., *Behavioral Mechanisms in Evolutionary Ecology,* 258–278. Chicago: University of Chicago Press.

Baerends, G. P. 1941. Fortpflanzungsverhalten und Orientierung der Grabwaspe *Ammophila compestris* Jur. *Tijdschrift voor Entomologie Deel* 84:68–275.

Baker, R. R. 1978. *The Evolutionary Ecology of Animal Migration.* New York: Holmes and Meier.

————. 1984. *Bird Navigation: The Solution of a Mystery?* New York: Holmes and Meier.

Becker, L. 1958. Untersuchungen über das Heimfindenvermögen der Bienen. *Zeitschrift fur vergleichende Physiologie* 41:1–25.

Bennett, A. T. D. 1996. Do animals have cognitive maps? *Journal of Experimental Biology* 199:219–224.

Berthold, P. 1991. Spatiotemporal programmes and genetics of orientation. In P. Berthold, ed., *Orientation in Birds,* 86–105. Basel: Birkhauser.

Brower, L. P. 1996. Monarch butterfly orientation: Missing pieces of a magnificent puzzle. *Journal of Experimental Biology* 199:93–103.

Byrne, R. W. 1982. Geographical knowledge and orientation. In A. W. Ellis, ed., *Normality and Pathology in Cognitive Functions,* 239–264. London: Academic.

Calder, W. A., III. 1984. *Size, Function, and Life History.* Cambridge, Mass.: Harvard University Press.

Cartwright, B. A., and T. S. Collett. 1983. Landmark learning in bees. *Journal of Comparative Physiology,* ser. A, *Sensory, Neural, and Behavioral Physiology* 151:521–543.

Cheng, K., T. S. Collett, A. Pickhard, and R. Wehner. 1987. The use of visual landmarks by honeybees: Bees weight landmarks according to their distance from the goal. *Journal of Comparative Physiology,* ser. A, *Sensory, Neural, and Behavioral Physiology* 161:469–475.

Churchland, P. S., and T.J. Sejnowski. 1991. *The Computational Brain.* Cambridge: MIT Press.

Collett, T. S. 1996. Insect navigation en route to the goal: Multiple strategies for the use of landmarks. *Journal of Experimental Biology* 199:227–235.

Collett, T. S., and J. Baron. 1994. Biological compasses and the coordinate frame of landmark memories in honeybees. *Nature* 368:137–140.

Collett, T. S., E. Dilmann, A. Giger, and R. Wehner. 1992. Visual landmarks and route following in desert ants. *Journal of Comparative Physiology,* ser. A, *Sensory, Neural, and Behavioral Physiology* 170:435–442.

Craig, P. C. 1973. Orientation of the sandbeach amphipod, *Orchestoidea corniculata.* *Animal Behaviour* 21:699–706.

256   Fred C. Dyer

Dickinson, J. A. 1994. Bees link local landmarks with celestial compass cues. *Natur-wissenschaften* 81:465–467.

Dingle, H. 1980. Ecology and evolution of migration. In S. A. Gauthreaux, Jr., ed., *Animal Migration, Orientation, and Navigation*, 1–101. New York: Academic.

———. 1996. *Migration: The Biology of Life on the Move.* Oxford: Oxford University Press.

Dyer, F. C. 1985. Nocturnal orientation by the Asian honeybee, *Apis dorsata. Animal Behaviour* 33:769–774.

———. 1987. Memory and sun compensation in honeybees. *Journal of Comparative Physiology*, ser. A, *Sensory, Neural, and Behavioral Physiology* 160:621–633.

———. 1991. Bees acquire route-based memories but not cognitive maps in a familiar landscape. *Animal Behaviour* 41:239–246.

———. 1994. Spatial cognition and navigation in insects. In L. A. Real, ed., *Behavioral Mechanisms in Evolutionary Ecology*, 66–98. Chicago: University of Chicago Press.

———. 1996. Spatial memory and navigation by honeybees on the scale of the foraging range. *Journal of Experimental Biology* 199:147–154.

Dyer, F. C., and J. A. Dickinson. 1996. Sun-compass learning in insects: Representation in a simple mind. *Current Directions in Psychological Science* 5:67–72.

Etienne, A. S., R. Maurer, and V. Séguinot. 1996. Path integration in mammals and its interactions with visual landmarks. *Journal of Experimental Biology* 199:201–209.

Fahrbach, S. E., and G. E. Robinson. 1995. Behavioral development in the honeybee: Toward the study of learning under natural conditions. *Learning and Memory* 2:199–224.

Fraenkel, G. S., and D. L. Gunn. 1940. *The Orientation of Animals.* Oxford: Oxford University Press.

Frisch, K. von. 1950. Die Sonne als Kompaß im Leben der Bienen. *Experientia* (Basel) 6:210–221.

Frisch, K. von. 1967. *The Dance Language and Orientation of Bees.* Cambridge, Mass.: Harvard University Press.

Frisch, K. von, and M. Lindauer. 1954. Himmel und Erde in Konkurrenz bei der Orientierung der Bienen. *Naturwissenschaften* 41:245–253.

Gallistel, C. R. 1990. *The Organization of Learning.* Cambridge, Mass.: MIT Press.

Görner, P., and B. Claas. 1985. Homing behaviour and orientation in the funnel-web spider, *Agelena labyrinthica.* In F. G. Barth, ed., *Neurobiology of Arachnids*, 275–297. Berlin: Springer-Verlag.

Gould, J. L. 1980. Sun compensation by bees. *Science* 207:545–547.

———. 1986. The locale map of honeybees: Do insects have cognitive maps? *Science* 232:861–863.

Griffin, D. R. 1952. Bird navigation. *Biological Reviews* 27:359–400.

Gwinner, E. 1996. Circadian and circannual programmes in avian migration. *Journal of Experimental Biology* 199:39–48.

Hartmann, G., and R. Wehner. 1995. The ant's path integration system: A neural architecture. *Biological Cybernetics* 73:483–497.

Hartwick, R. F. 1976. Beach orientation in talitrid amphipods: Capacities and strategies. *Behavioral Ecology and Sociobiology* 1:447–458.

Hoffmann, K. 1959. Die Richtungsorientierung von Staren unter der Mitternachtsonne. *Zeilshrift für vergleichende Physiologie* 41:471–480.

Jander, R. 1975. Ecological aspects of spatial orientation. *Annual Review of Ecology and Systematics* 6:171–188.

———. 1990. Arboreal search in ants: Search on branches. *Journal of Insect Behavior* 3:515–527.

Keeton, W. T. 1974. The orientational and navigational basis of homing in birds. In D. S. Lehrman, J. S. Rosenblatt, R. A. Hinde, and E. Shaw, eds., *Advances in the Study of Behavior,* Vol. 5, 47–132. San Francisco: Academic Press.

Kramer, G. 1950. Weitere analyse der faktoren, welche die zugaktivität des gekäftigten vogels orientieren. *Naturwissenschaften* 37:377–378.

Krebs, J. R., and A. Kacelnik. 1991. Decision making. In J. R. Krebs and N. B. Davis, eds., *Behavioural Ecology: An Evolutionary Approach,* 105–136. Oxford: Blackwell Scientific.

Kühn, A. 1919. *Die Orientierung der Tiere im Raum.* Jena, Germany: Gustav Fischer Verlag.

Land, M. F. 1981. Optics and vision in invertebrates. In H. Autrum, ed., *Handbook of Sensory Physiology,* Vol. VII/6B, 472–592. Berlin: Springer-Verlag.

Lehrer, M. 1996. Small-scale navigation in the honeybee: Active acquisition of information about the goal. *Journal of Experimental Biology* 199:253–261.

Leonard, B., and B. L. McNaughton. 1990. Spatial representation in the rat: Conceptual, behavioral, and neurophysiological perspectives. In R. P. Kesner and D. S. Olton, eds., *Neurobiology of Comparative Cognition,* 363–422. Hillsdale, N.J.: Erlbaum.

Lindauer, M. 1957. Sonnenorientierung der bienen unter der aequatorsonne und zur nachtzeit. *Naturwissenschaften* 44:1–6.

———. 1959. Angeborene und erlernte Komponenten in der Sonnenorientierung der Bienen. *Zeitschrift für vergleichende Physiologie* 42:43–62.

Loeb, J. 1918. *Forced Movements, Tropisms, and Animal Conduct.* Philadelphia: Lippincott.

Lohmann, K. J., and C. M. F. Lohmann. 1996a. Orientation and open-sea navigation in sea turtles. *Journal of Experimental Biology* 199:73–81.

———. 1996b. Detection of magnetic field intensity by sea turtles. *Nature* 380:59–61.

Loomis, J. M., R. L. Klatzky, R. G. Golledge, J. G. Cicinelli, J. W. Pelligrino, and P. A. Fry. 1993. Nonvisual navigation by blind and sighted: Assessment of path integration ability. *Journal of Experimental Psychology: General* 122:73–91.

McNaughton, B., C. A. Barnes, J. L. Gerrard, K. Gothard, M. W. Jung, J. J. Knierem, H. Kudrimoti, Y. Qin, W. E. Skaggs, M. Suster, and K. L. Weaver. 1996. *Journal of Experimental Biology* 199:173–185.

Mittelstaedt, H. 1985. Analytical cybernetics of spider navigation. In F. G. Barth, ed., *Neurobiology of Arachnids,* 298–316. Berlin: Springer-Verlag.

Müller, M., and R. Wehner. 1988. Path integration in desert ants, *Cataglyphis fortis.*

Proceedings of the National Academy of Sciences of the United States of America 85:5287–5290.

New, D. A. T., and J. K. New. 1962. The dances of honeybees at small zenith distances of the sun. *Journal of Experimental Biology* 39:279–291.

Pardi, L., and A. Ercolini. 1986. Zonal recovery mechanisms in talitrid crustaceans. *Bollettino di Zoologia* 53:139–160.

Pardi, L., and F. Scapini. 1985. Inheritance of solar direction finding in sandhoppers: Mass-crossing experiments. *Journal of Comparative Physiology,* ser. A, *Sensory, Neural, and Behavioral Physiology* 151:435–440.

Perdeck, A. C. 1958. An experiment on the ending of autumn migration starlings, *Sturnus vulgaris* L., and chaffinches, *Fringilla coelebs* L., as revealed by displacement experiments. *Ardea* 46:1–37.

Phillips, J. B. 1986. Two magnetoreception pathways in a migratory salamander. *Science* 233:765–767.

Phillips, J. B., and S. C. Borland. 1994. Use of a specialized magnetoreception mechanism for homing by the red-spotted newt *Notphathalmus viridescens*. *Journal of Experimental Biology* 188:275–291.

Phillips, J. B., and J. Waldvogel. 1988. Celestial polarized light patterns as a calibration reference for sun compass of homing pigeons. *Journal of Theoretical Biology* 131: 55–67.

Ringelberg, J., and B. J. G. Flik. 1994. Increased phototaxis in the field leads to enhanced diel vertical migration. *Limnology and Oceanography* 39:1855–1864.

Ristau, C. A., ed. 1991. *Cognitive Ethology: The Minds of Other Animals—Essays in Honor of Donald R. Griffin.* Hillsdale, N.J.: Erlbaum.

Scapini, F., and M. C. Mezzetti. 1993. Integrated orientation responses of sandhoppers with respect to complex stimuli-combinations in their environment. In *Proceedings of the 1993 Conference on Orientation and Navigation: Birds, Humans, and Other Animals.*

Schmidt-Koenig, K. 1961. Die Sonnenorientierung richtungsdressierter tauben in ihrer physiologischen nacht. *Naturwissenschaften* 48:110.

———. 1963. Sun compass orientation of pigeons upon displacement north of the arctic circle. *Biological Bulletin* 127:154–158.

Schmidt-Koenig, K., J. U. Ganzhorn, and R. Ranvaud. 1991. The sun compass. In P. Berthold, ed., *Orientation in Birds,* 1–15. Basel: Birkhauser.

Schmitt, D. E., and H. E. Esch. 1993. Magnetic orientation of honeybees in the laboratory. *Naturwissenschaften* 80:41–43.

Schöne, H. 1984. *Spatial Orientation.* Princeton, N.J.: Princeton University Press.

Stephens, D. W. 1993. Learning and behavioral ecology: Incomplete information and environmental unpredictability. In D. R. Papaj and A. C. Lewis, eds., *Insect Learning: Ecological and Evolutionary Perspectives,* 195–218. New York: Chapman and Hall.

Stephens, D. W., and J. R. Krebs. 1986. *Foraging Theory.* Princeton, N.J.: Princeton University Press.

Tinbergen, N., and W. van Kruyt. 1938. Uber die Orientierung des Bienenwolfes (*Philanthus triangulum* Fabr.) III. Die Bevorzugung bestimmter Wegmarken. *Zeitschrift für vergleichende Physiologie* 25:292–334.

Ugolini, A., and L. Pardi. 1992. Equatorial sandhoppers do not have a good clock. *Naturwissenschaften* 79:279–281.

Ugolini, H., and A. Pezzani. 1995. Magnetic compass and learning of y-axis (sea-land) direction in the marine isopod *Idotea baltica basteri*. *Animal Behaviour* 50:295–300.

Ugolini, A., and F. Scapini. 1988. Orientation of the sandhopper *Talitrus saltator* (Amphipoda, Talitridae) living on dynamic sandy shores. *Journal of Comparative Physiology*, ser. A, *Sensory, Neural, and Behavioral Physiology* 162:453–462.

Ugolini, A., F. Scapini, G. Beugnon, and L. Pardi. 1988. Learning in zonal orientation of sandhoppers. In G. Chelazzi and M. Vannini, eds., *Behavioral Adaptation to Intertidal Life*, 105–118. New York: Plenum.

Wallraff, H. 1991. Conceptual approaches to avian navigation systems. In P. Berthold, ed., *Orientation in Birds*, 128–165. Basel: Birkhauser.

———. 1996. Seven theses on pigeon homing deduced from empirical findings. *Journal of Experimental Biology* 199:105–111.

Wehner, R. 1981. Spatial vision in arthropods. In H. Autrum, ed., *Handbook of Sensory Physiology*, Vol. VII/6C, 287–616. Berlin: Springer.

———. 1982. Himmelsnavigation bei Insekten. Neurophysiologie und Verhalten. *Vierteljahresschrift der Naturforschenden Gesellschaft in Zürich* 5:1–132.

———. 1987. "Matched filters"—Neural models of the external world. *Journal of Comparative Physiology*, ser. A, *Sensory, Neural, and Behavioral Physiology* 161:511–531.

———. 1991. Visuelle navigation: Kleinstgehirn-strategien. *Verhandlungen der Deutschen Zoologischen Gesellschaft* 84:89–104.

Wehner, R., B. Michel, and P. Antonsen. 1996. Visual navigation in insects: Coupling of egocentric and geocentric information. *Journal of Experimental Biology* 199:129–140.

Wehner, R., S. Bleuler, C. Nievergelt, and D. Shah. 1990. Bees navigate by using vectors and routes rather than maps. *Naturwissenschaften* 77:479–482.

Wehner, R., and B. Lanfranconi. 1981. What do the ants know about the rotation of the sky? *Nature* 293:731–733.

Wehner, R., and M. Müller. 1993. How do ants acquire their celestial ephemeris function? *Naturwissenschaften* 80:331–333.

Wehner, R., and M. V. Srinivasan. 1981. Searching behavior of desert ants, genus *Cataglyphis* (Formicidae, Hymenoptera). *Journal of Comparative Physiology* 142:315–338.

Wehner, R., and S. Wehner. 1990. Insect navigation: Use of maps or Ariadne's thread? *Ethology, Ecology, and Evolution* 2:27–48.

Wiltschko, R. 1996. The function of olfactory input in pigeon orientation: Does it provide navigational information or play another role? *Journal of Experimental Biology* 199:113–119.

Wiltschko, R., and W. Wiltschko. 1981. The development of sun compass orientation in young homing pigeons. *Behavioral Ecology and Sociobiology* 9:135–141.

Wiltschko, R., D. Nohr, and W. Wiltschko. 1981. Pigeons with a deficient sun compass use the magnetic compass. *Science* 214:343–345.

Wiltschko, W., and R. Wiltschko. 1992a. Magnetic orientation and celestial cues in migratory orientation. In P. Berthold, ed., *Orientation in Birds,* 16–37. Basel: Birkhauser.

———. 1992b. Migratory orientation: Magnetic compass orientation of garden warblers (*Sylvia borin*) after a simulated crossing of the magnetic equator. *Ethology* 91: 70–79.

———. 1996. Magnetic orientation in birds. *Journal of Experimental Biology* 199: 29–38.

Wiltschko, W., R. Wiltschko, and W. T. Keeton. 1976. Effects of a "permanent" clock-shift on the orientation of young homing pigeons. *Behavioral Ecology and Sociobiology* 1:229–243.

Winston, M. L. 1987. *The Biology of the Honey Bee.* Cambridge, Mass.: Harvard University Press.

CHAPTER SEVEN

The Ecology and Neurobiology of Spatial Memory

DAVID F. SHERRY

7.1 INTRODUCTION

The study of spatial memory has served a central function in the development of a science of animal cognition for two reasons. The first reason is ecological: Most ecologically interesting behavior of animals has a spatial component. Dispersal, migration, territoriality, predator avoidance, mate search, nest site selection, provisioning young, foraging, and food storing all require animals to move through space and keep track of where they have been, where they are, and where they are going. The second reason is neurobiological: In 1978, O'Keefe and Nadel (1978) proposed that an ancient structure in the vertebrate brain, the hippocampus, is a dedicated processor of spatial information, a cognitive map. The hippocampal cognitive map became a paradigm for animal cognition—a cognitive system with more complex operating rules than those governing simple associations and implemented with a specialized neural architecture. The hippocampal cognitive map has had its critics, and current research presents a complicated picture; the idea, however, has provided an inescapable theoretical backdrop to most research on the neurobiology of spatial memory.

In this chapter, I will illustrate the ecological importance of spatial memory with a few examples of territoriality, mate search, foraging, and food storing. Next, I will describe some proposed mechanisms of spatial memory and discuss a few current topics in the neurobiology of spatial memory. This review will be selective, but I will chart the major trends in contemporary research on spatial memory. Dyer (this vol. chap. 6) examines spatial orientation from a different perspective—the adaptive design of navigational strategies.

7.2 THE BEHAVIORAL ECOLOGY OF SPATIAL MEMORY

How important is memory in spatial orientation? Some spatial-orientation tasks require no memory. In European blackcaps (*Sylvia atricapilla*), migratory direction is inherited (Berthold et al. 1992). When hand-raised blackcaps from two populations with autumn migratory directions that differed by 90°

were tested in captivity under controlled conditions, their autumn migratory directions corresponded to those of the respective parent population (Helbig et al. 1989). Migratory direction in one hand-raised group even included a 45° clockwise turn part way through the migratory period, which corresponded to the direction change made by the parent population as it rounded the eastern Mediterranean Sea and headed south into Africa (fig. 7.1). Without previous migratory experience (and therefore without memory of any previous migrations), these first-year blackcaps headed in the direction that was characteristic of their parent population's migratory path, and in one case, the birds were ready to make a right turn at the appropriate time.

Other aspects of spatial orientation, however, clearly depend more on experience, and a record of recent and sometimes long-past experience must be maintained to return home, revisit a newly discovered place, or adjust for changes in the environment.

7.2.1 Territoriality

A territory is a place that is defended for the exclusive use of the occupant (Davies and Houston 1984). A territory has a spatial extent and boundaries, and defense often occurs at prominent landmarks that define the limits of the territory. The benefits of the use of landmarks as territorial boundaries have been quantified by Eason et al. (1996). They plotted the territories defended by individually marked male cicada killer wasps (*Sphecius speciosus*) on a flat, grassy lawn. Males defended territorial boundaries by patrolling and chasing intruders. Next, 90-cm lengths of dowel were placed flat on the lawn such that the dowels were not aligned with existing territorial boundaries. The dowels rested slightly below grass level and were not used as perches by the wasps. By the following day, wasps had changed their territorial boundaries to coincide with the dowels. In another experiment, Eason et al. placed parallel pairs of dowels on the lawn and compared the time males spent in territorial defense at their dowel-marked and unmarked boundaries. Males spent 9.7% of the

---

Figure 7.1. The inheritance of migratory direction. Each compass circle shows the individual mean vectors, determined in Emlen cages, of hand-raised blackcaps from two populations with differing autumn migration routes as shown on the map. An Emlen cage records foot scratches made by a bird on the walls of the funnel-shaped cage. Statistical tests on the pattern of scratches are used to determine if activity is oriented and the mean bearing of oriented activity (Helbig 1991). Solid arrows in circles indicate means for well-oriented birds and dashed arrows indicate birds not significantly oriented during tests. For southwestern Germany, $n = 18$ and for eastern Austria, $n = 19$. There was a significant change in directions between September/October and November in birds from eastern Austria ($F_{2,35} = 26.42$, $P < .01$) but not in birds from southwestern Germany (from Helbig 1996; Helbig et al. 1989).

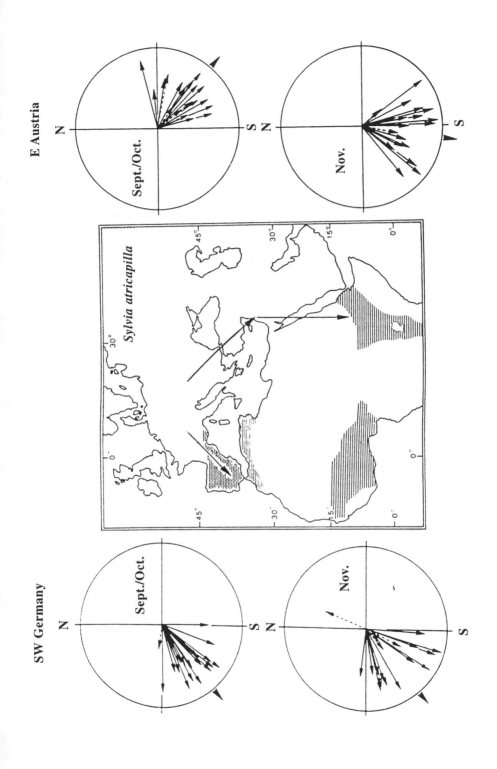

SW Germany

Sept./Oct.

Nov.

Sylvia atricapilla

E Austria

Sept./Oct.

Nov.

observation periods defending unmarked boundaries and only 1.4% of the observation periods defending boundaries marked with dowels. Territorial wasps thus changed the shapes of their territories to make use of landmarks at boundaries and benefitted by a reduction in the time spent defending marked boundaries.

Territorial boundaries are honored in both the breach and the observance by male song birds. Male white-throated sparrows (*Zonotrichia albicollis*) defend exclusive territories with song, but spend about 45% of their time outside territorial boundaries. They do not sing outside their own territory but instead move silently through the territories of neighbors (Falls and Kopachena 1994). This change in behavior at territory boundaries shows that the spatial location of boundaries is known to the birds and has a direct influence on behavior. The role of spatial memory in territoriality is further illustrated with experiments on birds' ability to associate the songs of their neighbors with the neighbors' territories (Falls 1982; Beecher et al., this vol. chap. 5). Male hooded warblers (*Wilsonia citrina*) responded relatively mildly to the played-back song of a neighbor when the song was broadcast from near the boundary with that neighbor. When the same song was played back from near another neighbor's boundary, territory holders responded much more strongly and approached quickly to within a few meters of the speaker; the birds then remained near the speaker during and after playback (Godard 1991). Playback of a familiar song from a new location was thus regarded as territory intrusion and produced a strong response from territory holders. Remarkably, comparable results were also obtained the following year when male hooded warblers returned to their territories after wintering in Central America and not hearing the songs of their neighbors for about 8 months (Godard 1991). Therefore, male hooded warblers not only learn the songs of their neighbors; the warblers associate their neighbors' songs with the boundaries of their neighbors' territories. The response to song depends on the place where the song originates (see Beecher et al., this vol. section 5.4).

Male songbirds are not the only territorial animals to associate the signals of their neighbors with spatial locations. The weakly electric fish *Gymnotus carapo* discriminates the electric discharges of neighbors and responds more aggressively and with a shorter latency when a familiar signal comes from a new location, the territory of another neighbor (McGregor and Westby 1992).

7.2.2 Mate Search

For some animals, like male songbirds, obtaining a mate requires defending a resource-rich territory and advertising occupancy. For other animals, finding a mate requires an active search. If a potential, but initially unreceptive, mate is found, there may be a reproductive advantage to returning to the same place at a later, more propitious time. The mating season of thirteen-lined ground squirrels (*Spermophilus tridecemlineatus*) lasts 1–3 weeks, and females are in estrus for about 4 days. Males compete for females in a polygynous scramble. Schwagmeyer (1994) removed females that were in estrus and found that males searched preferentially at sites where they had encountered the female on previous days. Males were not simply attracted by pheromones or other odors of the female because males did not find females that were experimentally displaced to new sites and did not search the sleeping burrow used most recently by the female (unless the female had been encountered there on a previous day). Observations showed that males did not follow regular routes during successive searches and avoided sites they had searched previously.

How important is spatial memory, in general, for locating mates? The importance of spatial memory varies, no doubt, with the mating system and the spatial distribution of potential mates (Davies 1991). When males mate polygynously and compete for mates by expanding their home ranges to include multiple female home ranges, selection can produce a difference in spatial memory between the sexes that favors males. Meadow voles (*Microtus pennsylvanicus*) are polygynous, but pine voles and prairie voles (*M. pinetorum* and *M. ochrogaster*) are monogamous. The home ranges of male meadow voles are larger than the home ranges of females, but home-range sizes do not differ between the sexes in prairie and pine voles (Gaulin and Fitzgerald 1986, 1989). Radiotelemetry data showed that range sizes of male meadow voles do not expand before sexual maturity or outside the breeding season (Gaulin and Fitzgerald 1989). Male meadow voles' spatial abilities are superior to those of females, as shown by males' performances in a variety of laboratory maze tasks; male and female pine and prairie voles' spatial abilities do not differ (Gaulin and Fitzgerald 1986, 1989).

It is possible that the superior spatial ability of males is not an endogenous component of the polygynous mating system of meadow voles; instead the superior spatial ability may be a consequence of spatial experience or level of activity. A series of experiments by Gaulin and his colleagues have shown, however, that sex differences in spatial experience and activity level cannot account for the observed sex difference in spatial ability (Gaulin et al. 1990; Gaulin and Wartell 1990). To determine if spatial ability is affected by experience, the maze performance of wild, captive adult prairie voles was compared with that of their lab-reared offspring. Wild captive adults had experienced 200-m^2 home ranges, and lab-reared offspring had been confined to small, plastic holding cages. No difference in maze performance was observed between the two groups (Gaulin and Wartell 1990). The similar performances of the groups were not simply the result of insensitive measurements of maze performance; significant differences in maze performance were observed between voles that were food deprived for 15 hours and voles that were food deprived for 24 hours, with the latter group showing better performance.

The hypothesis that sex or species differences in activity account for observed differences in maze performance was tested by recording the level of activity, measured with photocell-beam interruptions, of polygynous meadow voles and monogamous prairie voles of both sexes (Gaulin et al. 1990). As expected maze performance was not related to the observed activity level (male meadow voles performed better than females, and male and female prairie voles performed equally well). A statistical interaction between species and sex in maze performance was still found after controlling for activity level.

Organizational effects of hormones during development and activational effects at sexual maturity combine to produce the sex difference in spatial ability observed in meadow voles (Galea et al. 1996). As I will discuss in section 7.4, male and female meadow voles differ not only in home-range size and ability to solve laboratory spatial problems; the size of the male vole's hippocampus is larger. This sex difference does not occur in pine voles (Jacobs et al. 1990).

7.2.3 Foraging

Despite the explosion of interest during the 1970s and 1980s in optimal foraging, few foraging models explicitly considered how the spatial distribution of food influenced foraging decisions. The classic patch and prey models assumed only that food was distributed in patches and that encounter with prey was determined by an encounter-rate parameter (Stephens and Krebs 1986). Central-place models addressed the effect of distance to a patch on foragers that returned to the same central place after collecting prey, but these models did not address other properties of the spatial distribution of food. If the spatial properties of prey distribution or searches were considered, it was usually in simplified models showing, for example, that (1) directional search paths were more efficient than random search (Krebs 1978); (2) search areas were restricted after a forager encountered food; or (3) a forager may maximize renewal time of a resource by following a path around a continuous circuit (e.g. Davies and Houston 1981). Clearly the spatial distribution of food and memory for food location can have more complex effects on foraging decisions.

Healy and Hurly (1995) examined how foraging rufous hummingbirds (*Selasphorus rufus*) use memory for spatial locations. Many hummingbirds defend feeding territories to maintain exclusive access to enough flowers to meet their energy requirements. Because feeding depletes the flowers' nectar, hummingbirds can increase the rate at which they collect nectar by avoiding visits

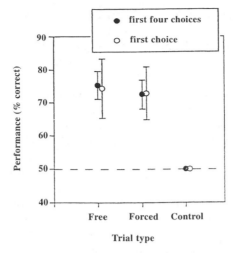

Figure 7.2. The accuracy of rufous hummingbird memory for the spatial locations of flowers. Four flowers, in an array of eight, had not been visited in the previous trial. Visits to these previously unvisited flowers were "correct" and are shown for trials during which birds' previous visits were either free or forced. Performance exceeded the control level during both free and forced trials (t tests, P < .05, n = 6). Performance on the first choice and for the mean of the first four choices did not differ. Error bars indicate the standard error of the mean, and the dashed line shows performance expected by chance (from Healy and Hurly, 1995).

to depleted flowers; rufous hummingbirds avoid revisiting depleted flowers by remembering which flowers they have visited. Healy and Hurly (1995) placed artificial flowers, each filled with 40 μl of 24% sucrose, in a circular array of eight flowers in hummingbird feeding territories. Birds were allowed to visit and deplete four of the eight flowers, and their subsequent flower visits were observed after intervals that ranged from 4 to 40 min. From 70% to 80% of the birds' next four visits were to nondepleted flowers; this outcome was observed if the birds were allowed to choose the four initial flowers to visit or if they were forced to visit four flowers determined randomly by the experimenters (fig. 7.2). The locations of particular flowers in the array were varied to control for memory of individual flowers, and additional tests showed the hummingbirds could not discriminate between nondepleted and depleted flowers without visiting them. Hummingbirds could, in principle, avoid revisits to depleted flowers by "trap lining," (i.e. following a consistent route from flower to flower). Healy and Hurly (1995) showed that rufous hummingbirds solved the revisiting problem not by trap lining but by remembering individual flower locations. Rufous hummingbirds did this either retrospectively, by remembering depleted flowers, or prospectively, by remembering which flowers had not been visited (Cook et al. 1985; Kesner and DeSpain 1988). In either case, memory was used to restrict visits to flowers that were not yet depleted.

The tendency of hummingbirds to avoid previously visited flowers, rather than return to them, depends on the birds' recent experience with the food source. In a subsequent experiment using artificial flowers that were not depleted after a single visit, birds were trained to search an array of four distinctively colored and patterned flowers until the one flower that contained sucrose was found (Hurly and Healy 1996). This flower was not depleted after a single visit, and hummingbirds learned to return to this flower and not other flowers in the array on subsequent visits. Nevertheless, birds still identified this flower by its spatial location and not its appearance as shown by a further experimental manipulation. Hurly and Healy (1996) were able to separate memory for the location of the baited flower from memory for its color and features by adapting an experimental design developed by Brodbeck (1994) (see section 7.3). After hummingbirds fed at the baited flower, the flower was exchanged with another flower in the array and the birds' subsequent choices were recorded. On their next visit, hummingbirds consistently returned to the location in the array where they had found nectar (now occupied by a different looking flower) instead of the flower with the matching color and pattern (now at a different location). Subsequent choices gave no indication of a preference for the flower with the matching color and pattern.

7.2.4 Food Storing

Few behaviors make such extraordinary demands on spatial memory as the retrieval of food from scattered caches. A variety of birds and mammals scatter hoard food and retrieve it by remembering the spatial locations of caches (Vander Wall 1990). Jays and nutcrackers store large numbers of acorns and pine nuts, respectively, and their failure to retrieve all that they store is an important means of seed dispersal for several species of oaks and pines (Tomback and Linhart 1990; Bossema 1979; Darley-Hill and Johnson 1981). Food-storing birds depend on their caches to survive the winter. With use of stored food to feed nestlings, food-storing birds are able to breed earlier than many other birds (Sherry 1985; Vander Wall 1990). Jays and nutcrackers (*Corvidae*), chickadees and tits (*Paridae*), squirrels (*Sciuridae*), and kangaroo rats (*Heteromyidae*) have all been observed to accurately remember the spatial locations of caches for periods ranging from a few days to many months (Kamil and Balda 1990; Sherry and Duff 1996; Shettleworth 1990; Jacobs and Liman 1991; Jacobs 1992).

Memory for spatial locations is not the only conceivable way to relocate hidden caches. Food-storing birds might encounter their caches by chance or place caches in sites that are likely to be searched again in the course of normal

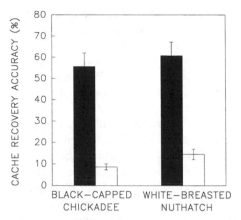

Figure 7.3. Memory for the spatial location of caches in black-capped chickadees and white-breasted nuthatches in captivity. Filled bars indicate observed cache-retrieval accuracy, open bars indicate accuracy expected by chance. Accuracy significantly exceeded that predicted by chance for both species (chickadees, $F_{1,14}$ = 43.93, P < .001; nuthatches, $F_{1,8}$ = 62.67, P < .001). There was no difference in accuracy between chickadees and nuthatches (chickadee data from Petersen and Sherry 1996; nuthatch data from Petersen and Sherry, unpublished).

foraging (Gibb 1960; Haftorn 1974; Källander 1978). Laboratory experiments have shown, however, that the performance of food-storing birds when searching for stored food was much better than that expected by chance (fig. 7.3), and food-storing birds showed no decrement in accuracy when their choice of cache sites was experimentally constrained (Kamil and Balda 1985; Shettleworth and Krebs 1982; Sherry et al. 1981). The birds did not follow a regular route when retrieving caches, and they did not retrace the route they used when making caches (Sherry 1984). When visual and olfactory cues from stored food were eliminated, birds showed no decline in accuracy, and when stored food was displaced a short distance away from the cache site, birds usually failed to find it (Bennett 1993; Cowie et al. 1981; Shettleworth and Krebs 1982). These results occurred because birds retrieved hoarded food by remembering the location of caches with respect to prominent nearby landmarks (Bossema 1979; Bennett 1993; Herz et al. 1994; Vander Wall 1982), perhaps supplemented with use of the sun compass for orientation (see section 7.3) (Balda and Wiltschko 1991; Wiltschko and Balda 1989). Food-storing rodents can detect buried caches by olfaction. These rodents, nevertheless, preferentially retrieved their own caches before those of others, which indicates that they too can remember the spatial locations of caches (Jacobs 1992; Jacobs and Liman 1991).

In the laboratory, black-capped chickadees (*Parus atricapillus*) were able to find their caches for at least 4 weeks after making them; after 4 weeks, their accuracy fell to that expected by chance (Hitchcock and Sherry 1990). Clark's nutcrackers (*Nucifraga columbiana*) found their caches in captivity at least 40 weeks after making them (Balda and Kamil 1992). In chickadees and tits in the wild, estimates of the interval between caching and cache retrieval range from several days (Cowie et al. 1991; Stevens and Krebs 1986) to several months (Brodin and Ekman 1994). Brodin and Ekman (1994) determined the cache-retrieval interval of willow tits (*Parus montanus*) by offering the birds seeds labelled with ^{35}S-cysteine, which is incorporated into the daily growth bars of feathers when the food is retrieved and eaten. By pulling a tail feather from each member of a marked population of willow tits to induce growth of a replacement feather and then collecting and analyzing the replacement feathers 2 months later, Brodin and Ekman (1994) obtained a record of the consumption of labelled food. Food was commonly retrieved and eaten 6–40 days after storage and, in some cases, after even longer periods of time. Because this interval exceeded the duration of memory for cache location in the laboratory, Brodin and Ekman (1994) proposed that long-term cache retrieval does not depend on memory mechanisms. Whether or not tits' memory for cache location can persist for 40 or more days in the field will, no doubt, be answered with further experimentation; however, it also seems possible that the duration of memory for cache location observed in captivity may underestimate the duration of this memory in the wild.

7.2.5 Conclusion

The ecological importance of spatial memory is widespread. A few cases in which spatial memory has not been well examined but seems potentially important are (1) poison-arrow frogs, *Dendrobates pumilio,* that remember, in all likelihood, the bromeliad leaf axils that hold their developing tadpoles (Weygoldt 1980), (2) brood parasites that visit and probably remember the locations of potential host nests (Sherry et al. 1993; Reboreda et al. 1996), and (3) migratory birds that show year-to-year nest-site fidelity (Ketterson and Nolan 1990). Although the ecological circumstances in which animals show memory for spatial locations is very diverse, the ways in which different species solve these problems are similar. In the next section, I will describe some of the more well-documented mechanisms that underlie memory for spatial locations.

7.3 MEMORY AND SPATIAL ORIENTATION

Some proposed mechanisms of spatial memory are well established, others have not been so unequivocally demonstrated, and some probably await discovery. A distinction can be made among *landmarks, global-reference systems, path integration,* and the *cognitive map.* Landmarks are stationary objects or surfaces that can be used to identify a location in space by its distance and direction from the landmark. With global-reference systems, celestial or geomagnetic information is used to determine a reference bearing (analogous to a compass bearing) and, in some cases, a position. Path integration is the use of idiothetic information, or information that is generated during motion, to determine distance and direction from the origin of the path. Idiothetic information about active displacement and rotation can be derived, at least in principal, from the vestibular system, proprioceptive feedback, optic flow, and efference copy of locomotor commands. Finally, the cognitive map usually refers to a representation of the spatial relations among locations that is constructed or inferred from partial knowledge of the spatial relations among places. This categorization is somewhat arbitrary. Landmarks can be important for correcting errors that accumulate during path integration, and global-reference systems are components of some cognitive map models. Nevertheless, these distinctions help describe the component parts of many models of spatial orientation. Dyer (this vol. chap. 6) examined the selective costs and benefits of these systems. I will deal with how the systems work. In the following sections, I will describe how landmarks and one global-reference system, the sun compass, contribute to memory of spatial locations. Path integration is discussed by Etienne et al. (1996), and discussions of the cognitive map can be found in Gallistel and Cramer (1996), Bennett (1996), and Nadel (1991).

7.3.1 Landmarks

Stable environmental landmarks provide the most readily identifiable (at least to humans) and easily manipulated sources of spatial information. The classic demonstration of landmark use is Tinbergen's (1932) homing experiment in the digger wasp, *Philanthus triangulum*. Tinbergen placed a circle of pine cones around an active nest and then either displaced the cones to surround a sham nest 30 cm away or returned the cones to their original positions around the active nest. As every student of ethology knows, returning wasps landed inside the circle of cones regardless of whether the cones surrounded the sham nest or the real nest. This simple experiment showed that wasps found their nest by remembering its location with respect to neighboring landmarks.

But how do landmarks specify a location in space? It seems obvious that the wasp always searched in "the same place" with respect to the circle of pine cones, but there is more than one way that landmarks can specify "the same place." For instance, did the wasp determine the nest position with use of the whole array of landmarks or just one particular landmark? In a very informal experiment, Tinbergen and Kruyt (1938) found that moving only a few landmarks was not sufficient to ensure identification; the whole array of landmarks had to be moved. Suzuki et al. (1980) designed an experiment in which rats were trained to find food on an eight-arm radial maze. A cylindrical black curtain surrounded the maze to eliminate use of extraneous landmarks, and the experimenters placed landmarks outside the maze beyond the ends of the radial arms. In experimental trials, rats were forced to visit three preselected maze arms (by blocking entrances to the other five arms) and then were confined briefly to the center of the maze. Then with all arms open, the rats' choices among previously visited and unvisited maze arms were observed. This method of providing forced choices before free choices, first used by Zoladek and Roberts (1978), controlled for the possibility that the rats may have changed the nature of the task—for example, by entering the most distinctive maze arms first. This method also allowed Suzuki et al. to manipulate landmarks during the time between the rats' forced and free choices. Landmarks were either rotated as a group by 180° or transposed randomly and reassigned to the maze arms. When landmarks were rotated 180°, rats entered the maze arms that were in the same position relative to the landmarks as the arms that were unvisited. That is, when the landmarks were rotated, the rats' choices rotated. When landmarks were transposed, animals chose randomly. This simple experiment illustrated how animals determine spatial location with use of landmarks. The rats used the array of landmarks as a unit to identify unvisited maze arms. They did not identify maze arms by the individual landmark at the end of the maze arm or by the nearest landmark. If this were the case, transposition of landmarks should have produced systematic choices of unvisited maze arms defined by their respective landmarks, because after transposition each maze arm still had a corresponding landmark. Landmarks were not used by rats in this experiment as individual beacons but as an array or configuration. This finding has been confirmed in a variety of contexts (Morris 1981; Kraemer et al. 1983; Olton and Collison 1979; Mazmanian and Roberts 1983; Spetch and Honig 1988).

Landmarks are used as a configuration and not as individual beacons because a single landmark usually cannot specify a location in space (apart from the spot occupied by the landmark) without additional information. A digger wasp may remember the distance her burrow entrance lies from a pine cone, but in what direction? Either a radially asymmetric landmark, additional landmarks, or another cue that polarizes the environment by providing a bearing (such as the north star or the earth's magnetic field) is required to determine the direction of the goal from the landmark. A pair of landmarks is sufficient if the landmarks can be distinguished or if the goal and the landmarks are collinear. Two identical landmarks reduce the number of possible locations of the goal to two points in space, and a third landmark defines the location.

Collet et al. (1986) showed that gerbils use landmarks as a configuration. Gerbils were trained to find hidden food in an array of three identical landmarks and then tested with various modifications and distortions of the land-mark array. Gerbils learned the distance and direction of the goal from each individual landmark but required multiple landmarks to locate the goal. For example, gerbils trained to search for food at the center of an equilateral trian-gle of identical landmarks searched in roughly the correct location if one land-mark was removed. This result incidentally indicated that the environment was polarized and must have contained some cue that provided a bearing; in this way, the gerbils determined the side of the triangle with which they were dealing. Without such a polarizing cue, the regions formerly inside and outside the triangle would have been unclear, and gerbils would have searched at two points, one on either side of the line connecting the remaining landmarks. If two landmarks were removed, gerbils regarded the remaining landmark as first one and then another of the known landmarks; gerbils searched at three points around the remaining landmark (fig. 7.4). Landmarks were learned individu-ally but used in a configuration. The experiments of Collett et al. showed, in addition, that gerbils used a variety of rules if landmarks provided conflicting information. Landmarks near a goal were weighted more than landmarks far from a goal. If some landmarks indicated one location and others did not, the location indicated with the most landmarks was chosen. If landmarks were distinct, gerbils learned both the geometrical properties of the landmark array and the individual features of landmarks to determine goal location.

Cheng (1986) showed that in some circumstances, however, landmark fea-tures are ignored altogether, and the geometrical properties of the landmark array are used alone to identify a goal. Cheng trained rats to locate food in a rectangular enclosure with use of distinctive tactile, olfactory, and visual cues in each corner of the enclosure. When the rats searched the enclosure during test trials with the food removed, rats tended to err at places that were at a 180° rotation from the correct site. The pattern of errors suggested that rats relied on the rectangular geometry of the enclosure more than the distinct features in each corner of the enclosure. Further experiments with rotations and transformations of distinct features showed that rats used features but principally when these features were in geometrically correct locations.

In rats, this distinction between the features of landmarks and their geomet-ric properties is relevant to a sex difference in spatial memory. Male rats, like male meadow voles, tend to perform better than females on spatial tasks. In rats, Williams et al. (1990) examined how hormonal effects during develop-ment may influence performance in the radial-arm maze. Males gonadecto-mized shortly after birth behaved like control females; females given estradiol benzoate, which has the masculinizing effect of the testosterone metabolite estradiol, behaved like control males. Control males and females given estra-diol benzoate consistently performed better than control females and gonadec-tomized males; most interestingly, the kind of information used by the former

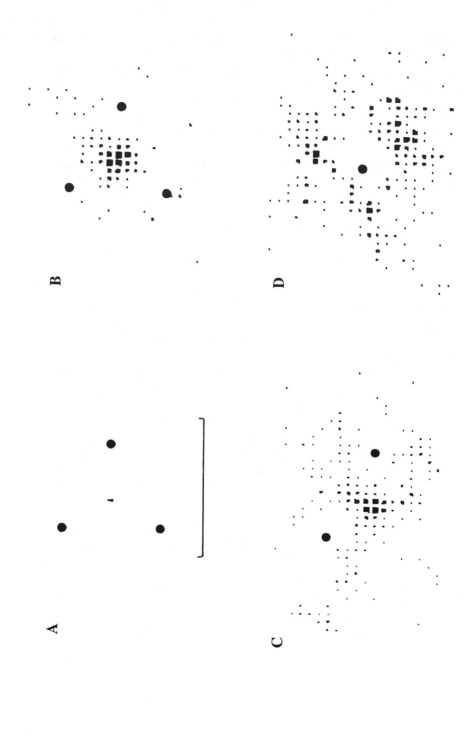

group differed from that used by the latter group. The performance of control males and females given estradiol benzoate was disrupted by changing the geometry of the enclosure that surrounded the maze, from rectangular to circular. Performance was unaffected by changes in landmarks outside the maze. Control females and gonadectomized males were affected by changing both the geometry of the enclosure and landmarks. Control females (and gonadectomized males) used more sources of information but performed less well than control males (and females given estradiol benzoate), which relied almost exclusively on the geometric properties of the region around the maze (Williams et al. 1990).

7.3.1.1 Food-Storing Birds

The scattered caches of food-storing birds are sometimes preferentially placed near conspicuous objects and edges instead of in more open areas without distinct features. Jays cache food more often near objects oriented with their long axis vertical, than near the same objects oriented with their long axis horizontal (Bossema 1979; Bennett 1993). These preferences likely arise because birds use landmarks to relocate their caches. One of the clearest demonstrations of the use of landmarks by food-storing birds was provided by Vander Wall (1982). Clark's nutcrackers displaced their searches for caches in the same direction and by the same distance as experimentally displaced landmarks. Herz et al. (1994) found that removal of landmarks immediately next to caches site had relatively little effect on the accuracy of searches by black-capped chickadees; however, removal of landmarks about 1 meter away from caches reduced search accuracy. Rotation of these more distant landmarks produced partial and inconclusive effects on searches, but a more recent experiment by Duff et al. (in press) showed that a 180° rotation of distant landmarks produced a nearly perfect 180° rotation of searches in chickadees.

The relative importance of the geometric relations and features of landmarks was determined by Brodbeck (1994) with a very elegant experiment on black-capped chickadees. Birds were trained to find food in one of four feeders placed on the walls of an aviary. Colored patterns on the feeders were unique

---

Figure 7.4. Use of landmarks by Mongolian gerbils. (a) Gerbils were trained to find food hidden at a site (triangle) at the center of a triangular array of identical landmarks (circles). Scale bar equals 1 meter. Search behavior is shown as time spent in each 11.0 × 13.3 cm cell, and cells are filled in proportion to the time spent searching. Data of three gerbils and from eight to 19 search trials per condition are shown. Search distributions are shown with (b) all landmarks present, (c) one landmark removed, and (d) two landmarks removed. Search was concentrated at a single site if two or three landmarks were present, but three different sites were searched if only one landmark was present (from Collett et al. 1986).

during each trial, and the four feeders were arranged in a different geometric array during each trial. Once birds located the baited feeder, they were removed from the aviary for a 5-minute retention interval; on return, food was available only in the previously baited feeder. The bird's task was to locate and remember for 5 minutes the feeder baited on this trial. Once birds had mastered this task, test trials were conducted to determine the information birds used to accomplish the task. During search trials, after the retention interval, birds returned to an array that had been displaced laterally along the aviary wall, and the baited feeder had been exchanged with another feeder in the array (fig. 7.5). With feeders in this configuration, the relative importance of the location of the feeder within the aviary, its location within the array, and the features of the feeder could be determined. This procedure is the same as that described for Hurly and Healy (1996) experiment with rufous hummingbirds, with the addition of a lateral displacement of the whole array. Brodbeck (1994) found a clear preference ranking in the chickadees' choices: location in the room, position within the array, and features (fig. 7.5). Brodbeck demonstrated that landmark features are not ignored or forgotten, but they are just not highly ranked among cues.

7.3.2 Global-Reference Systems

A global reference provides an animal with a spatial frame for orientation. Some global-reference systems allow the animal to determine its location and some merely provide a fixed reference bearing. The sun, stars, and the earth's magnetic field serve as global-reference systems. The sun's position provides a compass bearing if the animal has information about the time of day. The classic work of Kramer (1951) showed that starlings (*Sturnus vulgaris*) can use the sun's position to maintain a constant bearing in a small cage. The night sky, or more specifically the fixed point of rotation in the night sky, can also provide a compass bearing, as shown by Emlen (1970). Finally, experimental manipulations of the magnetic field have shown that birds can use the inclination of the earth's magnetic field (its deviation from horizontal)

Figure 7.5. Black-capped chickadees were trained to find food at one of four feeders and return to that feeder after a short retention interval. On probe trials (*top*), birds located food in one of the feeders (*C*); however, during the retention interval, the array was laterally displaced and another feeder was exchanged for the baited feeder. At testing (*bottom*), feeder *D* occupied the original location of the baited feeder, and feeder *A* occupied the original location of the baited feeder in the array. Most of the birds' first choices were to the original location of the baited feeder (*D*), and their second choices were the original array position (*A*). Their third choices were the feeder that matched in color and pattern (*C*). Results are shown with bar graphs. All three search distributions differed significantly from those predicted by chance (G-statistic, $P < .05$, $n = 4$) (from Brodbeck 1994).

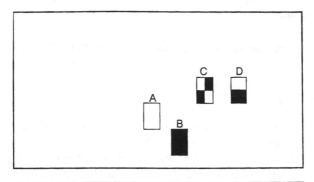

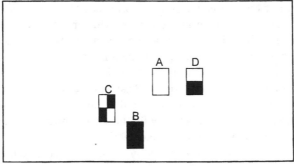

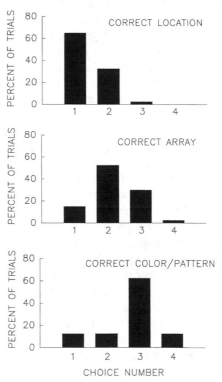

CORRECT LOCATION

CORRECT ARRAY

CORRECT COLOR/PATTERN

PERCENT OF TRIALS

CHOICE NUMBER

to derive a bearing and possibly a position as well (Wiltschko and Wiltschko 1993).

7.3.2.1 The Sun Compass

A variety of methods have been used to investigate animals' use of the sun compass. The most commonly used method of investigation is the clock-shift procedure. Because the azimuth of the sun changes during the day, the sun's position is useful only if combined with information about the time of day. The elevation of the sun is not used for orientation by animals. One way to demonstrate that the azimuth of the sun is used for orientation is to change the animal's estimate of time with a phase shift in its circadian rhythm. A phase shift should cause an error in orientation that is equal to the difference between the sun's azimuth at the animal's subjective time of day and the sun's actual azimuth at the local solar time. Phase advances and phase delays produce errors in bearing that can be measured in cages or in the field, and these errors are frequently as predicted with the sun-compass model (Schmidt-Koenig 1990; Wiltschko and Wiltschko 1993).

Much research on sun-compass use by vertebrates has assumed that solar information is most useful for long-range orientation because, at least initially, the animal may need only a correct heading. Insects use the sun compass to orient on a much smaller spatial scale (Dyer and Dickinson 1994), and recent work indicates that the sun compass may also serve a function in the small-scale orientation of food-storing birds. Wiltschko and Balda (1989) discovered that the clock-shift procedure caused pinyon jays (*Aphelocoma coerulescens*) to rotate their searches for cached food in a relatively small octagonal cage away from the actual cache locations and toward the sector of the cage predicted with use of the sun compass. Black-capped chickadees trained to find food in one sector of an octagonal cage exhibited similar changes in their direction of search (Duff et al., in press). Duff et al., however, also showed that the bearing derived with use of the sun compass is not sufficient for orientation in a small scale. Chickadees were trained to find food in one sector of a small octagonal cage. They were able to see surrounding landmarks through the mesh walls of the cage. In the presence of familiar landmarks, the clock-shift procedure produced a modest rotation of the mean search direction; however, in the presence of novel landmarks, the birds were not oriented in any direction. When information from familiar landmarks and the sun compass conflicted, the chickadees' search was clearly influenced by use of sun-compass information. Their search was probably a compromise between sun-compass and landmark information. However, when all landmarks were unfamiliar and the sun compass was the only source of directional information, the birds were disoriented. The explanation for this seemingly paradoxical result may be that birds form a "mosaic map" consisting of familiar areas (defined with local landmarks) that were linked with use of the sun compass or other global reference systems (suggested by Wiltschko and Wiltschko 1978). In familiar surroundings, the clock-shift procedure produces a change in orientation by means of its effect on the sun compass, as shown by Wiltschko and Balda (1989) and by Duff et al. (in press); however, in unfamiliar surroundings, a directional cue alone provides no information that the birds can use to locate food.

7.4 THE NEUROBIOLOGY OF SPATIAL MEMORY

There are many ecological contexts that necessitate use of spatial memory. Animals also use a variety of environmental sources of information to guide spatial orientation. Orientation, however, not only has functional utility; it also has underlying causes, and these causes are found in the nervous system. If function has influenced the evolution of spatial memory, then the same function has influenced the evolution of the nervous system. There is thus a direct link from ecological selection pressures on spatial orientation to spatial-memory processes to the neural implementation of spatial memory. If so, differences among animals' brains and nervous systems as consequences of adaptation to different ecological conditions would be expected. This idea is widely accepted for sensory processes. Evidence that animal cognition and its neural basis are ecological adaptations has only more recently become available. This evidence is comparative. In animals, similarities and differences among brain areas with spatial memory functions have been found that correspond to similarities and differences in uses of space.

7.4.1 The Hippocampus

As I mentioned (section 7.1), much research on the neurobiology of spatial memory has focused, in one way or another, on the hippocampus. In this section, I will describe briefly why the hippocampus has held center stage for so long, and then present some comparative research. In most mammals, the hippocampus is a curved C-shaped structure that surrounds the thalamus. In primates (including humans), the curve of the hippocampus descends into the dorsomedial aspect of the temporal lobe. In birds, the hippocampus extends along the dorsomedial surface of the forebrain for about one-third of its length. The hippocampus was named by the anatomist Giulio Cesare Aranzi (1530–1589), who saw a resemblance to the sea horse *Hippocampus* in the curving shape of the human structure (Lewis 1923) (fig. 7.6).

Birds and mammals have evolved independently for at least 310 million years, when the synapsid clade of reptiles, which gave rise to mammals, diverged from the diapsid clade, which gave rise to dinosaurs and thereafter

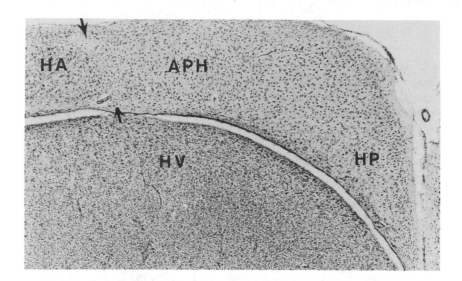

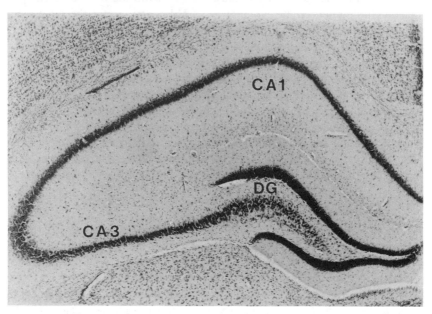

Figure 7.6. The hippocampuses of a white-breasted nuthatch (*top*) and a deer mouse (*bottom*). Both photomicrographs show coronal sections of the left hippocampus. The dorsal edge of the nuthatch hippocampus lies at the dorsal surface of the brain, and the deer mouse hippocampus is overlain by cortex. Abbreviations for nuthatch: APH = area parahippocampalis, HA = hyperstriatum accessorium, HP = hippocampus, HV = hyperstriatum ventrale. Arrows show the lateral boundary of the hippocampal region. Abbreviations for rat: CA1 and CA3 = subdivisions of the hippocampus proper, or Ammon's horn, DG = dentate gyrus. CA cell fields, the dentate gyrus, and associated structures such as the subiculum and entorhinal cortex make up the hippocampal formation. Deer mouse section courtesy of Tara Perrot-Sinal.

birds (Fastovsky and Weishampel 1996). Little anatomical or functional similarity between the hippocampuses of birds and mammals is expected, and as figure 7.6 shows, there is no gross anatomical similarity at all. Nevertheless, by embryological (Källén, 1962) and anatomical criteria (Casini et al. 1986; Erichsen et al. 1991; Krayniak and Siegel 1978; Szekely and Krebs 1996), the mammalian and avian hippocampuses appear to be evolutionary homologues; they are derived from the same medial forebrain structure in their most recent common ancestor.

O'Keefe and Nadel's (1978) theory of the function of the hippocampus is straightforward, and this idea has been rearticulated by Nadel (1991). The theory states that (1) the hippocampus is one of several brain structures that function in spatial memory; and (2) the primary function of the hippocampus is spatial memory (not spatial orientation or spatial behavior). The theory was derived with two key pieces of evidence: (1) lesions in the hippocampus produced spatial disorientation in rats; and (2) the rat hippocampus contained "place" cells, or cells that showed maximum firing rates when the animal was in a particular place (O'Keefe and Dostrovsky 1971). The theory states that the hippocampus deals with the input, storage, and retrieval of information from all sensory modalities that enable identification of places. The part of the theory that has been largely abandoned, even by its supporters, is the claim that the hippocampus is involved in long-term storage of memory. Various experiments have shown that the disruptive effects of hippocampal lesions are diminished if enough time is allowed to pass between learning and lesioning (Squire et al. 1984).

The principal debate about the function of the hippocampus concerns whether its characterization as a spatial-memory module is sufficiently broad to describe how the hippocampus actually functions. A variety of theories describe broader functions of the hippocampus. These proposed functions include declarative memory (Squire 1987), working memory (Olton et al. 1979), temporary memory storage (Rawlins 1985), and configural association (Sutherland and Rudy 1989). Lesions in the hippocampus disrupt the learning of relations among odor cues (Eichenbaum et al. 1988), and most of the deficits produced with hippocampal damage in primates and humans have no clear consequence on spatial functioning (Cohen and Eichenbaum 1993). A broader theory may indeed be required to characterize the function of the hippocampus. Species differences—in particular, the difference between primates and other mammals—are frequently ignored in theory construction but may mark important evolutionary divides in hippocampal function. The spatial function of the hippocampus may be more important in some taxa than others. Place cells and the more recently discovered head-direction cells (see the following) are located outside the hippocampus, and lesions in other brain regions can produce spatial deficits. In addition, researchers are focusing increasingly on the function of the neighboring cortex in the processing of hippocampal input and output (Nagahara et al. 1995). Whatever the final resolution of the hippocampal debate is, important correlations remain between spatial behavior and activity of single brain cells, between localized brain lesions and disruptions of spatial orientation, and between species differences in the uses of space and relative sizes of the hippocampus. These three topics are discussed in the following.

7.4.2 Head-Direction Cells

One of the most dramatic discoveries concerning the neural representation of space has been the identification and description of head-direction cells. First described by Ranck (1985), head-direction cells exhibit maximum firing rates when the head of the animal points in a particular horizontal direction. Notably, these cells lie outside the hippocampus in the postsubiculum, anterior thalamic nucleus, lateral dorsal thalamic nucleus, retrosplenial cortex, and striatum (Taube et al. 1996). Recordings of the electrical activity of head-direction cells in freely moving rats show that these cells fire maximally over a relatively narrow range of head directions (usually about 90°). The firing pattern is stable from day to day and is independent of the animal's behavior, its position in space, and roll or pitch attitude of the head (providing the head is pointed in the appropriate horizontal direction).

How do head-direction cells "know" where the head is pointing? Head-direction cells in the postsubiculum and anterior thalamic nucleus are highly sensitive to the location of landmarks in the rat's environment. Goodridge and Taube (1995) placed rats in a cylindrical enclosure with a single large landmark (a white card filling 100° of arc). The head direction that resulted in the maximum firing rate of head-direction cells was determined, and the card was removed. Some head-direction cells fired maximally with the same head direction in the absence of the card; other cells fired with the head rotated as much as 180° from its original orientation. When the card was returned, however, most of the cells that fired with a new head direction returned immediately to maximum firing with the head in the original orientation relative to the card. Of those cells that had fired with the original head orientation in the absence of the card, about half could be induced to fire with a new head orientation by reintroducing the card at a new position rotated 90°. These results showed that some head-direction cells maintain a stable firing rate with one head direction in the absence of landmarks, probably with use of idiothetic input; however, many head-direction cells fire with head directions that are relative to landmarks.

In another series of experiments, Taube and Burton (1995) found that when there was a conflict between landmark and idiothetic information, the rat's head direction that resulted in maximal firing of head-direction cells was determined primarily by landmarks. Idiothetic information could affect the firing rate of cells with head direction, but landmark information, when available, clearly dominated idiothetic information.

Taube and Burton (1995) also confirmed that head-direction cells respond to perceived direction and not merely prominent surrounding cues. The maximal firing rate of head-direction cells in the postsubiculum and anterior thalamic nucleus of the rat was determined in a cylinder with a single prominent landmark as described. The cylinder was connected by a short alley to a rectangular chamber with a single landmark in a different cardinal direction than the landmark in the cylinder. Head-direction cells did not usually fire with a new head direction after the rat left the cylinder, proceeded along the alley, and entered the rectangle (fig. 7.7). A maximal firing rate was established with head direction relative to a conspicuous landmark, and this firing rate with head direction was maintained in new surroundings with a new landmark.

7.4.3 Lesions of the Hippocampus

Lesions in the mammalian and avian hippocampus disrupt spatial orientation. Jarrard (1983) trained rats in the radial-arm maze and found that hippocampal lesions disrupted function of both spatial and working memory. Sherry and Vaccarino (1989) found that lesions in the hippocampus of black-capped chickadees did not disrupt food storing or the amount of searching for food caches; lesions instead reduced the accuracy of search to that predicted by chance. In a subsequent experiment, we found that performance of a nonstoring task that demanded memory for familiar spatial locations was disrupted by hippocampal lesions, while performance of a task that demanded memory for simple associations was not. Recent work by Hampton and Shettleworth (1996) showed the same kind of effect of hippocampal lesions with use of a very different task. Black-capped chickadees' memory for the location of stimuli on a touch screen was disrupted with hippocampal lesions, but memory for the colors of stimuli was not. These results show that lesions of the hippocampus do not disrupt all memory in black-capped chickadees; lesions disrupt memory of one kind and not others. In the terminology of neuropsychology, hippocampal lesions produce a dissociation in memory. Memory for spatial locations, whether places in a room or places on a touch screen, is impaired in birds with hippocampal lesions. Memory for other kinds of information, whether a cue that indicates food or a recently seen patch of color on a video monitor, is not.

Hippocampal lesions have effects in another natural setting: homing and orientation by pigeons. Homing pigeons show a complex pattern of deficits after hippocampal lesions. Experienced adult homing pigeons with hippocampal lesions released from a familiar site showed no deficit in homeward orien-

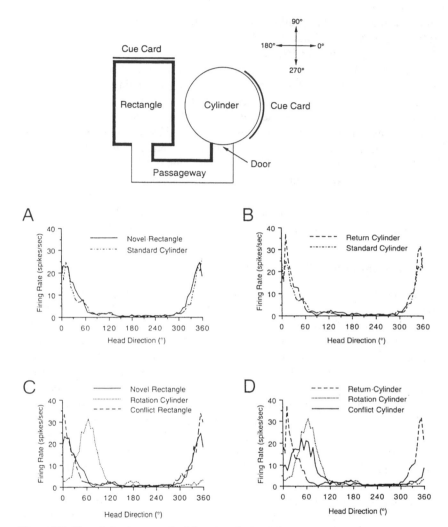

Figure 7.7. The relation between the firing rate of a head-direction cell and rat behavior are determined in the apparatus diagrammed (*top*). (*a*) The firing pattern that was established in the cylinder (standard cylinder) persisted when the rat traveled along the passageway and entered the rectangle for the first time (novel rectangle). (*b*) The cell maintained its firing pattern when the rat returned to the cylinder (return cylinder). (*c*) Rotation of the cylinder cue card by 90° counterclockwise caused an approximate 60° counterclockwise rotation in the head direction associated with maximal firing rate (rotation cylinder). The cell immediately fired maximally with the original head direction during a subsequent visit to the rectangle (conflict rectangle). (*d*) The cell fired maximally with the original head orientation when the rat returned to the cylinder (conflict cylinder). Data from novel-rectangle and return-cylinder trials are reproduced in *c* and *d* for comparison (from Taube and Burton 1995).

tation compared to that of controls (Bingman et al. 1984). Nevertheless, the pigeons with hippocampal lesions failed to return home. Such birds even failed to return home when released within sight of the loft. Thus, whatever mechanism these birds used to orient correctly at the release site (the possibilities include the sun compass, magnetic information, and release-site landmarks), this mechanism was not dependent on hippocampal function. What seems dependent on hippocampal function is the use of familiar landmarks near the home loft because lesioned pigeons manifested a deficit at this stage of homing. Other experiments, however, showed that homing from distant release sites is dependent on hippocampal function but in a different sense. An intact hippocampus is necessary for development of the ability to home from a distant release site although this ability, once acquired, becomes independent of hippocampal function (Bingman 1993).

7.4.4 Comparative Studies of the Hippocampus

The foregoing results show that the hippocampus and other parts of the brain function in memory of spatial locations. Does the brain, then, show adaptive specialization for spatial memory?

7.4.4.1 Food Storing

The hippocampus varies strikingly in size between species and, in some species, between the sexes. Much of this variation can be readily correlated with differences in the use of space between species and the sexes. The hippocampus of food-storing birds is as much as twice the size of that of birds with comparable brain and body sizes that do not store food (Sherry et al. 1989; Krebs et al. 1989). Within the food-storing parid and corvid families, there is variation in hippocampal size that correlates with the amount of food storing typically performed by different species (Hampton et al. 1995; Healy and Krebs 1992; Basil et al. 1996). Mexican chickadees (*Parus sclateri*) and bridled titmice (*P. wollweberi*), for example, both store food, although in controlled laboratory conditions, these birds store less than black-capped chickadees. In these birds, relative hippocampal size correlates with intensity of food storing (Hampton et al. 1995). Similar correlations are observed in European and North American food-storing corvids (Healy and Krebs 1992; Basil et al. 1996).

Comparable correlations occur among strains of pigeons (Rehkämper et al. 1988). Homing pigeons have a proportionally larger hippocampus than nonhoming strains. There seems to be no relation, however, between migration in passerines and relative hippocampal size (Sherry et al. 1989; Healy and Krebs 1991). This observation and that of Bingman, who described the effects of hippocampal lesions on homeward orientation in pigeons, suggest that the hippocampus may serve relatively little function in long-distance orientation of pigeons and passerines. Perhaps this is so because long-distance orientation is achieved with use of global-reference systems such as the earth's magnetic field, the sun compass, or other celestial cues. Bingman et al. (1984) did find that in pigeons the hippocampus functions in the use of familiar local cues near home. If the same is true in passerine migrants, the hippocampus may function in home recognition by birds that show year-to-year nest-site fidelity (Ketterson and Nolan 1990). Healy et al. (1994) found that age and migratory experience are related to hippocampal size in the garden warbler (*Sylvia borin*). Older birds have greater hippocampal volumes than younger birds, and birds with migratory experience have greater hippocampal volumes than inexperienced birds. Whether these age- and experience-dependent changes in hippocampal size indicate hippocampal involvement in the use of local landmarks, development of global-reference systems, or both, remains to be determined.

In mammals, food storing also influences relative hippocampal size (Jacobs and Spencer 1994). Merriam's kangaroo rats (*Dipodomys merriami*) are scatter hoarders, and the bannertail kangaroo rat (*D. spectabilis*) maintains and defends a larder. Relative hippocampal size is greater in Merriam's kangaroo rats than that in bannertail kangaroo rats perhaps because Merriam's kangaroo rats rely more on spatial memory to retrieve scattered food caches.

7.4.4.2 Mating Systems

Food storing and homing are not the only ecological factors that influence relative hippocampal size. Sex differences in hippocampal size have been discovered that are clearly related to sex differences in the use of space. Among the *Dipodomys* kangaroo rats described and the *Microtus* voles, there are species with monogamous mating systems and species with polygynous mating systems. I described (section 7.1) the polygynous mating system of meadow voles and the monogamous mating system of pine voles and their effects on spatial memory. In addition to having larger home ranges and better spatial abilities than females, male meadow voles have a larger hippocampus than females (fig. 7.8). No sex difference occurs in hippocampal size in monogamous pine voles, and the sexes do not differ in home-range size in the wild or spatial ability in the laboratory (Jacobs et al. 1990). Research on *Dipodomys* kangaroo rats confirmed the effect of mating system on hippocampal size. Merriam's and bannertail kangaroo rats mate polygynously, and during the breeding season, the home ranges of males are considerably larger than those of females. The hippocampus of males is larger than that of females in both species (Jacobs and Spencer 1994).

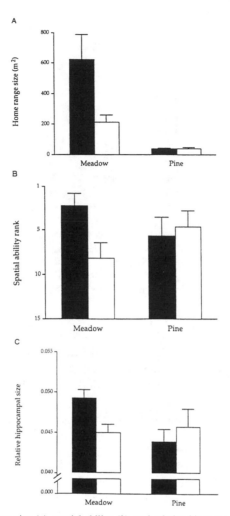

Figure 7.8. Home range size (*a*), spatial ability (*b*), and relative hippocampal size (*c*) in polygynous meadow voles and monogamous pine voles. Filled bars = males, open bars = females. (*a*) Home ranges were determined by telemetry for nine to 12 individuals of each sex in each species. Home-range size was greater in males than in females in meadow voles (*t* test, $P < .02$, $n = 21$) but not in pine voles (*t* test, $P < .66$, $n = 19$). (*b*) Rank order of maze performance for eight to 13 individuals of each sex in each species. A low rank indicates superior spatial ability. Note reversal of the y-axis scale. Male meadow voles performed better than female meadow voles (Mann-Whitney *U* test, $P < .02$, $n = 20$), but there was no difference between the sexes in pine voles (Mann-Whitney *U*-test, $P > .05$, $n = 21$). (*c*) The volume of the hippocampus is shown as a proportion of total brain volume in samples of 10 males and 10 females in each species. A significant interaction occurred between species and sex ($F_{1,38} = 4.61$, $P < .05$) (from Sherry et al. 1992).

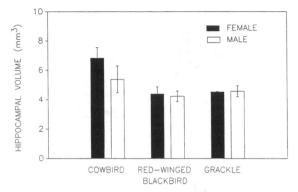

Figure 7.9. Hippocampal volume in male and female brown-headed cowbirds, red-winged blackbirds, and common grackles. Analysis of covariance, with telencephalon size as a covariate, indicated a significant sex difference in parasitic cowbirds ($F_{1,8}$ = 9.18, P < .02, n = 12) but not red-winged blackbirds ($F_{1,13}$ = 0.65, P < .5, n = 16). The sample size of common grackles (n = 4) was too small for statistical analysis (from Sherry et al. 1993).

7.4.4.3 Brood Parasitism

Cowbirds are brood parasites. In the best known species, the brown-headed cowbird (*Molothrus ater*), females lay about 40 eggs in the nests of various host species during an 8-week breeding period. Females lay at or before dawn, and spend the remainder of the morning searching for potential host nests in which to lay eggs on subsequent days. Female cowbirds have been reported to fly into shrubs and flush incubating birds from their nests, sit silently and watch nest building by potential hosts, and walk on the forest floor while scanning the canopy (Norman and Robertson 1975). Although not confirmed experimentally, it is likely that female cowbirds learn the locations of potential host nests and return to them after an interval of a day or two to lay eggs. Male cowbirds do not participate in the search for host nests in this species. The hippocampus of female cowbirds is significantly larger than that of males; this sex difference does not occur in closely related, nonparasitic icterine blackbirds (fig. 7.9) (Sherry et al. 1993).

Not all *Molothrus* cowbirds are parasitic. The shiny cowbird (*M. bonariensis*) is a generalist parasite like the brown-headed cowbird. The screaming cowbird (*M. rufoaxillaris*) is a specialist that uses one host, the bay-winged cowbird (*M. badius*). The bay-winged cowbird is not a nest parasite but a communal breeder with biparental care and helpers at the nest (Fraga 1991). Female shiny cowbirds search for host nests unassisted by males; however, male and female screaming cowbirds both have been reported to inspect the nests of their bay-winged cowbird hosts (Mason 1987). Reboreda et al. (1996) found that the hippocampus in the two parasitic species was larger (relative to the size of the telencephalon) than that in the nonparasite. In addition, a sex difference in relative hippocampal size that favored females occurred in shiny cowbirds but not in the other two species. This sex difference suggests that search for host nests by shiny cowbird females exerts an effect on relative hippocampal size.

Although all of these results indicate that search for host nests by female cowbirds has led to the evolution of greater hippocampal size, there are some contradictory data: The *obscurus* and *artemisiae* subspecies of the brown-headed cowbird exhibit no sex difference in relative hippocampal size (Uyehara and Narins 1992). Hippocampal size in *Molothrus ater obscurus* and *M. ater artemisiae* may be subject to selective pressures or constraints that do not influence hippocampal size in *Molothrus ater ater* or *M. bonariensis*.

7.5 CONCLUSIONS

Animals use spatial memory in a diverse set of ecological circumstances. Indeed, it is difficult to imagine many activities of mobile organisms that do not involve processing, directly or incidentally, some information about spatial location. The study of spatial memory is currently undergoing a sort of renaissance as a result of increased contact between field-based researchers concerned with navigation and orientation and laboratory-based researchers concerned with animal cognition. Some of the theoretical dividing lines that formerly demarcated mechanisms of long-distance navigation from those of local systems, such as landmark use, are vanishing. Future research may further integrate theoretical ideas. In addition, neurophysiological research has uncovered a variety of substrates in which cognitive mechanisms of spatial orientation may be implemented. Comparative work on one of these neurophysiological substrates, the hippocampus, has shown remarkable variation in neural structure that correlates with ecological variation in the use of space.

7.6 SUMMARY

Memory for spatial locations is an important component of behavior in a wide variety of ecological settings, including territoriality, mate search, foraging, and food storing. Animals defend territories, locate mates, find food, and retrieve stored food in part by remembering spatial locations. Various kinds of information can be used to identify and remember locations in space. The use of landmarks and the sun compass are two examples. Neural specializations that reflect the ecological importance of spatial memory can be found at the neuronal level (e.g. place cells, head-direction cells, and such structures as the avian and mammalian hippocampuses). Comparative analyses illustrate the relation between ecological selection pressures and evolutionary modification of spatial memory and its neural basis.

290   David F. Sherry

Acknowledgments

I would like to thank Andrew Check, Perri Eason, Bruce Falls, Peter McGregor, Bill Roberts, and Jeffrey Taube for much helpful discussion and information on spatial behavior and spatial memory. Thanks, too, to Reuven Dukas for his careful and patient editing and Nicky Clayton for her many helpful comments on the manuscript. Preparation of this paper was supported by the Natural Sciences and Engineering Research Council of Canada.

Literature Cited

Balda, R. P., and A. C. Kamil. 1992. Long-term spatial memory in Clark's nutcracker *Nucifraga columbiana. Animal Behaviour* 44:761–769.
Balda, R. P., and W. Wiltschko. 1991. Caching and recovery in scrub jays: Transfer of sun-compass directions from shaded to sunny areas. *Condor* 93:1020–1023.
Basil, J. A., A. C. Kamil, R. P. Balda, and K. V. Fite. 1996. Differences in hippocampal volume among food storing corvids. *Brain Behavior and Evolution* 47:156–164.
Bennett, A. T. D. 1993. Spatial memory in a food storing corvid. I. Near tall landmarks are primarily used. *Journal of Comparative Physiology,* ser. A, *Sensory, Neural, and Behavioral Physiology* 173:193–207.
———. 1996. Do animals have cognitive maps? *Journal of Experimental Biology* 199:219–224.
Berthold, P., A. J. Helbig, G. Mohr, and U. Querner. 1992. Rapid microevolution of migratory behaviour in a wild bird species. *Nature* 360:668–670.
Bingman, V. P. 1993. Vision, cognition, and the avian hippocampus. In H. P. Zeigler and H.-J. Bischof, eds., *Vision, Brain, and Behavior in Birds,* 391–408. Cambridge: MIT Press.
Bingman, V. P., P. Bagnoli, P. Ioalè, and G. Casini. 1984. Homing behavior of pigeons after telencephalic ablations. *Brain Behavior and Evolution* 24:94–108.
Bossema, I. 1979. Jays and oaks: An eco-ethological study of a symbiosis. *Behaviour* 70:1–117.
Brodbeck, D. R. 1994. Memory for spatial and local cues: A comparison of a storing and a nonstoring species. *Animal Learning and Behavior* 22:119–133.
Brodin, A., and J. Ekman. 1994. Benefits of food hoarding. *Nature* 372:510.
Casini, G., V. P. Bingman, and P. Bagnoli. 1986. Connections of the pigeon dorsomedial forebrain studied with WGA-HRP and ^{3}H-proline. *Journal of Comparative Neurology* 245:454–470.
Cheng, K. 1986. A purely geometric module in the rat's spatial representation. *Cognition* 23:149–178.
Cohen, N. J., and H. Eichenbaum. 1993. *Memory, Amnesia, and the Hippocampal System.* Cambridge: MIT Press.
Collett, T. S., B. A. Cartwright, and B. A. Smith. 1986. Landmark learning and visuospatial memories in gerbils. *Journal of Comparative Physiology,* ser. A, *Sensory, Neural, and Behavioral Physiology* 158:835–851.
Cook, R. G., M. F. Brown, and D. A. Riley. 1985. Flexible memory processing by

rats: Use of prospective and retrospective information in the radial maze. *Journal of Experimental Psychology: Animal Behavior Processes* 11:453–469.

Cowie, R. J., J. R. Krebs, and D. F. Sherry. 1981. Food storing by marsh tits. *Animal Behaviour* 29:1252–1259.

Darley-Hill, S., and W. C. Johnson. 1981. Acorn dispersal by the blue jay (*Cyanocitta cristata*). *Oecologia* 50:231–232.

Davies, N. B. 1991. Mating systems. In J. R. Krebs and N. B. Davies, eds., *Behavioural Ecology: An Evolutionary Approach,* 3d ed., 263–294. Oxford: Blackwell Scientific.

Davies, N. B., and A. I. Houston. 1981. Owners and satellites: The economics of territory defence in the pied wagtail, *Motacilla alba. Journal of Animal Ecology* 50: 157–180.

―――. 1984. Territory economics. In J. R. Krebs and N. B. Davies, eds., *Behavioural Ecology: An Evolutionary Approach,* 2d ed., 148–169. Sunderland Mass: Sinauer.

Duff, S. J., L. A. Brownlie, D. F. Sherry, and M. Sangster. In press. Sun compass orientation by black-capped chickadees (*Parus atricapillus*). *Journal of Experimental Psychology: Animal Behavior Processes.*

Dyer, F. C., and J. A. Dickinson. 1994. Development of sun compensation by honeybees: How partially experienced bees estimate the sun's course. *Proceedings of the National Academy of Sciences of the United States of America* 91:4471–4474.

Eason, P. K., and G. A. Cobbs. 1996. The effect of landmarks on territorial behavior. Animal Behavior Society, Flagstaff AZ. August 1996. Abstract 70.

Eichenbaum, H., A. Fagan, P. Mathews, and N. J. Cohen. 1988. Hippocampal system dysfunction and odor discrimination learning in rats: Impairment or facilitation depending on representational demands. *Behavioral Neuroscience* 102:331–339.

Emlen, S. T. 1970. Celestial rotation: Its importance in the development of migratory orientation. *Science* 170:1198–1201.

Erichsen, J. T., V. P. Bingman, and J. R. Krebs. 1991. The distribution of neuropeptides in the dorsomedial telencephalon of the pigeon (*Columba livia*): A basis for regional subdivisions. *Journal of Comparative Neurology* 314:478–492.

Etinne, A. S., R. Maurer, and V. Seguinot. 1996. Path integration in mammals and its interaction with visual landmarks. *Journal of Experimental Biology* 199:201–209.

Falls, J. B. 1982. Individual recognition by sounds in birds. In D. E. Kroodsma and E. H. Miller, eds., *Acoustic Communication in Birds,* Vol. 2, 237–278. New York: Academic.

Falls, J. B., and J. G. Kopachena. 1994. White-throated sparrow (*Zonotrichia albicollis*). In A. Poole and F. Gill, eds, *The Birds of North America* No. 128. Philadelphia: Academy of Natural Sciences.

Fastovsky, D. E., and D. B. Weishampel. 1996. *The Evolution and Extinction of the Dinosaurs.* Cambridge: Cambridge University Press.

Fraga, R. M. 1991. The social system of a communal breeder, the bay-winged cowbird *Molothrus badius. Ethology* 89:195–210.

Galea, L. A. M., M. Kavaliers, and K.-P. Ossenkopp. 1996. Sexually dimorphic spatial learning in meadow voles *Microtus pennsylvanicus* and deer mice *Peromyscus maniculatus. Journal of Experimental Biology* 199:195–200.

Gallistel, C. R., and A. E. Cramer. 1996. Computations on metric maps in mammals:

Getting oriented and choosing a multi-destination route. *Journal of Experimental Biology* 199:211–217.

Gaulin, S. J. C., and R. W. FitzGerald. 1986. Sex differences in spatial ability: An evolutionary hypothesis and test. *American Naturalist* 127:74–88.

Gaulin, S. J. C., and R. W. FitzGerald. 1989. Sexual selection for spatial-learning ability. *Animal Behaviour* 37:322–331.

Gaulin, S. J. C., R. W. Fitzgerald, and M. S. Wartell. 1990. Sex differences in spatial ability and activity in two vole species (*Microtus ochrogaster* and *M. pennsylvanicus*). *Journal of Comparative Psychology* 104:88–93.

Gaulin, S. J. C., and M. S. Wartell. 1990. Effects of experience and motivation on symmetrical-maze performance in the prairie vole (*Microtus ochrogaster*). *Journal of Comparative Psychology* 104:183–189.

Gibb, J. A. 1960. Populations of tits and goldcrests and their food supply in pine plantations. *Ibis* 102:163–208.

Godard, R. 1991. Long-term memory of individual neighbours in a migratory songbird. *Nature* 350:228–229.

Goodridge, J. P., and J. S. Taube. 1995. Preferential use of the landmark navigation system by head direction cells in rats. *Behavioral Neuroscience* 109:49–61.

Haftorn, S. 1974. Storage of surplus food by the Boreal Chickadee *Parus hudsonicus* in Alaska, with some records on the Mountain Chickadee *Parus gambeli* in Colorado. *Ornis Scandinavica* 5:145–161.

Hampton, R. R., D. F. Sherry, S. J. Shettleworth, M. Khurgel, and G. Ivy. 1995. Hippocampal volume and food-storing behavior are related in Parids. *Brain Behavior and Evolution* 45:54–61.

Hampton, R. R., and S. J. Shettleworth. 1996. Hippocampal lesions impair memory for location but not color in passerine birds. *Behavioral Neuroscience* 110:831–835.

Healy, S. D., E. Gwinner, and J. R. Krebs. 1994. Hippocampus size in migrating garden warblers: Effects of age and experience (abstract). *Journal für Ornithologie* 135:74.

Healy, S. D., and T. A. Hurly. 1995. Spatial memory in rufous hummingbirds (*Selasphorus rufus*): A field test. *Animal Learning and Behavior* 23:63–68.

Healy, S. D., and J. R. Krebs. 1991. Hippocampal volume and migration in passerine birds. *Naturwissenschaften* 78:424–426.

———. 1992. Food storing and the hippocampus in corvids: Amount and volume are correlated. *Proceedings of the Royal Society London*, ser. B 248:241–245.

Helbig, A. J. 1991. Experimental and analytical techniques used in bird orientation research. In P. Berthold, ed., *Orientation in Birds*, 270–306. Basel: Birkhäuser-Verlag.

———. 1996. Genetic basis, mode of inheritance, and evolutionary changes of migratory directions in palaearctic warblers (Aves: Sylviidae). *Journal of Experimental Biology* 199:49–55.

Helbig, A. J., P. Berthold, and W. Wiltschko. 1989. Migratory orientation of blackcaps (*Sylvia atricapilla*): Population-specific shifts of direction during the autumn. *Ethology* 82:307–315.

Herz, R. S., L. Zanette, and D. F. Sherry. 1994. Spatial cues for cache retrieval by black-capped chickadees. *Animal Behaviour* 48:343–351.

Hitchcock, C. L., and D. F. Sherry. 1990. Long-term memory for cache sites in the black-capped chickadee. *Animal Behaviour* 40:701–712.

Hurly, T. A., and S. D. Healy. 1996. Memory for flowers in rufous hummingbids: Location or local visual cues? *Animal Behaviour* 51:1149–1157.

Jacobs, L. F. 1992. Memory for cache locations in Merriam's kangaroo rats. *Animal Behaviour* 43:585–593.

Jacobs, L. F., and E. R. Liman. 1991. Grey squirrels remember the locations of buried nuts. *Animal Behaviour* 41:103–110.

Jacobs, L. F., and W. D. Spencer. 1994. Natural space-use patterns and hippocampal size in kangaroo rats. *Brain Behavior and Evolution* 44:125–132.

Jacobs, L. F., S. J. C. Gaulin, D. F. Sherry, and G. E. Hoffman. 1990. Evolution of spatial cognition: Sex-specific patterns of spatial behavior predict hippocampal size. *Proceedings of the National Academy of Sciences of the United States of America* 87:6349–6352.

Jarrard, L. E. 1983. Selective hippocampal lesions and behavior: Effects of kainic acid lesions on performance of place and cue tasks. *Behavioral Neuroscience* 97:873–889.

Källén, B. 1962. Embryogenesis of brain nuclei in the chick telecephalon. II. *Ergebnisse der Anatomie und Entwicklungsgeschichte* 36:62–82.

Källander, H. 1978. Hoarding in the rook Corvus frugilegus. *Anser* (suppl.) 3:124–128.

Kamil, A. C., and R. P. Balda. 1985. Cache recovery and spatial memory in Clark's nutcracker (*Nucifraga columbiana*). *Journal of Experimental Psychology: Animal Behavior Processes* 11:95–111.

———. 1990. Spatial memory in seed-caching corvids. In G. H. Bower, ed., *The Psychology of Learning and Motivation,* Vol. 26, 1–25. San Diego: Academic.

Kesner, R. P., and M. J. DeSpain. 1988. Correspondence between rats and humans in the utilization of retrospective and prospective codes. *Animal Learning and Behavior* 16:299–302.

Ketterson, E. D., and V. Nolan, Jr. 1990. Site attachment and site fidelity in migratory birds: Experimental evidence from the field and analogies from neurobiology. In E. Gwinner, ed., *Bird Migration,* 117–129. Berlin: Springer-Verlag.

Kraemer, P. J., M. E. Gelbert, and N. K. Innis. 1983. The influence of cue types and configuration upon radial-arm maze performance in the rat. *Animal Learning and Behavior* 11:373–380.

Kramer, G. 1951. Eine neue Methode zur Erforschung der Zugorientierung und die bisher damit erzielten Ergebnisse. In S. Hörstadius, ed., *Proceedings of the Tenth International Ornithological Congress, Uppsala,* 269–280. Uppsala: Almqvist & Wicksell.

Krayniak, P. F., and A. Siegel. 1978. Efferent connections of the hippocampus and adjacent regions in the pigeon. *Brain Behavior and Evolution* 15:372–388.

Krebs, J. R. 1978. Optimal foraging: Decision rules for predators. In J. R. Krebs and N. B. Davies, eds., *Behavioural Ecology: An Evolutionary Approach,* 23–63. Oxford: Blackwell Scientic.

Krebs, J. R., D. F. Sherry, S. D. Healy, V. H. Perry, and A. L. Vaccarino. 1989. Hippocampal specialization of food-storing birds. *Proceedings of the National Academy of Sciences of the United States of America* 86:1388–1392.

Lewis, F. T. 1923. The significance of the term *hippocampus*. *Journal of Comparative Neurology* 35:213–230.

Mason, P. 1987. Pair formation in cowbirds: Evidence found for screaming but not shiny cowbirds. *Condor* 89:349–356.

Mazmanian, D. S., and W. A. Roberts. 1983. Spatial memory in rats under restricted viewing conditions. *Learning and Motivation* 14:123–139.

McGregor, P. K., and G. W. M. Westby. 1992. Discrimination of individually characteristic electric organ discharges by a weakly electric fish. *Animal Behavior* 43:977–986.

Morris, R. G. M. 1981. Spatial localization does not require the presence of local cues. *Learning and Motivation* 12:239–260.

Nadel, L. 1991. The hippocampus and space revisited. *Hippocampus* 1:221–229.

Nagahara, A. H., T. Otto, and M. Gallagher. 1995. Entorhinal-perirhinal lesions impair performance of rats on two versions of place learning in the Morris water maze. *Behavioral Neuroscience* 109:3–9.

Norman, R. F., and R. J. Robertson. 1975. Nest-searching behavior in the brown-headed cowbird. *Auk* 92:610–611.

O'Keefe, J., and J. Dostrovsky. 1971. The hippocampus as a spatial map: Preliminary evidence from unit activity in the freely-moving rat. *Brain Research* 34:171–175.

O'Keefe, J., and L. Nadel. 1978. *The Hippocampus as a Cognitive Map.* Oxford: Clarendon.

Olton, D. S., J. T. Becker, and G. E. Handelmann. 1979. Hippocampus, space, and memory. *Behavioral and Brain Science* 2:313–365.

Olton, D. S., and C. Collison. 1979. Intramaze cues and odor trails fail to direct choice behavior on an elevated maze. *Animal Learning & Behavior* 7:221–223.

Petersen, K., and D. F. Sherry. 1996. No difference occurs in hippocampus, food-storing, or memory for food caches in black-capped chickadees. *Behavioral Brain Research* 79:15–22.

Ranck, J. B., Jr. 1985. Head direction cells in the deep cell layer of dorsal presubiculum in freely moving rats. In G. Buzsáki and C. H. Vanderwolf, eds., *Electrical Activity of the Archicortex,* 217–220. Budapest: Akadémiai Kiadó.

Rawlins, J. N. P. 1985. Associations across time: The hippocampus as a temporary memory store. *Behavioral and Brain Sciences* 8:479–496.

Rehkämper, G., E. Haase, and H. D. Frahm. 1988. Allometric comparison of brain weight and brain structure volumes in different breeds of the domestic pigeon, *Columba livia f. d.* (fantails, homing pigeons, strassers). *Brain Behavior and Evolution* 31:141–149.

Reboreda, J. C., N. S. Clayton, and A. Kacelnik. 1996. Species and sex differences in hippocampus size in parasitic and non-parasitic cowbirds. *Neuroreport* 7:505–508.

Schmidt-Koenig, K. 1990. The sun compass. *Experientia* 46:336–342.

Schwagmeyer, P. L. 1994. Competitive mate searching in thirteen-lined ground squirrels (Mammalia, Sciuridae): Potential roles of spatial memory. *Ethology* 98:265–276.

Sherry, D. F. 1984. Food storage by black-capped chickadees: Memory for the location and contents of caches. *Animal Behaviour* 32:451–464.

———. 1985. Food storage by birds and mammals. *Advances in the Study of Behaviour* 15:153–188.

Sherry, D. F., and S. J. Duff. 1996. Behavioral and neural bases of orientation in food-storing birds. *Journal of Experimental Biology* 199:165–172.

Sherry, D. F., M. R. L. Forbes, M. Khurgel, G. O. Ivy. 1993. Females have a larger hippocampus than males in the brood-parasitic brown-headed cowbird. *Proceedings of the National Academy of Sciences of the United States of America* 90: 7839–7843.

Sherry, D. F., L. F. Jacobs, and S. J. C. Gaulin. 1992. Spatial memory and adaptive specialization of the hippocampus. *Trends in Neurosciences* 15:298–303.

Sherry, D. F., J. R. Krebs, and R. J. Cowie. 1981. Memory for the location of stored food in marsh tits. *Animal Behaviour* 29:1260–1266.

Sherry, D. F., and A. L. Vaccarino. 1989. Hippocampus and memory for food caches in black-capped chickadees. *Behavioral Neuroscience* 103:308–318.

Sherry, D. F., A. L. Vaccarino, K. Buckenham, and R. S. Herz. 1989. The hippocampal complex of food-storing birds. *Brain Behavior and Evolution* 34:308–317.

Shettleworth, S. J. 1990. Spatial memory in food-storing birds. *Philosophical Transactions of the Royal Society of London,* ser. B, 329:143–151.

Shettleworth, S. J., and J. R. Krebs. 1982. How marsh tits find their hoards: The roles of site preference and spatial memory. *Journal of Experimental Psychology: Animal Behaviour Processes* 8:354–375.

Squire, L. R. 1987. *Memory and Brain.* New York: Oxford University Press.

Squire, L. R., N. J. Cohen, and L. Nadel. 1984. The medial temporal region and memory consolidation: A new hypothesis. In H. Weingartner and E. S. Parker, eds., *Memory Consolidation,* 185–210. Hillsdale, N.J.: Erlbaum.

Spetch, M. L., and W. K. Honig. 1988. Characteristics of pigeons' spatial working memory in an open-field task. *Animal Learning and Behavior* 16:123–131.

Stephens, D. W., and J. R. Krebs. 1986. *Foraging Theory.* Princeton, N.J.: Princeton University Press.

Stevens, T. A., and J. R. Krebs. 1986. Retrieval of stored seeds by marsh tits *Parus palustris* in the field. *Ibis* 128:513–525.

Sutherland, R. J., and J. W. Rudy. 1989. Configural association theory: The role of the hippocampal formation in learning, memory, and amnesia. *Psychobiology* 17: 129–144.

Suzuki, S., G. Augerinos, and A. H. Black. 1980. Stimulus control of spatial behavior on the eight-arm maze in rats. *Learning and Motivation* 11:1–18.

Székely, A. D., and J. R. Krebs. 1996. Efferent connectivity of the hippocampal formation of the zebra finch (*Taenopygia guttata*): An anterograde pathway tracing study using *Phaseolus vulgaris* leucoagglutinin. *Journal of Comparative Neurology* 368: 198–214.

Taube, J. S., J. P. Goodridge, E. J. Golob, P. A. Dudchenko, and R. W. Stackman. 1996. Processing the head-direction cell signal: A review and commentary. *Brain Research Bulletin* 40:477–484.

Taube, J. S., and H. L. Burton. 1995. Head-direction cell activity monitored in a novel environment and during a cue conflict situation. *Journal of Neurophysiology* 74: 1953–1971.

Tinbergen, N. 1932. Über die Orientierung des Bienenwolfes (*Philanthus triangulum* Fabr.). *Zeitscrhift für vergleichende Physiologie* 16:305–334.

Tinbergen, N., and W. Kruyt. 1938. Über die Orientierung des Bienenwolfes (*Philanthus triangulum* Fabr.) III. Die Bevorzugung bestimmter Wegmarken. *Zeitscrhift für vergleichende Physiologie* 25:292–334.

Tomback, D. F., and Y. B. Linhart. 1990. The evolution of bird-dispersed pines. *Evolutionary Ecology* 4:185–219.

Uyehara, J. C., and P. M. Narins. 1992. Sexual dimorphism in cowbird brains and bodies: Where does it end? *Proceedings of the Third International Congress of Neuroethology, Montreal* Abstract No. 146.

Vander Wall, S. B. 1982. An experimental analysis of cache recovery in Clark's nutcracker. *Animal Behaviour* 30:84–94.

———. 1990. *Food Storing in Animals*. Chicago: University of Chicago Press.

Weygoldt, P. 1980. Complex brood care and reproductive behavior in captive poison-arrow frogs, *Dendrobates pumilio* O. Schmidt. *Behavioral Ecology and Sociobiology* 7:329–332.

Williams, C. L., A. M. Barnett, and W. H. Meck. 1990. Organizational effects of early gonadal secretions on sexual differentiation in spatial memory. *Journal of Neuroscience* 104:84–97.

Wiltschko, W., and R. P. Balda. 1989. Sun compass orientation in seed-caching scrub jays (*Aphelocoma coerulescens*). *Journal of Comparative Physiology,* ser. A, *Sensory, Neural, and Behavioral Physiology* 164:717–721.

Wiltschko, W., and R. Wiltschko. 1978. A theoretical model for migratory orientation and homing in birds. *Oikos* 30:177–187.

———. 1993. Navigation in birds and other animals. *Journal of Navigation* 46:174–191.

Zoladek, L., and W. A. Roberts. 1978. The sensory basis of spatial memory in the rat. *Animal Learning and Behavior* 6:77–81.

CHAPTER EIGHT

Risk-Sensitive Foraging: Decision Making in Variable Environments

MELISSA BATESON AND ALEX KACELNIK

8.1 INTRODUCTION

Analysis of foraging has greatly benefited from the integration of evolutionary, ecological, and cognitive research. In this chapter, we focus on risk sensitivity, the area of foraging theory that concerns how and why animals respond to variability in food sources. Our intention is to show how optimality modeling and cognitive research can and need to operate in unison if we are to understand risk sensitivity. For many years, data collection and theoretical developments regarding risk sensitivity have progressed in parallel in the behavioral ecology and psychology literatures. The lack of interchange between these two disciplines reflects a difference in the types of questions asked and the types of explanations sought. In behavioral ecology, research has centered on theoretical predictions that under certain ecological conditions, natural selection should favor foragers that are sensitive to environmental variance. Conversely, research in psychology has been driven by animals' observed responses to variance and, more recently, by hypotheses about the information-processing mechanisms that underlie foraging. We demonstrate the necessity of both approaches and advocate a program of research that simultaneously considers evolutionary ecology and cognitive mechanisms.

8.1.1 What Is Risk Sensitivity?

We begin with three examples that introduce the phenomenon of risk sensitivity. In the first experiment, bumblebees (*Bombus edwardsi*) were allowed to forage on an array of two types of artificial flowers. One color of flower always contained a constant volume of 0.1 µl of nectar, and the other color of flower contained either 1 µl of nectar (10% of flowers) or no nectar (90% of flowers). Once the bees had experienced both flower types and presumably learned something about their different properties, the bees showed a strong preference for the flowers that contained the constant volume of nectar (Waddington et al. 1981). In the second experiment, thirsty pigeons (*Columba livia*) were given repeated choices between two keys. The keys were arranged such that pecks on one key led to water delivery after a fixed 15-second delay, and pecks on the other key led to water delivery after a variable delay averaging 15 seconds. After training, the birds preferred water delivery with the variable delay (Case et al. 1995). In the third experiment, yellow-eyed juncos (*Junco phaeonotus*) were given repeated choices between two feeding stations, one of which offered four millet seeds on every visit, and the other either one or seven seeds with equal probability. The birds' preferences depended on the ambient temperature: At 1°C, they preferred the variable option; however, at 19°C, they preferred the constant option (Caraco et al. 1990).

In all three experiments, the animals faced two options that offered equal average rates of gain. The animals' preferences appeared to be influenced by the variance in the rate of gain that differed between the two options. In all cases, the variable option was programmed such that it was impossible for the animals to know exactly the gain that would result from choosing this option; with the variable option, the animals could learn only the probability distribution of possible outcomes. In the foraging literature, this type of environmental variance is referred to as *risk,* and an animal with preferences that are affected by the variance in the rate of gain is called *risk sensitive.* When two options offer the same long-term rate of gain, animals that prefer the less variable option (e.g. Waddington's bumblebees) are *risk averse,* and animals that prefer the more variable option (e.g. the pigeons of Case et al.) are *risk prone.*

Thus in foraging theory, risk has a very specific meaning. Risk should not be confused with the problem faced by an animal that has incomplete information and knows neither the mean nor variance of the foraging options it faces; this problem is often referred to as uncertainty. Also, risk should not be confused with risk of predation, which some authors prefer to call danger.

8.1.2 Patterns of Risk Sensitivity

In a recent review (Kacelnik and Bateson 1996), we collated the results of 59 experimental studies of risk sensitivity in 28 animal species including insects, fish, birds, and mammals. In most of these studies, we found evidence for sensitivity to risk, and a number of consistent patterns emerged. A brief review of our findings follows.

Most of the variation in the direction of risk-sensitive preferences is explained by the component of rate of gain that is programmed to be risky. Because rate of gain is a function of the amount of food obtained and the time taken to obtain it, variation in either of these components can generate risk in the rate of gain. When risk is generated with variance in amount, as in the bee example described, animals are usually risk averse (fig. 8.1*a*). By contrast, when risk is generated with variance in the time associated with getting food or water, as in the pigeon example described, animals are almost universally risk prone (see fig. 8.1*a*).

In some species (for example juncos), the direction of preference appears to be dependent on the relationship between the energy needs of the animal and the average rate of energy gain available from the foraging options (the *energy budget*). Usually, if an animal's rate of gain exceeds its needs (i.e. the animal is on a *positive energy budget*), then risk aversion occurs. However, if the rate of gain is insufficient for the animal to meet its needs (i.e. the animal is on a *negative energy budget*), then risk proneness occurs (fig. 8.1*b*). This effect of the energy budget seems to be restricted to experimental situations in which risk is generated with variance in food amount. As yet, there is virtually no evidence that preference for variable time delays is affected by the energy budget; however, this lack of effect has not been thoroughly tested. There is also some evidence that a response to the energy budget may be dependent on the typical body weight of the species; lighter species appear more likely to respond to changes in the budget than heavier species (fig. 8.1*c*).

In this chapter, we will present a critical review of the main theories that have been proposed to account for risk sensitivity. We will concentrate on identifying which of the above findings the different theories can and cannot explain. However, before we proceed with these theories, we need to explain why risk sensitivity challenges classical optimal foraging theory, which until recently has provided the dominant framework for the modeling of animal decision making.

8.1.3 Classical Optimal Foraging Theory Predicts Indifference to Risk

In Charnov's (1976a, 1976b) two seminal optimality models, he assumed that foraging animals make decisions that maximize their rate of energy gain. In these models, rate of energy gain is computed as the ratio of the expected energy obtained from food to the expected time spent foraging. This form of rate is also known as *long-term rate,* or the *ratio of expectations* (Bateson and Kacelnik 1996). Thus,

$$\text{long-term rate} = \frac{\text{E\{energy obtained from food\}}}{\text{E\{time spent foraging\}}}, \qquad (8.1)$$

where E{ } designates an expectation or average. The choice of long-term rate is justified on the grounds that it makes sense to start with the assumption that natural selection should favor animals that maximize this currency. The

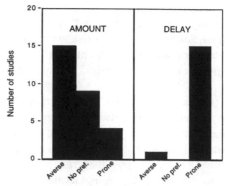

A

B

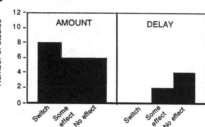

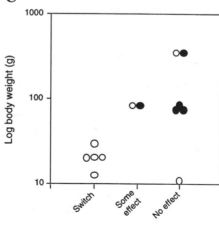

C

rationale is that both the survival and reproduction components of fitness have an obvious relationship with the long-term rate of energy gain. The more energy gained, the greater the probability of survival and the greater amount of energy that can be put into reproduction. Also, the less time spent feeding, the more time there is for other activities such as predator avoidance and reproduction. Therefore, animals may be expected to maximize their long-term rate of energy gain because, by doing so, they obtain the most energy per unit of time that is devoted to foraging (see Ydenberg, this vol. section 9.2, for further discussion of currencies).

As equation 8.1 shows, only expectations or averages enter the computation of long-term rate. Therefore, in most risk-sensitivity experiments in which the average amount and time are the same in the two options, long-term rate should be identical and foraging decisions should be unaffected by risk. However, the results summarized in figure 8.1 clearly show that considering only the average values of these variables is insufficient because risk has considerable effects on preference in most studies. Thus, maximizing long-term rate cannot be all that is involved in decision making. In the following section, we will introduce Jensen's (1906) inequality, a mathematical result that helps us understand why risk affects foraging decisions.

8.1.4 Jensen's Inequality

Let us start by imagining a function $y = f(x)$. Although it is mathematically irrelevant what x and y are, we will assume that y represents fitness and x represents the amount of food an animal gains from a foraging option. Thus, for example, the function allows us to determine how many units of fitness (y) a junco will obtain from four millet seeds (x). In the case of a risky foraging option in which there are two or more possible values of x (either one or seven seeds in the case of the juncos), we need to calculate the average fitness value of the option; however, there are two different ways to calculate the value. We can compute the average value of x (with use of the same notation described, $E\{x\}$) and use this value to obtain a single value of y (written $E\{y\} =$

Figure 8.1. The responses of animals to risk. Data are from 28 species of insects, fish, birds, and mammals. (a) The differential effects of variability in amount and delay on risk-sensitive preferences. (b) The effects of energy-budget manipulations on risk-sensitive preferences. "Switch" indicates that subjects switched from being risk averse on positive budgets to risk prone on negative budgets. "Some effect" indicates there was a similar, but smaller shift in preference as that of the switch group. "No effect" indicates that manipulations of energy budget did not change the direction or degree of preference. (c) The response of bird species to energy-budget manipulations versus body weight. Empty circles indicate studies examining risk in amount, and the filled circles indicate studies examining risk in delay (replotted from Kacelnik and Bateson 1996).

$f(E\{x\})$). Or we can compute the y value obtained with each value of x and take the average of the resulting values of y (written $E\{y\} = E\{f(x)\}$). In other words, we can either average the values of x before applying the function or average the values of y obtained after applying the function to each value of x.

If $f(x)$ is a linear function, then averaging may be performed at either stage because the outcomes will be equal: $E\{f(x)\} = f(E\{x\})$. However, if $f(x)$ is nonlinear, then the stage at which averaging occurs matters because *Jensen's inequality* applies, and $E\{f(x)\} \neq f(E\{x\})$. Specifically, if $y = f(x)$ is increasing and decelerating, then calculating $E\{x\}$ first and applying the function to the result produces a higher value than calculating the average of the y values obtained after applying the function (i.e. $E\{f(x)\} < f(E\{x\})$). By contrast, if $y = f(x)$ is increasing and accelerating, the opposite is true, and $E\{f(x)\} > f(E\{x\})$.

If we make the assumption that fitness is the ultimate currency of all decisions, then the above results produce the following consequences for animals that must choose between fixed and variable foraging options with equal long-term rates of gain. In cases in which the fitness function is linear, risk sensitivity is not predicted because both options offer equal fitness. However, if the fitness function is nonlinear and animals average the fitness they obtain after applying the function, then risk sensitivity will occur, and the direction will depend on the shape of the function. If $y = f(x)$ is increasing and decelerating, then risk aversion is predicted; if $y = f(x)$ is increasing and accelerating, then risk proneness is predicted (fig. 8.2).

Thus, risk sensitivity is predicted if two conditions are met. First, the function that relates the amount of food gained to fitness must be nonlinear. We will refer to this condition as the *nonlinearity condition*. Second, averaging of outcomes from variable options must be performed with the fitness values obtained after the fitness function has been applied. We will refer to this condition as the *averaging condition*. Classical optimal-foraging models do not predict risk sensitivity because average long-term rate of gain is computed by averaging outcomes before the fitness function is applied (i.e. the averaging condition is not met).

Most existing models of risk sensitivity rely on Jensen's inequality. The models differ, however, in a number of respects that can obscure this underlying link. A major difference concerns the identity of the function $y = f(x)$. We introduced Jensen's inequality with use of the function to relate energy gain to fitness. However, alternative explanations of risk sensitivity focus on different functions. The alternative explanations also differ in the biological justifications they provide for the nonlinearity and averaging conditions. We divide the explanations into two types. The explanations presented in sections 8.2–8.4 are called *functional* because they assume that risk sensitivity has

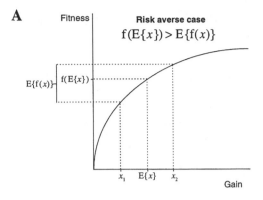

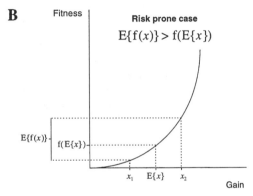

Figure 8.2. Graphics demonstrating Jensen's inequality. (*a*) f(*x*), increasing and decelerating, leads to risk-averse behavior. (*b*) f(*x*), increasing and accelerating, leads to risk-prone behavior.

evolved because of its fitness consequences. By contrast, the explanations in sections 8.5 and 8.6 are called *mechanistic* because they explain risk sensitivity as arising from basic psychological processes such as associative learning and perception. As Tinbergen (1963) made clear, these are logically distinct kinds of explanations, both of which are needed to fully understand behavior. Every behavioral phenomenon has a functional and mechanistic explanation; therefore, one type of explanation does not substitute for the other. Thus, we expect functional and mechanistic models of the same phenomenon to lead to identical predictions about behavior. We shall see, however, that differences in the background and motivation of behavioral ecologists and psychologists—who are responsible for the functional and mechanistic models, respectively—result in models that address different aspects of risk sensitivity and make very different predictions about behavior.

8.2 Functional Explanation 1: Long-Term Rate Maximization

In section 8.1.3, we argued that risk sensitivity is not explained by classical optimal-foraging models that assume maximization of long-term rate. However, this is only the case if the long-term rate of gain is equal in the fixed and variable options of a risk-sensitive foraging experiment. In this section, we suggest that there are a number of reasons why the long-term rate of gain experienced by a forager may differ from that intended by the experimenter.

In most risk-sensitivity studies, a tacit assumption is made that there is a linear relationship between the characteristics of the foraging option programmed by the experimenter and what is experienced by the forager. Thus for example, it is assumed that if one millet seed results in a net energy intake of 1 calorie, then seven seeds will provide 7 calories. However, the programmed amount of food may not have a linear relationship to the amount ingested by the animal, and the programmed time delay before food is obtained may not have a linear relationship to the delay experienced by the animal. It is also possible that the gross amount of food ingested by the animal may not have a linear relationship to the net energy derived from it. Because it is impossible for an animal to average quantities not yet experienced, the averaging condition is bound to be met and Jensen's inequality will apply if any of the functions we described is nonlinear. We, therefore, have a potentially trivial explanation for some cases of risk sensitivity.

The different techniques for generating risk can affect the shape of the function that relates what is programmed to what is experienced by the forager. Of the studies that generate risk by varying the amount of reward, most manipulate either the number of similarly sized food items, the duration of access to a food hopper, or the volume or concentration of nectar delivered. Although it is usually claimed that the average amounts obtained from constant and variable options are equal, evidence is rarely presented. In many cases, it is possible that animals may not consume the entire reward available or may not consume rewards of different sizes at the same rate. In cases in which amount is controlled by the time of access to food, the linear relationship assumed between programmed access time and the amount of food taken is usually not verified. It has been demonstrated that the amount of grain consumed by pigeons feeding from a hopper is an increasing, but decelerating function of the time of hopper access (Epstein 1981) (fig. 8.3). Jensen's inequality predicts that given an increasing, decelerating function, pigeons should be risk averse if they are offered a choice between two foraging options with the same average time of hopper access but with different variances in the length of access. In order to eliminate this possible source of risk-sensitivity, linearity between programmed and experienced reward needs to be proved in each study. For

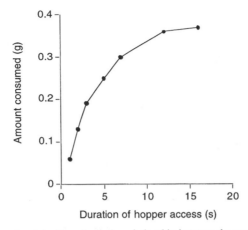

Figure 8.3. An example of nonlinearity in the relationship between the programmed duration of hopper access and the amount of grain consumed by pigeons. The nonlinearity is probably produced by two factors: the maximum amount of food the bird can retrieve from the hopper during one episode of reinforcement and the decreasing accessibility of grain as the pigeon eats (replotted from Epstein 1981). This graph shows that if a pigeon is given a choice between a fixed reinforcement of a 10-second duration versus a risky reinforcement equally likely to be 5 or 15 seconds in duration the bird maximizes its long-term rate of food intake by choosing the fixed option.

example, Reboreda and Kacelnik (1991) reported risk sensitivity in starlings (*Sturnus vulgaris*) despite the establishment of a linear relationship between time of hopper access and grams of food consumed in this species.

Experimenters that explicitly claim to vary the amount of food almost always also vary the time taken to acquire the food; large rewards usually take longer to deliver, handle, and consume than small rewards. This source of variability is usually ignored because it will not cause differences in the long-term rate of gain that is available from a fixed and variable option if the function that relates the reward size to handling time is linear. However, if there is a nonlinear relationship between the reward size and handling time, then differences in long-term rates can result because of Jensen's inequality. For example, Possingham et al. (1990) argued that a bumblebee presented with a choice of a flower type that offers a fixed 3 µl of nectar versus a variable flower type that offers 1 or 5 µl should be risk prone if the bee is maximizing its long-term rate of energy gain. This prediction is based on the observation that in bumblebees there is an increasing, but decelerating relationship between the volume of nectar taken from a flower and the time spent on the flower (fig. 8.4) (Hodges and Wolfe 1981). The effect of this relationship is that—although, on average, the volume of nectar taken from the fixed and variable flowers will be identical—the average time taken to get the nectar

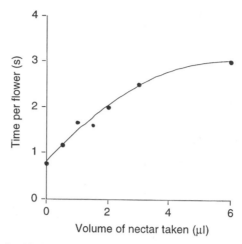

Figure 8.4. The relationship between volume of nectar taken and handling time per flower in the bumblebee, *Bombus appositus* (replotted form Hodges and Wolfe 1981). Bees extract larger volumes of nectar proportionately more quickly; initially, more of the hairy tongue comes into contact with the nectar, and lapping is more efficient (Harder 1986).

from the variable flowers will be less than that taken with the fixed flowers. Hence, the variable flowers will provide a higher long-term rate of nectar intake than the fixed flowers. This effect of handling time ought to be important in animals such as bees because handling takes up a substantial portion of total foraging time. However, the argument presented does not help us understand risk sensitivity in bees because the reported cases show risk aversion (Real 1981; Waddington et al. 1981; Real et al. 1982) and not risk proneness (we will return to the problem of why bees are sometimes risk averse in section 8.5). Despite this specific problem, the general point remains that nonlinearities in handling time can generate differences in the long-term rate of intake that is provided by fixed and variable food sources (which ostensibly should provide the same rate).

As with food amount, the delay to obtain food can also be controlled in a number of ways. Most studies use schedules of reinforcement, in which reward is programmed on the basis of either the number of responses performed (*ratio schedules*), the amount of time elapsed (*time schedules*), or a combination of these (*interval schedules*). With time schedules, the timing of reward is independent of the behavior of the animal; however, with interval or ratio schedules, reward is contingent on the animal completing the required responses. Thus, a programmed interval or ratio need not have a linear relationship to the actual delay experienced by the animal. For example, Ha et al. (1990) used fixed and variable ratio schedules to examine risk sensitivity in grey jays (*Persoreus canadensis*) and reported risk proneness. However, to assess the source of risk sensitivity in this experiment it must be established if the rate of responding is constant over the range of programmed ratios. If, for instance, animals worked faster when the ratio was higher, then the average time taken to complete a variable ratio would be less than the time taken to complete a fixed ratio with the same average value. Thus, animals that appear to be risk sensitive with respect to the programmed ratios may simply be rate maximizers when experienced times are taken into account.

The examples we have discussed all relate to nonlinearities in the functions that relate the parameters programmed by the experimenter to those experienced by the forager. Fewer data exist on the shape of the function that relates the gross rate of energy gain to the net rate of energy gain because obtaining this information requires measurements of energy expenditure. However, it is theoretically feasible that this function may also be nonlinear. For instance, in the case of honeybees that collect nectar by hovering from one flower to the next, the cost of staying aloft per unit time increases with the load already collected; successive equal volumes of nectar collected have different net values to the bee (Schmid-Hempel et al. 1985; see also Ydenberg, this vol. section 9.2).

The explanations for risk sensitivity discussed in this section may seem uninteresting because they do not require anything beside long-term rate maximizing; thus, these explanations are consistent with the classical optimal-foraging theory. In many experimental studies, however, risk is programmed without consideration for the shape of the functions that relate what is programmed to what is experienced, and nonlinearity in these functions must be acknowledged as a candidate explanation for some reports of risk sensitivity.

From the ecological perspective that examines how resource characteristics affect the exploitation of the resource by foragers, the cause of risk sensitivity is irrelevant. What matters is that foragers may treat food sources that provide the same average quantity of food differentially depending on the variance in food quantity. This differential treatment could have consequences for the evolution of food species. For example, in the only risk-sensitive foraging study that used a natural food source, Cartar (1990) showed that bumblebees' choices between two flower species were influenced by variance differences in the rate of gain associated with the species. Such behavior should place selective pressures on the distribution of nectar between flowers because a plant can theoretically attract more pollinators with simply a change in this distribution.

8.3 FUNCTIONAL EXPLANATION 2: RISK-SENSITIVE FORAGING THEORY

Risk-sensitive foraging theory is the main, functionally based framework developed to explain why animals are sensitive to risky food sources. This theory focuses on the function that relates the energy gained, as the result of a foraging decision, to fitness and provides biological justifications for both the nonlinearity and the averaging conditions.

8.3.1 The Energy-Budget Rule

The simplest and best known risk-sensitive foraging model (Stephens 1981) concerns a small endotherm, such as a bird, that has to attain a minimum reserve threshold by dusk to survive the night. This threshold provides the necessary nonlinear relationship between rate of gain and fitness. If the animal has more than the threshold level of reserves, it survives until the next day; however, if the animal has less, it dies and has a fitness of zero. Consider a bird faced with two foraging options that differ only in the variance of the expected number of seeds available per reward (for example the junco experiment described in section 8.1.1). If the fixed option offers a rate of gain that is sufficiently high to take the bird above the threshold (i.e. it is on a positive-energy budget), then the bird should be risk averse; however, if the rate is not sufficiently high (i.e. the bird is on a negative energy budget), then the bird's only chance of survival is to be risk prone and gamble on the variable option that gives an above average rate of gain (fig. 8.5). These options have been summarized in the *daily energy-budget rule,* which states that a forager on a positive budget should be risk averse and a forager on a negative budget should be risk prone (Stephens 1981). Note that "budget" is not equivalent to the animal's absolute level of reserves; budget refers to the relationship between an animal's needs and its potential rates of gain. A hungry animal close to starvation could be on a positive budget if it is currently facing a rich food source. By a similar argument, a well-fed animal could be on a negative budget. If everything else is assumed to be constant, however, a subject with low reserves has higher needs and is therefore more likely to be on a negative budget than a counterpart with high reserves.

Many investigators have sought to test the energy-budget rule directly by experimentally manipulating the energy budget of the subjects. Relatively few studies, however, have produced convincing support for the theory. In our recent review, we examined 24 studies that manipulated energy budget: In 18 of these studies, risk sensitivity was investigated when amount was variable. Time was variable in five studies, and both amount and time were variable in one study. Among the experiments with variable amount, 14 showed a shift toward risk proneness when energy budgets were reduced and risk aversion when energy budgets were increased; although, only eight of these studies showed a complete switch in preference between significant risk proneness and significant risk aversion (Caraco et al. 1980; Caraco 1981; Caraco 1983; Barnard and Brown 1985; Moore and Simm 1986; Caraco et al. 1990; Young

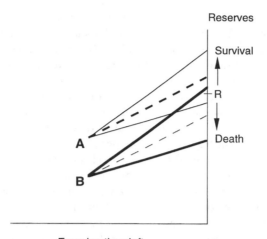

Figure 8.5. Graphic demonstration of the logic that underlies the energy-budget rule. *A* and *B* represent two foragers that differ in their energy reserves. Both have to make one foraging decision before nightfall and must choose between a fixed option (*broken line*) that gives a certain amount of food or a risky option (*solid line*) that gives with equal probability a small or large amount of food. *R* represents the level of reserves needed by dusk if the foragers are to survive the night. The optimal choices for *A* and *B* are different: *A* is on a positive budget and ensures its survival by choosing the fixed option; however, *B* is on a negative budget, and its only chance of survival is provided with the variable option. Optimal choices for each forager are shown in bold.

et al. 1990; Croy and Hughes 1991). Despite this apparent support for the energy-budget rule, we suspect that the real number of failures to obtain the predicted switch in preference with budget manipulations is actually greater; studies that fail to reject the null hypothesis (lack of effect of budget) are harder to publish and probably seldom submitted for publication. Some studies that show failure may be salvaged for publication if a convincing post hoc argument can be made for why a switch in preference was not predicted. A possible example is Barkan's (1990) study of black-capped chickadees (*Parus atricapillus*), which examined two very different rates of intake; the continuing risk aversion of the birds despite apparent budget changes was justified on the basis of careful calculations that showed birds were, in fact, always on positive budgets.

Of the few budget-manipulation experiments with variability in delay, all have failed to demonstrate convincing shifts in preference. Ha and colleagues (Ha et al. 1990; Ha 1991) tried unsuccessfully to induce a preference switch in gray jays; we failed also to show a preference switch in starlings (Bateson and Kacelnik, 1997). Similar failure was met by Case et al. (1995) with use of water as a reward for pigeons and manipulation of water budgets. A study of rats by Zabludoff et al. (1988) showed some evidence for a preference switch; the rats became risk prone as their body weights dropped from 85% to 75% of their free-feeding weights. This study, however, is difficult to interpret because the decrease in body weight is confounded by an increase in the variance of the more variable option; other studies have shown increasing risk proneness with increasing variance in delay to reward. Reboreda and Kacelnik (1991) found that individual starlings that were less efficient at using hopper access time (and as consequence got smaller food rewards) were significantly more risk prone than more efficient birds. This observation agrees with the prediction that needier animals should be more risk prone. However, because the amount of food the birds obtained was not manipulated experimentally, this evidence is only correlational. Thus, there is little direct evidence that energy budget affects risk sensitivity when the delay to receive food is variable. Instead, risk proneness with variable delays seems universal.

Differences in body weights of the various species of birds and mammals used in risk-sensitivity studies may explain the species that respond to budget manipulations and those that do not. Smaller species may be more likely than large species to be subject to selection for short-fall minimization, and thus, smaller species may be more likely to show energy budget-associated switches in risk sensitivity. In support of this prediction, studies of larger species of birds such as pigeons (Hamm and Shettleworth 1987; Case et al. 1995), jays (Ha et al. 1990; Ha 1991), starlings (Bateson and Kacelnik 1997), and rats (Leventhal et al. 1959; Kagel et al. 1986) have failed to find support for the energy-budget rule. In an attempt to reduce the impact of phylogenetic confounds (such as differences in metabolic rate) between birds and mammals, we analyzed the relationship between body weight and the effects of budget on risk sensitivity in birds only. In small birds, we found that changes in budget were more likely to cause appropriate switches in foraging preferences (fig. 8.1c). A closer inspection of the data, however, reveals that all studies with variability in delay have been conducted on larger species such as pigeons, jays, and starlings. Thus, it is not presently clear if the lack of an effect of budget in these studies is due to the component of rate of gain that was varied or the body weights of the experimental species because these two variables are confounded.

It is interesting to speculate if differences in energy budget were responsible for the differences in responses to amount and delay variability observed (fig. 8.1a). When food amount is variable, some of the variation in behavior is potentially attributable to the energy budget. Among the risk-averse animals, some were certainly on positive budgets (Caraco 1982; Stephens and Ydenberg 1982; Caraco and Lima 1985; Tuttle et al. 1990); however in one study, the subjects were probably on negative energy budgets (Wunderle et al. 1987).

Among the risk-prone animals, the shrews of Barnard et al. (1985) were probably on negative energy budgets. In the other three studies (Essock and Rees 1974; Young 1981; Mazur 1985), pigeons were maintained at 80% of their free-feeding weights and could also have been operating on negative budgets. However, the difficulty of explaining the variation is exemplified with the studies on pigeons. Despite the similar procedures employed and the maintenance of all birds at 80 or 85% of their free-feeding weights, two studies showed risk aversion (Menlove et al. 1979; Hamm and Shettleworth 1987) and three showed indifference (Staddon and Innis 1966; Essock and Rees 1974; Hamm and Shettleworth 1987) (in addition to the studies that showed risk proneness mentioned). This variation cannot be explained with any obvious differences in procedure that may have resulted in budget differences.

If delay is variable, animals are almost universally risk prone (the only exception has been shown in the concurrent schedule study of Rider [1983] in which a questionable measure of preference was used) (see Kacelnik and Bateson 1996). Given that most studies were performed by psychologists with pigeons maintained at body weights as low as 75% of free-feeding body weights, it is possible that subjects may have been on negative energy budgets. However, this seems unlikely because pigeons are relatively large birds that can be maintained at 75%–80% of their free-feeding weights for long periods of time. Moreover, given that most of the daily ration is generally received in the experiment, the rate of intake experienced must be sufficient to result in a positive energy budget.

Three studies (Logan 1965; Reboreda and Kacelnik 1991; Bateson and Kacelnik 1995a) directly compared responses to variability in amount and delay under the same conditions of energy budget. All showed that animals (rats in the first study and starlings in the other two) were risk averse when variability was in amount but risk prone when variability was in delay. These results suggest that the difference in response to variability in amount and delay is not attributable to energy-budget differences.

8.3.2 Beyond the Energy-Budget Rule

The energy-budget rule is insufficient to explain the patterns of risk sensitivity reported in the literature. Most experimental tests have focused on the energy-budget rule; however, this rule is not a universal prediction of all risk-sensitive foraging models. Stephens's (1981) original risk-sensitive foraging model inspired the creation of a number of variant models that explore the modification of his various original assumptions. It is not our intention to give an exhaustive review of the theory because an excellent review is available (McNamara and Houston 1992); however, we want to present the current level of sophistication and complexity of the theory's predictions.

One of the most important constraints in Stephens's model is that the forager is allowed to make only a single foraging decision and is then required to stick with this choice for the rest of the day. This constraint led to the criticism that risk proneness would be very rare in nature because it would occur only if the forager's probability of dying was over 50% that day (Krebs et al. 1983). However, if the forager is allowed to make sequential decisions that can vary according to its current state, then risk proneness becomes far less dangerous and therefore more likely. A risk prone forager with a run of good luck that creates a positive trajectory can capitalize on this luck by switching to risk aversion instead of chancing the creation of a negative budget again (Houston and McNamara 1982; McNamara 1983; McNamara 1984). Whether a single- or sequential-choice model is more realistic for a given foraging situation will depend on the degree to which an animal commits itself when it makes a choice. Single-choice models may be more appropriate to large-scale habitat choices, and sequential-choice models will be appropriate to the modeling of prey choices within a habitat. Most experiments on risk sensitivity probably approximate the sequential-choice model more closely due to their small-spatial scale and the large number of choices per session.

A second assumption of Stephens's model is that the only way to die is by failing to meet the critical level of reserves by nightfall. The possibility of starving while foraging is not included, even though for small mammals (such as the shrew) starvation during foraging is a very real danger. If the model is modified so that a forager is assumed to forage continuously to maintain reserves above a lethal level, then it can be shown that the optimal policy is always to be risk averse, if the mean net gain from a foraging choice is positive (Houston and McNamara 1985). Houston and McNamara (1985) combined in a single model the possibility of death (by allowing reserves to fall below a lethal level) during foraging with the need to build reserves to survive the night. In this case, the optimal policy was a compromise between those results in the separate models: A forager is risk averse at all levels of reserves except for a wedge-shaped region in the reserves-versus-time-of-day space near dusk when it is optimal for animals on negative budgets to be risk prone so that the required level of reserves may be achieved (fig. 8.6a)

A third limitation of Stephens's original model (and of all the others mentioned) is that the optimality criterion is restricted to maximization of probability of survival. There may be animals, particularly insects, that have the option of using energy for immediate reproduction. McNamara et al. (1991) modeled the situation in which reproduction occurs above a certain level of reserves and causes a reduction in reserves. They showed that the policy that maximizes lifetime reproductive success is different from one that minimizes mortality. In these reproduction models, risk proneness can occur at high levels of reserves because there are conditions in which a risk-prone decision could take a for-

(a)

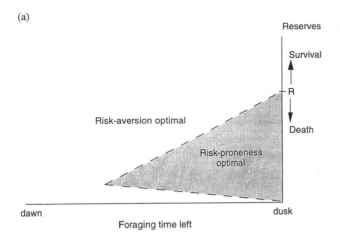

(b)

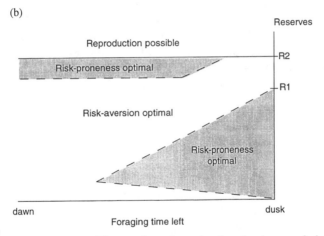

Figure 8.6. Optimal policy. (*a*) Diagram shows the wedge-shaped region near dusk in which it is optimal to be risk prone. *R* is the level of reserves required to survive the night (redrawn from Houston and McNamara 1988). (*b*) Diagram shows the effect of introducing the possibility of reproduction: If reserves are above the *R2*, then reproduction is possible. This second threshold gives rise to risk proneness at high levels of reserves in addition to the wedge-shaped region (see McNamara et al. 1991).

ager over a threshold above which it can reproduce (fig. 8.6*b*). Given certain parameters, it can be shown that as reserves increase, the optimal policy changes successively from risk aversion (to escape an immediate lethal level) to risk proneness (to have a chance of meeting a daily requirement) back to risk aversion (when the requirement can be reached but there is no chance of

314 Melissa Bateson and Alex Kacelnik

reproduction) and finally back to risk proneness again (when this offers the possibility of reproduction).

A final deficiency of the models discussed is that they only consider variance in amount of food; yet, clearly variability in the time associated with obtaining food can also produce risk in the rate of intake. Zabludoff et al. (1988) modeled choice between a foraging option in which there was a fixed delay to food and one in which the delay had the same mean but was variable. In this situation they showed that if the forager must reach a critical level of reserves by nightfall, then the energy-budget rule describes the optimal policy. However, McNamara and Houston (1987) showed that, under some circumstances, the effects of variability in amount and time are not equivalent. Variability in amount and time affect variance in rate of intake; however, unlike variable amounts, variable delays additionally introduce variance in the number of choices that can be made because the length of the delay affects the amount of time remaining.

The message from our brief overview of risk-sensitive foraging theory is that, as Houston and McNamara stressed (Houston 1991; McNamara et al. 1991; McNamara and Houston 1992), there is no single model of risk-sensitive foraging and no single prediction. Notably, the energy-budget rule is not a universal prediction of risk-sensitive foraging theory. This complexity has a number of ramifications for how the risk-sensitive foraging theory can be tested.

8.3.3 Problems with Testing Risk-Sensitive Foraging Theory

Failure to find evidence supporting the energy-budget rule is inconclusive because it is always possible that the wrong model has been tested. Even if the correct model can be chosen for a given situation, there still will be too many unknowns to predict quantitatively how an animal should respond to a given manipulation. For example, the juncos of Caraco et al. (1990) switched from an average of 60% preference for variability in a cold regime to an average of 37% preference for variability in thermoneutral conditions; however, risk-sensitive foraging theory predicts only the direction of the switch, and not the magnitude of the switch.

An additional problem with testing of risk-sensitive foraging theory (and most purely functional models) is that assumptions must be made about the mechanisms an animal uses to assess variables such as the energy budget or rate of intake. Although the many different manipulations tried can theoretically modify energy budgets, modification does not mean that a shift in budget will necessarily be registered by the animal; registration of the shift depends on the proximate mechanism used by the animal to assess its budget. For example, in the natural habitat, ambient temperature may be easy to measure and provide a reliable correlate of energy requirements. An animal, however, that has evolved a rule-of-thumb to change its risk preference with use of ambient temperature may fail to register a change in energy requirements induced by other means, such as a period of restricted access to food. At present, we do not know how animals assess their energy budgets; a failure to demonstrate an effect of an energy-budget shift on preference does not imply that the energy-budget rule will not predict behavior in the wild. A similar argument can be applied to the assessment of variance in rate of gain. These problems are only partially solved with experimentation in the field (e.g. Cartar 1991) because under natural conditions it is very difficult to control and measure the experience of individual foragers and thus a different set of problems are introduced.

From this discussion, it should be clear that testing the ecological validity of risk-sensitive foraging theory will be exceedingly difficult by rejecting predictions of specific models; the rejection of a model can always be attributed to some cause other than the general validity of the theory. Given this difficulty, we must rely on confirmation of predictions of risk-sensitive foraging models. If a theory makes original and counterintuitive predictions that are later confirmed, we gain confidence in the theory. The energy-budget rule emerges as the best candidate for a prediction that is currently unique to risk-sensitive foraging theory. We suggest that a sound empirical demonstration of this rule would provide strong evidence for the theory. The demonstration by Caraco et al. (1990) of the predicted switch in juncos may provide such evidence (Houston and McNamara 1990), but wide replication of their results is needed.

Another specific prediction of risk-sensitive foraging theory is that all risk-sensitive foraging models necessitate an unpredictable environment instead of just a variable environment. A starving junco is predicted to prefer a risky option because the possibility of a run of large seven-seed rewards will create a positive trajectory. A variable, but predictable option in which the junco receives one and seven seeds alternately offers no benefits over a fixed option that always yields four seeds because the total number of seeds obtained is the same. Thus, we can predict that if the observed preferences have evolved for the reasons proposed by risk-sensitive foraging theory and psychological mechanisms tightly match fitness consequences, then animals should be risk sensitive in the context of unpredictable variance but not predictable variance. We recently tested this idea in starlings. We reached the tentative conclusion that the birds appeared not to treat predictable and unpredictable variable delays differently in the manner predicted by risk-sensitive foraging theory (Bateson and Kacelnik, 1997). However, this negative result may be another example of a difficulty already mentioned. The prediction may have failed because predictable variance is uncommon in the natural world; animals may have evolved a general rule to treat all variability as risk regardless of its predictability or unpredictability.

8.4 Functional Explanation 3: Time Discounting

It is possible that not all cases of risk sensitivity will be explained with the same theory, and this may be particularly true for the difference in response to variability in amount and time. We will discuss a completely different functional explanation for why animals may be sensitive to variance in food delay based on the probability of being interrupted while foraging. We will introduce the argument by considering a choice between an immediate reward and another reward of the same magnitude that is delivered later. After this we will investigate the effects of adding variance to the time before a reward is delivered.

8.4.1 Interruptions and Temporal Discounting

Animals are expected to prefer immediate rewards over delayed rewards for a variety of reasons; however, we will focus on the effects of the probability of losing a reward because of an unpredictable interruption, such as the arrival of a predator or a change in weather, during the delay. If a predicted reward has a continuous chance of being lost before consumption, the longer the delay, the greater the cumulative probability that the reward does not arrive. Thus, a delayed reward will have a lower expected value than an immediate one because the value of a delayed reward is its immediate value multiplied by the probability that it materializes. Immediate rewards should also be preferred over delayed rewards because the time saved by accepting an immediate reward can be used to search for the next reward. This consideration, however, only should apply if an animal is making a sequence of foraging decisions and should not apply to an isolated choice. For clarity, we will restrict our discussion to the effects of the probability of interruption on an isolated choice and return to the additional considerations posed with sequential choices in section 8.5.

The value of a delayed reward, v, will be a function of the immediate value of the reward and its delay and can be written $v = f(r, d)$ where r is the value of the reward if it is obtained immediately and d is the delay from the point of choice; $f(r, d)$ is known as the *discounting function,* and there is much theoretical and empirical literature on the shape of this function (e.g. Myerson and Green 1995). If we assume that (1) the reason underlying discounting is probability of loss by interruption; (2) interruptions occur as a Poisson process with rate α (i.e. the chances of the reward being lost are constant per unit of waiting time); and (3) the rate of interruption, α, is constant for rewards of different magnitude, then the probability of getting a reward

d seconds in the future is $\exp(-\alpha d)$. Therefore, if the subject is fully informed of all the parameters of the problem, then the discounting function $f(r, d) = r\exp(-\alpha d)$. This is a linear function of reward amount and a negative exponential (i.e. decreasing and decelerating) function of delay. This prediction has been made independently in the literature on economics (Samuelson 1937) and foraging (Kagel et al. 1986; McNamara and Houston 1987).

We will now examine the effects of introducing inexact knowledge of reward arrival. This situation can be treated as mathematically equivalent to the existence of variability in the expected delay to reward. Consider a fixed option in which a reward of fixed magnitude, r, is promised for delivery after a fixed delay, d_f, and a variable option in which a reward also of fixed magnitude, r, is promised after a delay of either $d_f - \delta$ or $d_f + \delta$ with equal probability. The expected delay in the variable option is $d_v = 0.5 \, [(d_f - \delta) + (d_f + \delta)] = d_f$. The discounted value of the fixed option is $v_f = f(r, d_f)$, but the subject may employ different algorithms to assign value to the variable option. The subject may use either

$$v_{v1} = f(r, d_v) \tag{8.2a}$$

or

$$v_{v2} = 0.5[f(r, d_f - \delta) + f(r, d_f + \delta)]. \tag{8.2b}$$

In equation 8.2a, the subject discounts reward value with use of the expectation of the delay; in equation 8.2b, the subject averages the discounted value of each possible outcome. Thus, we discriminate between the order in which averaging and attributing value are performed. If the averaging of the two possible delays is performed first and the discounting function is applied later, we have equation 8.2a. If the discounting function is applied to each possible delay and the averaging is performed later, we have equation 8.2b. We encountered the same distinction when we introduced Jensen's inequality (section 8.1.4). We are, therefore, familiar with the various outcomes: If the discounting function is linear, then these two equations produce the same result and $v_f = v_{v1} = v_{v2}$. If the discounting function is decreasing and accelerating, then $v_f = v_{v1} > v_{v2}$. If the discounting function is decreasing and decelerating, then $v_f = v_{v1} < v_{v2}$. If the assumptions involved in predicting the discounting function are valid and the function is a negative exponential, then the third case applies and the value computed for the variable option according to equation 8.2b will be higher than that of the fixed option. In the case of a subject facing an isolated choice between a fixed and variable delay option, it is possible to prescribe equation 8.2b, because this equation maximizes the expected fitness consequences of the decision. To illustrate the effect of variability in delay, let $\alpha = 0.1$ and compare a fixed delay of 5 seconds with a variable delay that is either 1 or 9 seconds with equal probability. In the fixed option, the probability of getting the reward is $\exp(-0.5) = 0.606$. The corresponding probability in the variable option is $0.5 [\exp(-0.1) + \exp(-0.9)] = 0.656$. Thus, we predict that animals should be risk prone if faced with an isolated choice between a fixed and variable delay to food when unpredictable interruptions have exerted an important selective pressure on the evolution of decision making.

In the case of an isolated choice between a fixed and variable amount of reward, the discounting function is assumed to be linear with respect to reward amount. Therefore, under the current theory of time discounting, variability in amount per se should have no effect on preference. If, however, r and d are correlated (as is the case if large food items take longer to handle than small ones), then variability in amount can affect preference by means of the variability's effects on delay (Caraco et al. 1992). We will return to this issue in section 8.5.

8.4.2 Time Discounting Is Not Exponential but Hyperbolic

There have been several attempts to explore how humans discount single delayed rewards. In a recent study, Myerson and Green (1995) gave subjects a choice between two notional sums of money: r_1 was delivered immediately, and r_2 was delivered after a delay, d. By systematically varying r_1 for each pair of values of r_2 and d, they found the value of r_1 at which each subject switched preference from one option to the other. This procedure gave estimates of the relative values of the immediate and the delayed rewards from which the shape of the time-discounting function, $f(r, d)$, could be derived. Like other similar studies, this experiment showed that the shape of the discounting function is hyperbolic and not exponential; for a given absolute increase in d, the fraction of value lost decreased as d increased. By contrast, with an exponential function, the proportional reduction in value with an increase in d is independent of the absolute value of d (fig. 8.7). Therefore, the empirically determined discounting function is inconsistent with the function predicted with our assumptions. We suggest that time discounting cannot be explained as a consequence of the probability of loss by interruption unless additional assumptions are invoked.

Nevertheless, the function that most closely fits the available data is

$$v = r/(a + kd), \tag{8.3}$$

where a and k are constants; a is small and k is close to $1s^{-1}$. In equation 8.3, value is also a linear function of reward and a decreasing decelerating function of delay. Therefore, whatever the reasons there are for devaluing delayed rewards, an isolated reward that is predicted to occur after a variable delay ought to be preferred to a reward that is predicted after a fixed delay of its arithmetic

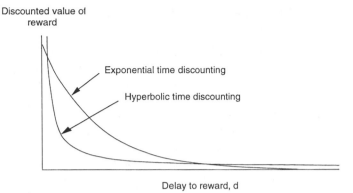

Figure 8.7. Exponential and hyperbolic time discounting functions. Both functions are decreasing and decelerating; however, their shapes differ.

mean. However, we have not been able to find empirical tests of this specific prediction.

8.4.3 Inadequacies of Purely Functional Accounts

All the functional explanations of risk sensitivity we have considered assume that animals can calculate accurate, unbiased estimates of parameters, such as their long-term rate of gain and the associated variance. In optimal-foraging models the tacit assumption is made that if adaptive behavior in animals necessitates accurate estimates of these parameters, then natural selection will provide the animals with the necessary cognitive equipment to obtain these estimates. When explaining subtle details of behavior, however, there are many reasons why a strictly adaptationist approach is insufficient.

In the following sections, we will discuss what is known about how animals process information while foraging. We will show how some of these findings may account for risk sensitivity with the generation of nonlinearities in the function that relates what is actually available to the subjective value the forager assigns to this. The explanations we will present differ from those in previous sections; risk sensitivity will be interpreted as a consequence of basic cognitive mechanisms instead of a direct functional response to external conditions. By promoting the development of mechanistically based models, we do not imply an anti-optimality stance. Behavioral mechanisms must also be the product of natural selection, and we hope that future work may help us understand why these mechanisms—which we regard as constraints in the context of the current discussion—have evolved. We believe natural selection is the main agent of evolutionary change, and we hypothesize that cognitive mechanisms, when viewed in a broader context, can be viewed as adaptations.

8.5 Mechanistic Explanation 1: Short-Term Rate Maximization

Many experiments on nonhuman animals have examined the effects on choice of the length of delay to reward. It is tempting to generalize the ideas presented in the previous section and treat animal experiments as equivalent to the human experiments on time discounting. There are, however, problems with extrapolating from work in humans to that in animals; the preferences of animals can only be tested after training, which involves repeated exposure to the options. Repeated exposure introduces two differences between work in animals and that in humans. First, animals learn the probability of interruptions in the experimental conditions; because interruptions are typically not programmed, the subject can learn they do not occur. Therefore, whereas learned probability of interruption could influence the choices of humans who are instructed verbally to make an isolated choice, learned probability of interruption is less likely to explain animals' choices. Second, subjects learn to experience a sequence of choices; therefore, immediate rewards are more valuable than delayed rewards because the time saved with a short delay can be used to pursue the next reward. Thus, in animal experiments, a delay to reward can be viewed as a loss of foraging opportunity (Stephens and Krebs 1986).

From a functional point of view, an appropriate measure of value given a sequence of choices is long-term rate of gain (equation 8.1), which is the currency of classical optimal-foraging models. Notice the similarities between equations 8.1 and 8.3. The numerator of both functions is a measure of the immediate value of a food reward, and the denominator of both functions includes the time associated with acquiring the reward. Given that time spent foraging in equation 8.1 includes the delay to reward, d, long-term rate can be written as

$$\text{Long-term rate} = \frac{\text{E\{energy obtained from food\}}}{\text{E\{delay to food\} + E\{other time\}}}, \qquad (8.4)$$

where "other time" refers to all periods in the foraging cycle other than the delay between choice and reward, such as intertrial intervals, postreward handling times, and travel times. Note that, like in equation 8.3, long-term rate drops hyperbolically with the delay to food. Long-term rate maximizing is therefore formally identical to hyperbolic time discounting. Animal experiments have demonstrated hyperbolic time-discounting functions (e.g. Mazur 1987; Rodriguez and Logue 1988). Therefore, if animals are offered a sequence of choices between two rewards—one immediate and another delayed—their choices are basically compatible with those in long-term rate maximizing. This fact has often been overlooked; authors have invoked time discounting when, in reality, rate considerations came into play. The two approaches could be combined because interruptions during a series of choices will affect the magnitude of the numerator and denominator in equation 8.4; although, an interruption may cause the loss of a reward, the interruption will release time to search for the next reward.

If variability is introduced and an animal must choose between rewards after fixed and variable delays of the same mean duration, the problem of whether to average first and assign value later or vice versa arises again. Unlike a human that faces an isolated choice, a maximizer of long-term rate should calculate averages first and assign value later as in equation 8.2a; this algorithm correctly characterizes the ratio of expected gain over expected time. As a consequence, this forager would be insensitive to risk as discussed in section 8.1.3. However, many experiments have demonstrated that animals, given sequences of choices, prefer variable over fixed delays with the same means (Herrnstein 1964; Davison 1972; Gibbon et al. 1988; Reboreda and Kacelnik 1991; Bateson and Kacelnik 1995a); the results are compatible with those from animals that assign value first and average later, as depicted with equation 8.2b. *Short-term rate* or the *expectation of the ratios* of amount to time is maximized (Bateson and Kacelnik 1995a, 1996):

$$\text{short-term rate} = \text{E}\left\{\frac{\text{energy obtained from food}}{\text{time spent foraging}}\right\}. \qquad (8.5)$$

Note that the difference between short- and long-term rates is in the position of averaging: for long-term rate, averaging precedes the computation of the ratio of gain over time, whereas for short-term rate, averaging follows the computation of the ratio. "Short" and "long" are often mistakenly interpreted as referring to the "memory window"—the amount of previous experience on which a rate estimate is based.

Evidence for maximizing short-term rate comes from experiments in which animals had to determine when values of fixed- and variable-delay options were equivalent. For example, we performed a titration experiment in which starlings chose between fixed (initially 20 seconds) and variable (2.5 or 60.5 seconds with 50% probability) delays to food. The birds initially preferred the variable option, and we reduced the fixed delay to 5.61 seconds before the birds showed indifference between the two options (fig. 8.8) (Bateson and Kacelnik 1996). At this *indifference point*, short-term rate was approximately equal for the fixed and variable options; the hypothesis that the birds use this currency to value the options was supported.

Further evidence for maximizing short-term rate comes from studies of bumblebees, which are risk averse if there is variability in the volume of nectar in flowers (Real 1981; Waddington et al. 1981). Harder and Real (1987) showed this finding is compatible with bees maximizing short-term rate. In

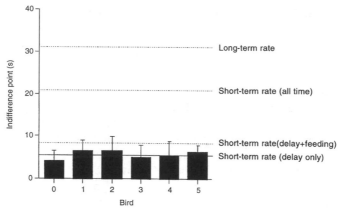

Figure 8.8. Indifference points derived from a titration experiment with six starlings. Bars show the mean (+1 standard deviation) of the value of the fixed delay for which the birds were indifferent between the fixed and variable options. The variable option had two delay times (2.5 and 60.5 seconds) that occurred with equal probability. Lines show the predictions of the six currencies under consideration (only one line appears for long-term rate because the three predictions are identical). The data are significantly different from the predicted data (one sample t tests, $P < .05$) with the exception of expectation of ratios (short-term rate, shown as a solid line) that was calculated with only the delay to reward (from Bateson and Kacelnik 1996).

bees, the correlation between nectar volume, r, and handling time, d, results in the appearance of r in the numerator and denominator of $f(r, d)$; an increasing, but decelerating function that relates the volume of nectar taken to net-intake rate from each decision results (fig. 8.9). This relationship does not lead to risk sensitivity if bees are maximizing long-term rate because volumes are averaged before the computation of rate. However, if bees are maximizing short-term rate, then risk aversion is predicted owing to Jensen's inequality. The magnitude of this effect of variability in reward amount will depend on the proportion of total foraging time that is spent handling the food item. In bees, this time is long; however, in many bird studies, the effects of handling time are likely to be overshadowed by longer delays to reward and intertrial intervals. Thus, variability in reward amount is unlikely to lead to detectable risk aversion even if there is a correlation between reward amount and handling time. In bees, risk aversion may be indicative of maximization of short-term rate given the measured correlation between nectar volume and handling time. In bees, risk aversion is abolished if the amount of reward is controlled with nectar concentration and not nectar volume; this occurs probably because there is no longer a correlation between r and d (fig. 8.9) (see Banschbach and Waddington 1994; Waddington 1995).

In animals, maximizing short-term rate is puzzling because an animal that

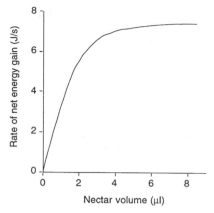

Figure 8.9. Function relating nectar volume to rate of net energy gain derived from the empirically established biomechanics of nectar extraction in bumblebees (replotted from Harder and Real 1987). The nonlinearity results from a correlation between nectar volume and handling time, (although Harder and Real assumed this function to be linear instead of nonlinear as shown in fig. 8.4 for the purposes of deriving this function).

maximizes this currency fails to account correctly for the loss of foraging opportunity and makes choices inconsistent with maximization of long-term rate.

8.5.1 Which Time Intervals Matter?

All time spent foraging including time lost traveling, pursuing, and handling prey items causes loss of foraging opportunity; this is why the denominators in equations 8.1 and 8.4 are the expectations of the sum of all time spent foraging. However, in operant experiments, the delay between the choice of a foraging option and a reward has a much greater impact on the value of an option than similar time intervals after the reward (Snyderman 1987). A reward associated with a short choice-reward delay and a long postreward delay is preferred over a reward associated with two delay lengths that are reversed (such that the choice-reward delay is long and the post-reward delay is short). The strong impact of choice-reward delays is also found in studies in which one of the options has variable delays. For example, in the starling experiment described (Bateson and Kacelnik 1996), there were two additional times in the foraging cycle—the reward time (or handling time) and the intertrial interval. We calculated the indifference points predicted with the six algorithms that result from applying either short- or long-term rates; each calculated with three combinations of time intervals (choice-reward delay only, choice-reward delay plus handling time, and choice-reward delay plus handling time plus the intertrial interval). The indifference point was very close to that predicted with short-term rate maximization calculated without only the intertrial interval (fig. 8.8). This apparently maladaptive behavior can be interpreted as follows: In the wild, animals never experience time delays between obtaining food and being able to make their next foraging decision. A wild starling is free to decide its next action the moment prey consumption is complete. Thus, starlings may fail to attend to the intertrial intervals because these time intervals do not occur in nature.

To those familiar with the foraging literature, it may seem strange that intertrial intervals have little impact on decision making when tests of the Marginal Value Theorem (Charnov 1976b) indicate that travel time between patches affects patch-leaving decisions (e.g. Kacelnik 1984). This apparent anomaly disappears when one understands that travel times are more analogous to choice-reward delays than intertrial intervals; the travel time *follows* the decision to leave a patch. Like choice-reward delay, travel time is between the forager's decision and its next reward (in the case of the Marginal Value Theorem, encounter with the next patch). In the laboratory, analogues of this foraging paradigm have led to the same conclusions as those from work with individual rewards. Kacelnik and Todd (1992) found that patch-residence time is shorter if travel time is variable than if travel time is fixed when average travel time is constant. This result contradicts maximization of long-term rate; however, this result is consistent with the birds paying attention to the intervals between decisions and their consequences. With this observation, we are strongly reminded of the theme of this book: Ecological models are inadequate without reference to the cognitive processes of animals in nature.

In summary, if animals are faced with sequences of choices, they maximize short-term rate. The only time intervals used to compute this rate are those between the choice and reward. If there is no risk in the choices, this currency results in behavior equivalent to long-term rate maximizing. If there is risk, however, maximizing short-term rate can result in risk proneness when variability is in delay or risk aversion when variability is in reward amount (if there is a correlation between the reward amount and the handling time). Identifying short-term rate as the algorithm that best approximates the currency used by animals to attribute value to foraging options does not explain risk sensitivity. We have not yet identified a functional or mechanistic basis for the use of short-term rate. In the remainder of this section (8.5), discussion is devoted to some possible bases.

8.5.2 The Costs of Short-Term Rate Maximizing

The cognitive mechanisms that produce maximization of short-term rate in laboratory studies may produce behavior that maximizes something close to long-term rate in the natural environment. If this is the case, there would be no need to explain why animals use a rate algorithm that results in suboptimal behavior in the laboratory. Maximization of short-term rate was initially rejected by optimal-foraging theorists because animals that used it appeared to make grossly maladaptive foraging decisions (Gilliam et al. 1982; Turelli et al. 1982; Possingham et al. 1990). In most experiments designed to separate the predictions of maximizing short and long-term rates, the discrepancy between the predictions was deliberately accentuated. Discrepancy between predictions is generated by programming very high variance in the variable-delay option. The lower the variance in delay, however, the smaller the difference in behavior predicted with maximizing short- and long-term rates. At the limit, when there is no variance in delay, short- and long-term rates are formally identical. Therefore, to determine the cost of maximizing short-term rate for a forager in its natural environment, we need to know the natural variability in the time it takes the animal to obtain prey items. If there is little variation in the natural environment, then the currency used will make little functional difference; even if the animal is maximizing short-term rate, behavior will be similar to that with maximization of long-term rate. To test this idea, measurements were made of the distribution of interprey intervals experienced by starlings that were foraging in natural pastures (Bateson and Whitehead 1996). These intervals were highly variable. Calculations proved that, given the observed level of variability, a maximizer of short-term rate would suffer a substantially lower daily food intake than a maximizer of long-term rate. Thus, at least in this case, maximizing short-term rate cannot be defended on the grounds that it generates adaptive behavior in the natural environment. We must look elsewhere for an explanation for why animals maximize short-term rate.

8.5.3 Associative Learning and Maximizing Short-Term Rate

In this section, we suggest that maximizing short-term rate can be explained mechanistically by considering the processes animals use to learn about causal relationships in the environment. We argue that the basic mechanisms of associative learning constrain animals' abilities to estimate the rate of food intake they are experiencing, and animals, as a result, behave suboptimally in some circumstances and display risk sensitivity.

In animal studies of risk-sensitivity, the foraging options are usually identified with stimuli such as differently colored flowers or pecking keys. When preferences are tested, the subjects choose among the stimuli instead of the rewards. It is therefore crucial to understand the process subjects use to attribute value to the stimuli, which have no worth before training because they are arbitrarily assigned to the foraging options. The training of subjects in foraging experiments follows a protocol that is analogous to that used in psychological studies of conditioning. In a conditioning experiment, the subject is exposed to a neutral stimulus; after some time delay, the subject experiences a meaningful event (such as receipt of food) that has fitness consequences. The originally neutral stimulus is called the conditioned stimulus, or CS, and the meaningful event is called the unconditioned stimulus, or US. For the animal, the truly important event is the arrival of the US, and most of the changes in the animal's psychological state occur at this time. During training, the value of the CS changes every time an US is experienced. A variety of factors are known to affect the rate of acquisition and the asymptotic level of the value attributed to a particular CS (for a review see Dickinson 1980). We will be concerned with only two of these factors first, the delay between the onset of the CS and delivery of the US and, second, the value of the US.

In the following two sections, we will present models that produce maximization of short-term rate as a result of the conditioning process. The first model uses the effects of the CS–US delay on learning to provide a mechanistic explanation for risk proneness with delay. The second model uses the effects of the US value on learning to provide a mechanistic explanation for risk aversion with amount.

8.5.3.1 Risk Proneness in Birds and Mammals and the CS–US Delay

Since Pavlov's work early this century, it has been known that, usually, rewards presented shortly after the onset of the CS strengthen the association between the CS and US more efficiently than rewards presented after a longer delay. Also, the function that relates the speed of learning to the length of the CS–US delay is nonlinear and appears to be approximately hyperbolic for any given intertrial interval (for data in pigeons see Gibbon et al. 1977). We can explain these findings mechanistically as follows. If a naive subject is exposed to a stimulus (such as a red light [CS]), the stimulus leaves a trace in working memory that decays with time so that the strength of the memory is inversely proportional to the time elapsed since the occurrence of the stimulus. If a meaningful event (the US) then occurs, the subject attributes some value to the previously neutral stimulus it remembers and the magnitude of the change in value is proportional to the strength of the memory of the CS (fig. 8.10).

Compare what happens if a subject is trained with (1) a CS followed by a fixed delay to the US and (2) a different CS followed by a variable delay to the US. In the case of the fixed delay, the subject attributes value in proportion to its memory of the CS every time the US occurs, and this value is always the same. In the case of the variable delay, the value attributed to the CS in each trial depends on the length of the CS–US delay: If the delay is short, the change in value will be large; if the delay is long, the change in value will be smaller. Note that this mechanism of learning imposes value assignment

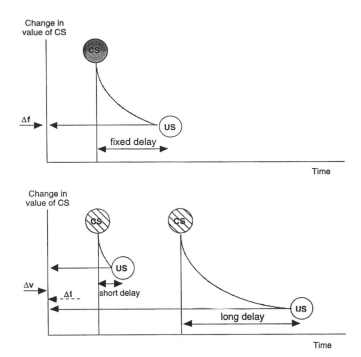

Figure 8.10. Model of how associative learning accounts for risk proneness if variability is in delay. *Upper panel:* the CS–US delay is fixed and results in a fixed change in value of Δf each time the US is experienced. *Lower panel:* the CS–US interval is variable (short or long with equal probability). The average change in the value of CS is Δv. If the function that relates the CS–US interval to the change in value of the CS in nonlinear, then Δv > Δf. The CS associated with the variable delay should be preferred over the CS associated with the fixed delay after similar exposure to the two options, because the CS associated with the variable delay will have a higher value.

before value averaging. If the function that relates the change in value to the CS–US delay is hyperbolic (as the literature suggests), then over a series of trials a CS followed by a fixed delay will acquire less value than a CS followed by a variable delay that has the same average length (another case of Jensen's inequality).

Consider what will occur if the subject, during the course of training, is allowed to choose between a CS associated with a fixed delay and one associated with a variable delay of the same average length. We predict that the subject will prefer the CS associated with the variable delay because this CS has a higher value, which reflects a stronger association with the US. We therefore can explain risk proneness as a product of the way animals learn if variability is in delay to reward. We can also explain why the CS–US delay, which is equivalent to the choice-reward delay, should be particularly important to foraging animals. However, this idea needs further development because we do not know if risk-sensitive preferences that result from training can translate into the stable long-term preferences in the risk-sensitivity literature. Our explanation does not substitute for an optimality analysis; our explanation merely changes the optimality question asked. We are left to explain why a learning mechanism of this type has evolved.

8.5.3.2 Learning and Risk Aversion in Bees

In this section, we describe an artificial neural network that is designed to simulate how bumblebees learn the amount of nectar gained from flowers of different colors. Artificial neural networks are an increasingly popular tool for modeling the processes involved in learning and decision making. These models illustrate how a nervous system, composed of simple units, can produce seemingly complex behavior (Dukas, this vol. section 1.5). The network we describe is of interest because it produces risk aversion if there is variability in nectar volume. Risk sensitivity occurs as a result of how the volume of nectar obtained from a flower (i.e. the value of the US) affects the value attributed by the bee to flowers of that color (i.e. the CS).

In bumblebees, there is an increasing, decelerating function that relates the volume of nectar in a flower to the net rate of energy gain derived from the nectar (fig. 8.9). Montague et al. (1995) incorporated this finding into a simple neural network model. They assumed that the output, $r(t)$, of a reward neuron, R, is equal to the net energy derived from a flower (figure 8.11). Thus, there is an increasing, decelerating function that relates volume of nectar taken to $r(t)$. Neuron R is connected to neuron P by a nonmodifiable synapse such that neuron R affects the output of P $[=\delta(t)]$ directly. Because of Jensen's inequality, flowers that contain a fixed volume of nectar cause a higher average value of $r(t)$ than flowers that contain variable volumes with a mean that equals the fixed volume. Neuron P also receives input by way of modifiable synapses from three neurons that carry sensory input about the colors of flowers currently in the visual field (Y = yellow, B = blue, N = neutral). The strength of the synapses W_Y, W_B, and W_N is modified according to the following learning rule. If the bee finds food, the change in weight that occurs at a particular synapse is proportional to the activation of that sensory neuron in the previous time step (i.e. how much of that color is in the visual field before the bee landed on the flower) and the current output of P $[\delta(t)]$, which is controlled entirely by $r(t)$ at the time of reward and reflects the net energy gain). Thus, flower colors associated with fixed volumes of nectar result in higher synaptic weights than flowers associated with variable volumes (with Jensen's inequality).

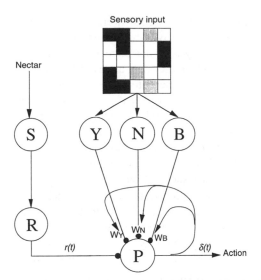

Figure 8.11. Architecture of a neural net that produces risk-averse behavior (redrawn from Montague et al. 1995). P is a simple linear unit that receives convergent sensory information and is hypothesized to correspond to the real bee neuron VUMmx1, which delivers information about reward during classical conditioning experiments (Hammer 1993). At the time (t) the bee gets a volume of nectar, $r(t)$ takes on the value derived from the utility curve shown in figure 8.9. The outputs of neurons N, Y, and B represent changes in the percentage of neutral, yellow, and blue colors in the visual field. These neurons influence the output of neuron P through weights W_N, W_Y, and W_B. The latter two weights are modifiable and W_N is fixed. The output of P, $\delta(t)$, controls the heading of the bee. As the model bee changes its heading above the array of flowers, the activity of these neurons change. Changing the weights with each encounter with a flower enables representation of information about the predictive relationship between sensory input and the amount of nectar obtained.

During the searching phase of foraging, the bee is not obtaining nectar and the output of P, $\delta(t)$, is equal to the weighted sum of the visual inputs $[r(t) = 0]$. This output controls the bee's probability of reorientation. If the bee is primarily looking at a flower color with a high synaptic weight, then $\delta(t)$ will be high and the bee will continue in the same direction. If the bee is looking at a flower color with a low weight, then $\delta(t)$ will be lower and random reorientation will occur. The higher synaptic weights associated with flowers containing fixed volumes inhibit random reorientation and thus promote approaches to these flower types. By contrast, the lower synaptic weights associated with variable-volume flowers result in random reorientation and make approach to variable flowers less likely. This simple model provides a mechanistic explanation for risk aversion in bumblebees. The behavior of the model closely resembles the risk aversion seen in experiments with bumblebees presented with different colored flowers associated with fixed or variable volumes of nectar (Real 1981; Waddington et al. 1981; Real et al. 1982).

Note that if the variability is in the concentration instead of the volume of nectar, the model does not predict risk aversion because the function that relates concentration of nectar to net rate of energy gain is linear. In bees, recent results have shown insensitivity to risk if variability is in nectar concentration (Banschbach and Waddington 1994; Waddington 1995).

The learning models we described in this section and section 8.5.3.1 have three crucial features in common. First, both models assume that environmental variability (the CS–US delay or the volume of the US) affects the speed of learning about the CSs associated with the fixed and variable options. Second, in both models, learning is assumed to occur each time an US is encountered, and attribution of value to the CS occurs before the averaging of value. Third, both models assume that the strength of the association of a particular CS to reward affects the probability that the forager will choose, or move toward, this CS. The combination of these features results in maximization of short-term rate and consequent risk sensitivity.

8.6 MECHANISTIC EXPLANATION 2: PERCEPTUAL ERROR

We have not made any reference to the accuracy and precision with which environmental information is processed. In this section, we will first examine the types of errors animals make when they estimate environmental parameters, such as the length of time intervals and the size of food rewards. We will then consider how these errors could affect animals' responses to fixed and variable food sources and the role these errors may have in risk sensitivity.

8.6.1 Weber's Law and the Scalar Property

In the first example, we consider the accuracy and precision of starlings' memories for time intervals. The birds were given a task that simulated foraging in patches (Kacelnik et al. 1990; Brunner et al. 1992); the patch was a standard operant pecking key. Each patch contained an unpredictable number of food items (between zero and four) that could be obtained with pecking at the key. Once the patch was exhausted, the bird had to travel to the next patch by flying between two perches. Within the patch, the food items were delivered on a fixed-interval schedule; the first peck made by the bird after the programmed interval resulted in a food reward. No signal was given when the final food item in the patch was delivered, and therefore the birds could only detect patch exhaustion by timing the intervals between successive food items. When arriving at a patch, the bird would peck at the key, and its pecking frequency increased toward the time of completion of each interval. If a reward was not delivered because the patch was exhausted, pecking rate declined. The time corresponding to the highest frequency of pecks, the *peak time,* provided an indication of how accurately the birds estimated the fixed interval. The variance in their peak times provided an indication of how precise their estimates were. This experiment was repeated with several different fixed intervals to examine how accuracy and precision varied with the length of the fixed interval. The birds' peak times were close to the fixed interval and showed the birds could measure times accurately. However, the standard deviations of their peak times were proportional to the fixed intervals and showed that precision is inversely proportional to the length of the time interval being estimated (fig. 8.12*a*).

In a second experiment (Bateson and Kacelnik 1995b), we examined accuracy and precision of starlings' estimates of food amounts. The birds faced a choice between two options indicated with colored pecking keys; each key was associated with a fixed quantity of food. To choose the option that offered the greater amount, the birds had to remember the quantity of food associated with each option. One option was a standard, and in the other, adjusting option, the quantity available was altered systematically from trial to trial. The amount of food available in the adjusting option was increased until the bird detected that the adjusting option yielded more food than the standard; then food yielded by the adjusting option was reduced until the bird detected that the adjusting option yielded less food than the standard. The variation in quantity of reward with the adjusting option therefore gave a measure of the accuracy of discrimination between the standard and the adjusting option. If the bird had perfect memory for all amount sizes, then the quantity of food available in the adjusting option would have ranged between one unit above and below the number of units available in the standard. If the bird only detected a difference when the two rewards differed by at least 3 units, then the adjusting option would have ranged within three units above and below the standard. We examined the range of values in the adjusting option for two differently sized standard values. In both cases, the range of values seen in the adjusting option centered around the same value as the standard value; therefore the birds had accurate memories for the size of food rewards. However, if the size of the standard was increased threefold, the range in the magnitude of the adjusting option increased by a similar factor; therefore, precision was proportional to the magnitude of the reward being remembered (fig. 8.12*b*).

Two properties apply to starlings' memories of time intervals and amounts of food. First, memories seemed to be centered around the true physical value experienced; the assumption that starlings are able to learn the foraging parameters accurately was justified. Second, even if the foraging parameter had no variance, starlings' estimates were imprecise, and most important, the standard deviations of their estimates were proportional to the value of the parameter being estimated. The phenomenon of proportional standard deviations is some-

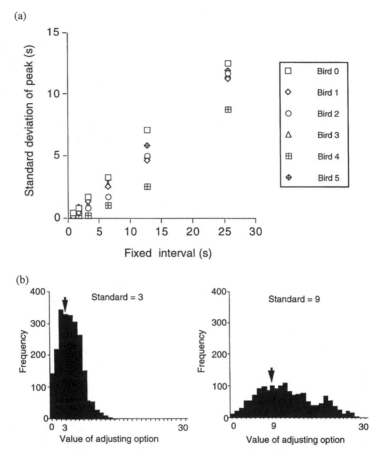

Figure 8.12. Demonstrations of Weber's law in starlings. (*a*) Memory for time. Graph shows the standard deviation of the peak time as a function of the fixed interval. Note the linear relationship between the standard deviation and the fixed interval: Short times are remembered more precisely than long times (replotted from Brunner et al. 1992). (*b*) Memory for amount of food. The range of values taken by the adjusting option with two different standards (three and nine units of food). Note the greater variability when the standard is nine units: Small quantities are remembered more precisely than large quantities (after Bateson and Kacelnik 1995b).

times known as the *scalar property*, which is a strong form of Weber's law. Weber's law is a very widespread phenomenon in discrimination and has been demonstrated for a number of different sensory modalities in many species including rats, pigeons, and humans. We will now examine how the scalar property interacts with environmental variance to give a mechanistic explanation for why animals are generally risk averse if variability is in amount and risk-prone if variability is in delay.

8.6.2 Scalar Expectancy Theory

Reboreda and Kacelnik (1991) (see also Bateson and Kacelnik 1995a) developed a model of risk sensitivity that is based on an information-processing model of timing known as scalar-timing theory (Gibbon et al. 1984; Gibbon et al. 1988). They assumed that the memory formed for a fixed time interval or amount of food can be modeled as a normal distribution with a mean that is equal to the real value of the stimulus and a standard deviation that is proportional to the mean (as suggested by the data presented). The memory representation for a variable option is formed from the sum of the memory distributions of its components. Thus, a food source that offers three or five seeds with equal probability will be remembered as the sum of the memory distributions of three and five seeds. Similarly, a stimulus followed by either a 3 or 5 second delay to food will be remembered as the sum of the memories of these two intervals. The memory representations of variable options will be positively skewed because the memory for the smaller stimulus has a smaller standard deviation than the memory for the large stimulus (fig. 8.13).

It is assumed that subjects choose between two foraging options by (1) retrieving a sample from their memory for each available option; (2) comparing these samples; and (3) choosing the option that offers the more favorable sample (bigger reward or shorter delay). According to the previous description of memory representations, memory distributions for variable options are skewed to the right; however, memory distributions for fixed options with the same mean are symmetric around the true mean. If the subject makes a choice by picking a pair of samples—one from the variable and one from the fixed memory distribution—then in more than half of the comparisons the sample from the memory for the variable option will be smaller than the sample from the memory for the fixed option. Thus, the model predicts that an option that offers variability in delay to food should be chosen more often than a fixed alternative because the memory for the variable option will more often yield the shorter sample. An option that offers variability in amount of food should be chosen less often than a fixed alternative again because the memory for the variable option will more often yield a smaller sample. The difference in choice arises because options that offer short delays are good and should be preferred; options that offer small amounts are bad and should be avoided.

The major prediction of this model agrees with experimental observations because animals are usually risk averse if variability is in amount and risk prone if variability is in delay (fig. 8.1a) (Gibbon et al. 1988; Reboreda and

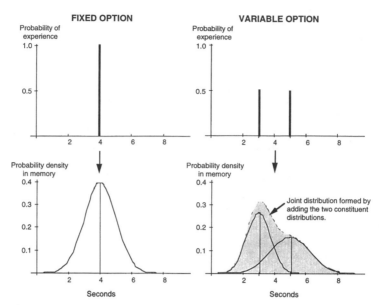

Figure 8.13. The scalar timing theory information-processing model. *Upper panels:* the experienced distribution of outcomes (delays or amounts) in fixed (*left*) and variable (*right*) options. *Lower panels:* the distributions that are assumed to be formed in memory as a result of the experiences (*upper panels*). Note the skew in the memory distribution for the variable option that results from the differing accuracies with which the constituent stimuli are represented (reproduced from Bateson and Kacelnik 1995a).

Kacelnik 1991; Bateson and Kacelnik 1995a). The model is also attractive because it predicts partial preferences, unlike the other models. The model, however, fails in some respects. First, although the qualitative agreement of data with the model is strong, there are quantitative discrepancies. For instance, the model predicts that animals should be indifferent between a fixed and variable option if the fixed option is equal to the *geometric mean* of the two alternative possibilities (amounts or delays) in the variable option (Bateson and Kacelnik 1995a) (i.e. the square root of the product of the two alternative possibilities in the variable option). Subsequent research, however, has demonstrated that, for delays at least, indifference occurs at approximately the *harmonic mean* (i.e. the outcome of maximizing short-term rate) (Bateson and Kacelnik 1996; Mazur 1984, 1986). Second, the model does not address the observed effects of energy budget on preference: It predicts universal risk proneness if variability is in delay and risk aversion if variability is in amount. Third, the model does not address what animals should choose if the two foraging options differ in the variance of both amount and delay simultaneously. A more recent version of scalar timing (Brunner et al. 1994) deals with some of these problems; however, further developments are required to address all aspects of the data available in the risk-sensitivity literature.

8.7 CONCLUSIONS

The theories we have described rely on two types of explanations: functional (or evolutionary) arguments that consider the circumstances under which risk-sensitive behavior can be adaptive and mechanistic arguments that consider how risk sensitivity may emerge as a result of the cognitive processes used by animals to perceive, learn, and remember information about the environment. Functional and mechanistic approaches are complementary forms of explanation in animal behavior because all adaptive behavior must be implemented in some way (Tinbergen 1963). However, as we have presented with many examples, these alternative forms of explanation often produce different predictions about the details of risk-sensitive behavior. Different predictions occur because different approaches have been taken by behavioral ecologists and psychologists. Functional hypotheses are driven by the assumption that risk-sensitive behavior is adaptive. By contrast, mechanistic hypotheses indicate that risk sensivitity is a consequence of the cognitive mechanisms that underlie behavior. Although these mechanistic hypotheses are based on observations of how animals actually behave, these hypotheses do not address the adaptive importance of the mechanisms they postulate or the risk-sensitive behavior they predict. In the future, we hope to see a fusion of these two approaches with production of models that give equal weight to cognitive and evolutionary considerations.

By accounting for empirical observations of risk sensitivity, we found that all of the approaches have advantages and disadvantages, which strongly suggest that an integrated approach is the way forward. Risk-sensitive foraging theory is the only theory that can explain why the direction of risk-sensitive preferences should be affected by energy budget. However, this theory does not explain the many failures to obtain shifts in risk preference or why animals should more often be risk averse with amount but risk prone with delay. The theory is difficult to test conclusively because it has many variants. To make precise predictions about how manipulations will affect behavior, it is necessary to know the current status of an animal precisely. One effect of these uncertainties is that only qualitative predictions are being tested. For instance, the energy-budget rule predicts exclusive preferences; however, even the best examples of shifts in preference are only partial shifts in the predicted directions.

Time discounting that arises from the probability of interruption potentially explains why animals should be risk prone if variability is in delay. However, time discounting fails quantitatively because with its simplest applications in isolated choices, it predicts exponential time discounting, and there is clear evidence that humans discount time hyperbolically. Also models that are based solely on time discounting do not address the effects on choice of variability in amount.

Maximizing short-term rate can account for the hyperbolic time discounting observed in animals and the consequent risk proneness if variability is in delay. However, assuming that animals maximize this currency does not explain all the data. The currency can only accommodate risk aversion with variable amounts if there is a positive correlation between amount and the handling time. Also maximizing short-term rate does not address the effects of energy budget. The major problem in maximizing short-term rate is explaining its functional and mechanistic bases. We suggested that maximizing short-term rate may emerge as a consequence of the basic mechanisms of associative learning that can be defended as adaptive in a context that is broader than that of risk sensitivity.

Scalar-timing theory can explain partial preferences and why animals are risk prone with variable delays and risk averse with variable amounts. However, the theory sometimes fails quantitatively and does not address the effects of energy budget.

Thus, whether risk-sensitive foraging preferences are largely the products of direct selection driven by the fitness consequences of variable food sources or whether they are consequences of selection for efficient cognitive mechanisms in broader contexts are unanswered problems. Of course, it is possible that the phenomena we categorize as risk sensitivity may only be superficially related; these phenomena may be the products of different underlying mechanisms, or they may have evolved under different selective pressures (in which case they will never be explained with a single unifying theory). Whatever the answers will be, it is unquestionable that satisfactory explanations for risk sensitivity will include ecological and cognitive considerations. This conclusion is not specific for risk sensitivity; however, its truth is particularly well exemplified with this area of research—perhaps because the field has attracted such extensive data collection and optimality modeling.

8.8 SUMMARY

Animals presented with foraging options that offer the same average rate of gain but differ in variance generally show strong preferences. This widespread behavioral phenomenon is known as risk sensitivity. Risk-sensitivity has implications for the ecological interaction of animals with their environments and the cognitive mechanisms that underlie foraging behavior. Several factors affect the direction of risk-sensitive preferences: the component of rate that is variable (amount or time), the energy needs of the subjects, and possibly the typical body weight of the species. We reviewed explanations for these data arising from functional, descriptive, and mechanistic arguments. Animals can respond to environmental variance in ways that enhance fitness, and there is some evidence for use of adaptive policies such as the energy-budget rule. However, some well-documented cognitive processes also generate risk-sensitive behavior that does not appear to be directly adaptive. The broader adaptive importance of these cognitive processes remains to be established; however, we believe that cognitive processes cannot be ignored in any form of foraging research.

ACKNOWLEDGMENTS

We thank the Wellcome Trust for financial support (grants 45243/Z/95 and 461/Z/95), NATO Collaborative Research Grant (920308), and the many people with whom we have discussed the ideas presented in this chapter.

LITERATURE CITED

Banschbach, V. S., and K. D. Waddington. 1994. Risk-sensitivity in honeybees: No consensus among individuals and no effect of colony honey stores. *Animal Behaviour* 47:933–941.
Barkan, C. P. L. 1990. A field test of risk-sensitive foraging in black-capped chickadees (*Parus atricapillus*). *Ecology* 71:391–400.
Barnard, C. J., and C. A. J. Brown. 1985. Risk-sensitive foraging in common shrews (*Sorex araneus* L.). *Behavioral Ecology and Sociobiology* 16:161–164.
Barnard, C. J., C. A. J. Brown, A. Houston, and J. M. McNamara. 1985. Risk-sensitive foraging in common shrews: An interruption model and the effects of mean and variance in reward rate. *Behavioral Ecology and Sociobiology* 18:139–146.
Bateson, M., and A. Kacelnik. 1995a. Preferences for fixed and variable food sources: Variability in amount and delay. *Journal of the Experimental Analysis of Behaviour* 63:313–329.
———. 1995b. Accuracy of memory for amount in the foraging starling (*Sturnus vulgaris*). *Animal Behaviour* 50:431–443.
———. 1996. Rate currencies and the foraging starling: The fallacy of the averages revisited. *Behavioral Ecology* 7:341–352.
———. 1997. Starlings' preferences for predictable and unpredictable delays to food: Risk-sensitivity or time discounting? *Animal Behaviour* 53:1129–1142.
Bateson, M., and S. C. Whitehead. 1996. The energetic costs of alternative rate currencies in the foraging starling. *Ecology* 77:1303–1307.
Brunner, D., J. Gibbon, and S. Fairhurst. 1994. Choice between fixed and variable delays with different reward amounts. *Journal of Experimental Psychology: Animal Behavior Processes* 20:331–346.
Brunner, D., A. Kacelnik, and J. Gibbon. 1992. Optimal foraging and timing processes

in the starling, *Sturnus vulgaris:* Effect of inter-capture interval. *Animal Behaviour* 44:597–613.

Caraco, T. 1981. Energy budgets, risk, and foraging preferences in dark-eyed juncos (*Junco hyemalis*). *Behavioral Ecology and Sociobiology* 8:213–217.

———. 1982. Aspects of risk-aversion in foraging white-crowned sparrows. *Animal Behaviour* 30:719–727.

———. 1983. White-crowned sparrows (*Zonotrichia leucophrys*) foraging preferences in a risky environment. *Behavioral Ecology and Sociobiology* 12:63–69.

Caraco, T., W. U. Blanckenhorn, G. M. Gregory, J. A. Newman, G. M. Recer, and S. M. Zwicker. 1990. Risk-sensitivity: Ambient temperature affects foraging choice. *Animal Behaviour* 39:338–345.

Caraco, T., A. Kacelnik, N. Mesnik, and M. Smulewitz. 1992. Short-term rate maximization when rewards and delays covary. *Animal Behaviour* 44:441–447.

Caraco, T., and S. L. Lima. 1985. Foraging juncos: Interaction of reward mean and variability. *Animal Behaviour* 33:216–224.

Caraco, T., S. Martindale, and T. S. Whittam. 1980. An empirical demonstration of risk sensitive foraging preferences. *Animal Behaviour* 28:820–830.

Cartar, R. V. 1990. A test of risk sensitive foraging in wild bumblebees. *Ecology* 72: 888–895.

Case, D. A., P. Nichols, and E. Fantino. 1995. Pigeon's preference for variable-interval water reinforcement under widely varied water budgets. *Journal of the Experimental Analysis of Behavior* 64:299–311.

Charnov, E. L. 1976a. Optimal foraging: Attack strategy of a mantid. *American Naturalist* 110:141–151.

———. 1976b. Optimal foraging: The marginal value theorem. *Theoretical Population Biology* 9:129–136.

Croy, M. I., and R. N. Hughes. 1991. Effects of food supply, hunger, danger, and competition on choice of foraging location by the fifteen-spined stickleback, *Spinachia spinachia* L. *Animal Behaviour* 42:131–139.

Davison, M. C. 1972. Preference for mixed-interval versus fixed-interval schedules: Number of component intervals. *Journal of the Experimental Analysis of Behavior* 17:169–176.

Dickinson, A. 1980. *Contemporary Animal Learning Theory.* Cambridge: Cambridge University Press.

Epstein, R. 1981. Amount consumed as a function of magazine-cycle duration. *Behavior Analysis Letters* 1:63–66.

Essock, S. M., and E. P. Rees. 1974. Preference for and effects of variable as opposed to fixed reinforcer duration. *Journal of the Experimental Analysis of Behavior* 21: 89–97.

Gibbon, J., M. D. Baldock, C. Locurto, L. Gold, and H. S. Terrace. 1977. Trial and intertrial durations and autoshaping. *Journal of Experimental Psychology: Animal Behavior Processes* 3:264–284.

Gibbon, J., R. M. Church, S. Fairhurst, and A. Kacelnik. 1988. Scalar Expectancy Theory and choice between delayed rewards. *Psychological Review* 95:102–114.

Gibbon, J., R. M. Church, and W. H. Meck. 1984. Scalar timing in memory. In J.

Gibbon and L. Allan, eds., *Timing and Time Perception,* 52–77. New York: New York Academy of Sciences.

Gilliam, J. F., R. F. Green, and N. E. Pearson. 1982. The fallacy of the traffic policeman: A response to Templeton and Lawlor. *American Naturalist* 119:875–878.

Ha, J. C. 1991. Risk-sensitive foraging: The role of ambient temperature and foraging time. *Animal Behaviour* 41:528–529.

Ha, J. C., P. N. Lehner, and S. D. Farley. 1990. Risk-prone foraging behaviour in captive grey jays *Perisoreus canadensis. Animal Behaviour* 39:91–96.

Hamm, S. L., and S. J. Shettleworth. 1987. Risk aversion in pigeons. *Journal of Experimental Psychology: Animal Behavior Processes* 13:376–383.

Hammer, M. 1993. An identified neuron mediates the unconditioned stimulus in associative olfactory learning in honeybees. *Nature* 366:59–63.

Harder, L. D. 1986. Effects of nectar concentration and flower depth on flower handling efficiency of bumblebees. *Oecologia* 69:309–315.

Harder, L. D., and L. A. Real. 1987. Why are bumblebees risk-averse? *Ecology* 68: 1104–1108.

Herrnstein, R. J. 1964. Aperiodicity as a factor in choice. *Journal of the Experimental Analysis of Behavior* 7:179–182.

Houston, A. I. 1991. Risk-sensitive foraging theory and operant psychology. *Journal of the Experimental Analysis of Behavior* 56:585–589.

Houston, A. I., and J. M. McNamara. 1982. A sequential approach to risk-taking. *Animal Behaviour* 30:1260–1261.

———. 1985. The choice of two prey types that minimizes the probability of starvation. *Behavioral Ecology and Sociobiology* 17:135–141.

———. 1988. A framework for the functional analysis of behaviour. *Behavioral and Brain Sciences* 11:117–163.

———. 1990. Risk-sensitive foraging and temperature. *Trends in Ecology and Evolution* 5:131–132.

Jensen, J. L. 1906. Sur les fonctions convexes et les inequalites entre les valeurs moyennes. *Acta Mathematica* 30:175–193.

Kacelnik, A. 1984. Central place foraging in starlings (*Sturnus vulgaris*). I. Patch residence time. *Journal of Animal Ecology* 53:283–299.

Kacelnik, A., and M. Bateson. 1996. Risky theories: The effects of variance on foraging decisions. *American Zoologist* 36:402–434.

Kacelnik, A., D. Brunner, and J. Gibbon. 1990. Timing mechanisms in optimal foraging: Some applications of scalar expectancy theory. In R. N. Hughes, ed., *Behavioural Mechanisms of Food Selection,* 63–81. Berlin: Springer-Verlag.

Kacelnik, A., and A. I. Todd 1992. Psychological mechanisms and the Marginal Value Theorem: Effect of variability in travel time on patch exploitation. *Animal Behaviour* 43:313–322.

Kagel, J. H., L. Green, and T. Caraco. 1986. When foragers discount the future: Constraint or adaptation? *Animal Behaviour* 34:271–283.

Kagel, J. H., D. N. MacDonald, R. C. Battalio, S. White, and L. Green. 1986. Risk-aversion in rats (*Rattus norvegicus*) under varying levels of resource availability. *Journal of Comparative Psychology* 100:95–100.

Krebs, J. R., D. W. Stephens, and W. J. Sutherland. 1983. Perspectives in optimal foraging. In A. H. Brush, and G. A. Clark, eds., *Perspectives in Ornithology,* 165–216. New York: Cambridge University Press.

Leventhal, A. M., R. F. Morell, E. J. Morgan, and C. C. Perkins. 1959. The relation between mean reward and mean reinforcement. *Journal of Experimental Psychology* 59:284–287.

Logan, F. A. 1965. Decision making by rats: Uncertain outcome choices. *Journal of Comparative and Physiological Psychology* 59:246–251.

Mazur, J. E. 1985. Probability and delay of reinforcement as factors in discrete-trial choice. *Journal of the Experimental Analysis of Behavior* 43:341–351.

———. 1987. An adjusting procedure for studying delayed reinforcement. In M. L. Commons, J. E. Mazur, J. A. Nevin, and H. Rachlin, eds., *Quantitative Analyses of Behavior: The Effects of Delay and of Intervening Events on Reinforcement Value,* 55–73. Hilsdale, N.J.: Erlbaum.

McNamara, J. M. 1983. Optimal-control of the diffusion-coefficient of a simple diffusion process. *Mathematics of Operations Research* 8:373–380.

———. 1984. Control of a diffusion by switching between 2 drift-diffusion coefficient pairs. *Siam Journal on Control and Optimisation* 22:87–94.

McNamara, J. M., and A. I. Houston. 1987. A general framework for understanding the effects of variability and interuptions on foraging behaviour. *Acta Biotheoretica* 36:3–22.

———. 1992. Risk-sensitive foraging: A review of the theory. *Bulletin of Mathematical Biology* 54:355–378.

McNamara, J. M., S. Merad, and A. I. Houston. 1991. A model of risk-sensitive foraging for a reproducing animal. *Animal Behaviour* 41:787–792.

Menlove, R. L., H. M. Inden, and E. G. Madden. 1979. Preference for fixed over variable access to food. *Animal Learning and Behavior* 7:499–503.

Montague, P. R., P. Dayan, C. Person, and C. J. Sejnowski. 1995. Bee foraging in uncertain environments using predictive Hebbian learning. *Nature* 377:725–728.

Moore, F. R., and P. A. Simm. 1986. Risk-sensitive foraging by a migratory bird (*Dendroica coronata*). *Experientia* 42:1054–1056.

Myerson, J., and L. Green. 1995. Discounting of delayed rewards: Models of individual choice. *Journal of the Experimental Analysis of Behavior* 64:263–276.

Possingham, H. P., A. I. Houston, and J. M. McNamara. 1990. Risk-averse foraging in bees: A comment on the model of Harder and Real. *Ecology* 71:1622–1624.

Real, L. A. 1981. Uncertainty and pollinator-plant interactions: The foraging behavior of bees and wasps on artificial flowers. *Ecology* 62:20–26.

Real, L. A., J. Ott, and E. Silverfine. 1982. On the trade-off between mean and variance in foraging: An experimental analysis with bumblebees. *Ecology* 63:1617–1623.

Reboreda, J. C., and A. Kacelnik. 1991. Risk sensitivity in starlings: Variability in food amount and food delay. *Behavioral Ecology* 2:301–308.

Rider, D. P. 1983. Choice for aperiodic versus periodic ratio schedules: A comparison of concurrent and concurrent-chains procedures. *Journal of the Experimental Analysis of Behavior* 40:225–237.

Rodriguez, M. L., and A. W. Logue. 1988. Adjusting delay to reinforcement: Comparing choice in pigeons and humans. *Journal of Experimental Psychology: Animal Behavior Processes* 14:105–117.

Samuelson, P. A. 1937. A note on the measurement of utility. *Review of Economic Studies* 4:155–161.

Schmid-Hempel, P., A. Kacelnik, and A. I. Houston 1985. Honeybees maximize efficiency by not filling their crop. *Behavioral Ecology and Sociobiology* 17:61–66.

Snyderman, M. 1987. Prey selection and self-control. In M. L. Commons, J. E. Mazur, J. A. Nevin, and H. Rachlin, eds., *Quantitative Analyses of Behavior,* 283–308. Hillsdale, N.J.: Erlbaum.

Staddon, J. E. R., and N. K. Innis. 1966. Preference for fixed vs. variable amounts of reward. *Psychonomic Science* 4:193–194.

Stephens, D. W. 1981. The logic of risk-sensitive foraging preferences. *Animal Behaviour* 29:628–629.

Stephens, D. W., and J. R. Krebs. 1986. *Foraging Theory.* Princeton, N.J.: Princeton University Press.

Stephens, D. W., and R. C. Ydenberg. 1982. Risk aversion in the great tit. In D. W. Stephens, ed.; *Stochasticity in Foraging Theory: Risk and Information,* Ph.D. diss., Oxford University.

Tinbergen, N. 1963. On aims and methods of ethology. *Zeitschrift fur Tierpsychologie* 20:410–433.

Turelli, M., J. H. Gillespie, and T. W. Shoener. 1982. The fallacy of the fallacy of the averages in ecological optimization theory. *American Naturalist* 119:879–884.

Tuttle, E. M., L. Wulfson, and T. Caraco. 1990. Risk-aversion, relative abundance of resources, and foraging preferences. *Behavioral Ecology and Sociobiology* 26:165–171.

Waddington, K. D. 1995. Bumblebees do not respond to variance in nectar concentration. *Ethology* 101:33–38.

Waddington, K. D., T. Allen, and B. Heinrich. 1981. Floral preferences of bumblebees (*Bombus edwardsii*) in relation to intermittent versus continuous rewards. *Animal Behaviour* 29:779–784.

Wunderle, J. M., M. Santa-Castro, and N. Fletcher. 1987. Risk-averse foraging by bananaquits on netative energy budgets. *Behavioral Ecology and Sociobiology* 21:249–255.

Young, J. S. 1981. Discrete-trial choice in pigeons: Effects of reinforcer magnitude. *Journal of the Experimental Analysis of Behavior* 35:23–29.

Young, R. J., H. Clayton, and C. J. Barnard. 1990. Risk sensitive foraging in bitterlings, *Rhodeus sericus:* Effects of food requirement and breeding site quality. *Animal Behaviour* 40:288–297.

Zabludoff, S. D., J. Wecker, and T. Caraco. 1988. Foraging choice in laboratory rats: Constant vs. variable delay. *Behavioural Processes* 16:95–110.

CHAPTER NINE

Behavioral Decisions about Foraging and Predator Avoidance

RONALD C. YDENBERG

9.1 INTRODUCTION

Everyone has watched a foraging bumblebee. Flying to an inflorescence, she usually alights on the bottom floret and works her way upward, sometimes visiting one or just a few florets, sometimes systematically probing each floret before moving to the next inflorescence. Hidden carefully somewhere in the vicinity is her home colony, with a queen, other workers, and a group of larvae whose growth and development are dependent on the proficiency with which she and her sister workers deliver nectar and pollen. Her sensory apparatus and nervous system developed under the direction of genes she inherited from her ancestors and are designed to enable exploitation of the flowers that supply essential resources. Ultimately, the reproductive success of the colony—the number of daughter queens and drones produced—depends on her foraging performance. I refer readers to Heinrich (1978) for a highly enjoyable and informative account of bumblebee biology.

The flight of the bumblebee poses important challenges; the senses and nervous system must deal with the formidable problems of proficient foraging in an everchanging and hazardous environment. To understand the bumblebee's experience, we must view the world with a bee's eye. The importance of adopting this perspective was recognized almost a century ago by physiologists such as Jakob von Uexküll—who was the first to promote the idea that an anthropomorphic viewpoint precludes understanding another species' behavior (see Thorpe 1979). Jakob von Uexküll taught that each species experiences the world in a unique way and the experience is determined with sensory abilities. This idea was very important in the development of ethology, and its practitioners laid great emphasis on von Uexküll's notion of the *Umwelt*—the sensory world in which each species lives.

While an appreciation of each species' unique Umwelt is important, generalizations are also required to build a broad understanding, which can be applied to many species. During the last thirty years, behavioral ecologists have developed a number of theoretical ideas that consider general properties of behavior. In a foraging context, much of the emphasis has been on analyzing decisions (Kacelnik and Krebs 1991). The fusion of simple general models from behavioral ecology with ethological ideas, such as that of Umwelt, and techniques from experimental psychology is currently creating the new discipline of cognitive ecology (Real 1993).

In this chapter, I will introduce some current ideas in cognitive ecology that are relevant to understanding foraging theory. I first will consider the structure of foraging models and the distinct foraging processes of feeding and provisioning. I next will explore the role of information in these models, and I hope to indicate how the study of foraging may help the development of cognitive ecology. I will conclude the chapter by considering the function of predation risk in foraging models.

9.2 FORAGING THEORY

9.2.1 The Function of Foraging Theory in Cognitive Ecology

The most basic assumption in foraging theory is that natural selection has led to the evolution of proficient foraging. As a result, the focus in foraging theory is on fitness costs and benefits (Stephens and Krebs 1986). The decision mechanisms involved are rarely specified. An important goal in cognitive ecology is to expand our understanding of these mechanisms; there are two ways in which a well-developed foraging theory—with its emphasis on ultimate factors—may enable this endeavor. First, foraging theory suggests what factors are important and, therefore, the sorts of things foragers ought to evaluate as they forage. Further, the models suggest how measurements may be integrated to form a view of the world that is relevant to the forager.

Second, foraging theory may help illuminate details of the mechanisms involved. When predictions on the basis of foraging models do not match experimental measurements, there are two possible explanations. The first explanation is that the prediction is wrong because some important factor was ignored in the model or the theory was not applicable.

The other and more interesting explanation is that the design of the experiment revealed some mechanism that caused the animals to perform "incorrectly." This is of direct interest because the nature of the "error" indicates the workings of the cognitive mechanisms used by the forager to make foraging decisions. For example, errors in estimating or recalling time intervals are implicated with the "unwillingness" of foragers to wait as long as foraging models predict they should. Such errors are assumed to arise from properties of the timing or memory mechanisms (Bateson and Kacelnik, this vol. section 8.6).

In other examples, investigators compared the observed and expected performances of foragers to derive insights into mechanisms. For example, Pyke (1981) observed hummingbirds foraging on inflorescenses of varying complexity. He constructed vertical inflorescenses with 12 florets arranged linearly, in a simple spiral, or in a more complex pattern that resembled real *Ipomopsis* inflorescences—a favorite forage plant. Pyke measured hummingbirds' performances during visits to an inflorescence by counting the number of revisits to florets that had been visited already (and were therefore empty). Surprisingly, birds performed best on the natural array and most poorly on the linear array. Pyke concluded that the mechanism used by the birds to decide which floret to fly to next (the "movement rule") had evolved so that birds performed well in the complex arrays of natural patches of flowers; poor performance on simple arrays was a side effect. Although this side effect is of little or no consequence to wild hummingbirds who live and forage in complex habitats, this side effect may be very important to cognitive ecologists who study hummingbird foraging on simple arrays in the laboratory (e.g. Sutherland and Gass 1995).

Cheverton (1983) made a similar observation to that of Pyke, but he reached a very different conclusion. He found that bumblebees performed poorly on simple floral arrays in the laboratory and concluded that the bees used evolved simple-movement rules (in this case, involving handedness) to solve difficult foraging problems. He suggested that the benefits of a simple rule used to decide quickly where to move next outweigh the costs of taking more time to make a better decision. Obviously, this conclusion necessitates knowing what "better" means; foraging theory may contribute to this knowledge.

9.2.2 Foraging Models

9.2.2.1 Terminology

In the following, I set out the basic parts of a foraging model. To be of use in cognitive ecology, a model must be flexible enough to include a wide diversity of foraging behavior yet simple enough that generalizations can be derived.

A number of terms are important and must be defined and used carefully. In the context of cognitive ecology, the central tenet of foraging theory is that selection (natural, sexual, or artificial) has acted on the structure and functioning of *decision mechanisms.* These mechanisms integrate information from the sensory organs and internal indicators of state (e.g. hunger) to allow choice among the possible alternative behaviors in a foraging situation (called *tactics* or *options*). Certain variants of the decision mechanism are assumed to have been favored by selection because they lead to behavior that best achieves a *strategy* (or objective or design feature), such as maximum rate of energy intake, minimum time spent foraging to reach a requirement, maximum delivery, minimum probability of an energy shortfall, or a host of others.

The investigator's measure of foraging performance or proficiency is called a *currency*. The currency is often referred to as a surrogate measure for fitness. In fact in much literature, there is hardly a distinction among currency, strategy, and fitness. I feel, for two reasons, that it is useful to reserve the term "currency" for the investigator's measures of foraging performance. First, different performance measures may be differently or even oppositely related to different fitness components (Salant et al. 1996). For example, the amount of food stored may increase with the amount of energy that a hoarding animal expends to fill its larder. The animal's survival, however, may decline with energy expenditure because of increased exposure to predators. The amount of food stored and survival both depend on energy expenditure, and both are positively related to fitness. The forager, however, cannot maximize the amount of food stored and survival simultaneously; the forager must balance two opposing demands. The investigator, however, may gain some insight by ignoring one of these demands and working to understand just how the other is related to fitness. Measuring fitness properly requires many details of population structure and density dependence not available in most foraging studies (Mylius and Diekmann 1995); I believe it is naive to assume that currencies and fitness are generally closely related.

Moreover, this assumption is not necessary. For example, Alexander (1982) successfully studied traits such as bone structure without making an explicit fitness assumption. In the case of femurs, he used the thickness of the femoral wall relative to femoral diameter as the performance measure. He assumed that this measured attribute was acted on by natural selection to minimize the weight necessary to bear a given load (the strategy). This assumption is different and less restrictive than the assumption that proposes such femurs maximize fitness.

Second, even if one could specify a currency directly related to fitness, there is no reason to suppose that the mechanisms used by animals to evaluate and measure their performance are similar to those used by investigators. In fact, the mechanisms are likely to be very different. Field biologists use stopwatches, but the foragers they study employ a cognitive timing mechanism that seems to function differently (Bateson and Kacelnik, this vol. section 8.6). This distinction between the investigator's measure of the animal's experience and the animal's own experience is especially important in cognitive ecology; with this distinction, a much better understanding of the cognitive and other mechanisms that animals use in foraging situations will be gained.

9.2.2.2 Adaptations and Constraints

As explained, the existence of an optimum is a deduction that is based on knowing how natural selection works; an optimum is not a manifestation of blind faith. There are undoubtedly many cases in which behavior is poorly matched to the environment; however, to discover these cases, we must define well and poorly matched. The term "optimality" has been widely associated with foraging theory; however, this term perhaps is best reserved for reference to the usually mathematical procedure of determining which behavioral options are most proficient. Used in this way, the term "optimal forager" refers to a forager that uses the mathematical techniques of optimality to determine which behavioral option is best in a specific situation; the term does not imply a perfectly performing forager. Animals do not use stopwatches; likewise, animals do not use differential calculus to solve foraging problems. Optimality is merely an investigator's technique that may help illuminate the biology.

A paper by Marrow et al. (1996) illustrates how, when used well, optimality does not lead merely to wildly speculative adaptationist tales, but can be used to explore both adaptations and constraints. Marrow et al. studied the reproduction of Soay sheep (*Ovus aries*), an ancient domesticated breed now feral on the St. Kilda archipelago of Scotland. In this harsh environment, the population fluctuates greatly and mortality is high every three or four years. Marrow et al. (1996) used an optimality approach to model the reproductive decisions of Soay ewes. In spring, each model ewe chose to not reproduce at all, have a single lamb, or have twins. Reproduction carries mortality risks, and lambs born near the population peak are less likely to survive than those borne near the trough of the cycle. With use of field data from a long-running study on the island, the models that made reproductive decisions with information about body condition and cycle stage (i.e. population density) performed best. Ewes should have twins when in good condition and at low and moderate population densities; however, ewes should behave more conservatively when in poor condition and as the population peak approaches. While the data show that females are sensitive to their own condition when making reproductive decisions, they do not make use of population density information—although use of this information would be simple. Somehow, ewes are constrained from using population density information. Marrow et al. (1996) suggested that perhaps ancient human selection for high fertility has made the breed less conservative than it ought to be in this environment. This sheep example shows how optimality can be used in the investigation of alternative behaviors, the nature of proficiency, and the mechanisms used to make behavioral decisions.

9.2.3 Foraging Processes

By adopting von Uexküll's idea of Umwelt, we acknowledge that every species' foraging situation is different and that a single strategy is unlikely to be universally applicable. In the following sections, I will outline a general framework that attempts to recognize some of this diversity. The emphasis is on general properties instead of particular models. The aim is to gain insight

FORAGING SITUATIONS

FEEDING   PROVISIONING
                 - delivery
                 - self-feeding

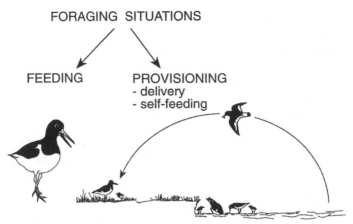

Figure 9.1. Classification of foraging situations that are exemplified with oystercatchers studied by Bruno Ens (1992). While *feeding,* animals capture prey for their own immediate consumption; oystercatchers are feeders for most of the year. While *provisioning,* a forager *delivers* some of the prey captured, obtaining the metabolic power for this activity by eating the other prey (*self feeding*); oystercatchers become provisioners during the breeding season.
Some parental oystercatchers must fly to deliver prey to inland nests; however, others have territories along the salt marsh edge and can take their broods to the foraging area. For these oystercatchers, provisioning does not involve a flight to a nest, but they give some of the captured prey to their chicks (delivery) and eat some themselves (self feeding). Thus, provisioning is not the same as central-place foraging.

into the kinds of information that are relevant, and how these kinds of information may be integrated to make foraging decisions.

First, the foraging context must be identified as *provisioning* or *feeding* (fig. 9.1). These are distinct foraging processes (Ydenberg et al. 1994). The difference is that feeders consume prey, but provisioners deliver a resource—food, water, or nesting material—to their offspring, nest mates, or a storage site. Provisioners must obtain the metabolic power for delivery by consuming some of the prey captured (usually on the spot, see Kacelnik 1984) or by searching for and consuming different foods (see Waite and Ydenberg 1994a), a process termed self feeding.

The difference can be appreciated by considering a situation that involves the delivery of a resource other than food. Male black wheatears (*Oenanthe leucura*) have evolved, apparently by sexual selection (Soler et al. 1996), the strange habit of delivering thousands of small stones to their nest site. Many social insects such as paper wasps deliver water to the colony (Kasuya 1982) on hot days. Water is collected at puddles and other sources and transported in the wasp's crop to the nest, where it is regurgitated and evaporated to cool the nest. A wasp must power the delivery of the water with energy obtained by self feeding; in doing so, the wasp must somehow take account of the time and energy needed for self feeding. To subtract the energy spent delivering the water from the water delivered does not make sense. This is not because we measure energy and water in different units (we could measure the weight of food needed to provide the energy, and the weight of the water in milligrams). The subtraction does not make sense because the water and energy are used differently. The same is true when food is delivered.

Second, provisioning need not involve delivery to a central place (fig. 9.1) (Orians and Pearson 1979). The essence of provisioning is that some or all of the resource is not for the provisioner's immediate consumption; the resource is delivered to an offspring, a nest mate, or a hoard, or the resource is given to another individual at the site where it is captured. (Ydenberg [1994] also considered provisioning situations in which food was consumed and later fed to offspring as a processed product, as in lactation.)

Feeding and provisioning are bound to be experienced differently by foragers—not least because a delivered resource is not consumed and, hence, does not affect hunger. While feeding or self feeding, a forager can assess its need for more food with feedback mechanisms that occur in the stomach, gut, or blood. However, provisioners must rely on very different feedback mechanisms to inform them that a demand has been satisfied. Parent songbirds respond to the level of begging by their nestlings (Price and Ydenberg 1995). Water-foraging honeybees apparently measure the eagerness with which workers in the hive unload them to determine how much more water is needed (Kühnholz 1994).

9.2.4 Currencies

I now turn to currencies. How should investigators measure foraging performance? Two main currencies are of interest. The currency called "rate" is the amount of energy gained per unit time and is defined as follows: the gross rate-of-energy capture with use of tactic i is b_i, the energy-expenditure rate is o_i, and so the net rate of gain is $b_i - c_i$. "Efficiency" is defined as the amount of energy gained per unit energy expenditure. (We could also differentiate between the gross (b_i/c_i) and net $([b_i - c_i]/c_i)$ efficiencies, but the latter is the same as $(b_i/c_i) - 1$ and efficiency usually refers to the former measure.) Other currencies have been used in foraging studies, but these currencies are mostly variants of rate and efficiency (see Ydenberg et al. 1994) or they are not very general (e.g. energy content per unit volume nectar for hummingbirds) (Montgomerie et al. 1984).

Most simple foraging models use the currency "net rate of energy intake," and the strategy "maximize the net rate of energy intake while foraging," and treat them virtually identically. Efficiency has been ignored almost entirely. But which quantity describes better the behavior of foragers?

Table 9.1
Studies that Contrast Rate and Efficiency Predictions for Foraging Behavior

Study	Behavior/Species Tested	Result
Feeding studies		
Juanes and Hartwick 1990	Prey choice/Dungeness crab	Efficiency
Welham and Ydenberg 1988	Patch time/Ring-billed gull	Intermediate[a]
Montgomerie et al. 1984	Nectar concentration/ Ruby-throated hummingbird	Efficiency[b]
Provisioning studies		
Kacelnik 1984	Load size/Starling	Rate[c]
Schmid-Hempel et al. 1985	Load size/Honeybee	Efficiency
Schmid-Hempel 1987	Load size/Honeybee	Efficiency
Wolf and Schmid-Hempel 1990	Load size/Honeybee	Efficiency
McLaughlin and Montgomerie 1990	Flight speed/Lapland longspur	Efficiency
Welham and Ydenberg 1993	Flight speed/Black terns	Efficiency
Waite and Ydenberg 1994a, 1994b	Load size/Grey jay	Rate, Efficiency[d]

Note: For each situation, the currency that best predicts the observed behavior is given.

[a] Gulls may have been provisioning.

[b] The currency used (net energy gain per unit volume nectar consumed) was close but not identical to efficiency. Nevertheless, the results approached more closely efficiency predictions instead of rate predictions.

[c] Results were close to the predictions of both currencies.

[d] Results were close to rate predictions in autumn, but shifted in the direction of efficiency predictions in winter.

9.2.4.1 Rate versus Efficiency: Tests

Studies that have examined the behavior of foragers in relation to the predictions of both rate and efficiency currencies are summarized in table 9.1. There have been relatively few studies of feeding (as opposed to provisioning) behavior. Juanes and Hartwick (1990) examined the prey-size selection of Dungeness crabs (*Cancer productus*) feeding on clams. Rate and efficiency currencies made very different predictions; with the former, a preference for large prey was predicted, and with the latter, a preference for small prey was predicted. The experiments clearly supported efficiency maximizing. A review of many other prey-choice experiments in the literature (Juanes 1992) showed that the preference for smaller prey (and hence efficiency) is a general pattern among decapods, although few of these studies made quantitative predictions.

There are more provisioning studies available. McLaughlin and Montgomerie (1990) and Welham and Ydenberg (1993) found that efficiency was a better predictor of the flight speeds of provisioning Lapland longspurs (*Calcarius lapponicus*) and black terns (*Chlidonias niger*), respectively. Finally, Schmid-Hempel and his colleagues studied the behavior of honeybees (*Apis mellifera*) and found good agreement with efficiency predictions and a marked divergence from rate predictions in three separate studies (Schmid-Hempel et al. 1985; Schmid-Hempel 1987; Wolf and Schmid-Hempel 1990).

While this survey of studies reveals that rate and efficiency are useful to the investigator as measures of foraging performance in different studies, this survey, as yet, tells us little about what a forager experiences or how it evaluates its own performance and integrates information to choose a behavior.

9.2.5 Strategies

How is it that rate and efficiency sometimes work as measures of foraging performance? These currencies have somewhat different properties. With an efficient tactic, energy is obtained cheaply in terms of energy expenditure; however, with this tactic, there is no direct account of the time involved. With rate-maximizing tactics, energy is obtained quickly. Foragers evidently choose rate-maximizing tactics under some circumstances and efficient tactics under other circumstances. What are these circumstances?

Most "classical" foraging theories assume that the foraging strategy is to maximize the long-term net rate of energy intake. This strategic objective always predicts rate-maximizing behavior and is therefore not supported with the data. I will consider two different possible strategies in the models I will outline. The forager maximizes the total net gain while foraging (denoted G_f) and the total net daily gain (G_d). These net gains seem very similar; however, they are not identical if any part of the day is spent resting because during rest energy expenditure continues but there is no intake. These models will show how, depending on the situation, foragers may choose a behavioral option (tactic) that is efficient or one that is quick; as a result, the investigator measures either rate of efficient behavior.

9.2.6 Constraints

The fundamental constraints in any foraging situation concern the time and energy available. Imagine that the total time available is T, and the forager has a maximum daily assimilable energy intake of K (Weiner 1992). If the forager's total intake reaches K before T expires, then the forager must rest for the remainder of the day. This situation is called "energy limitation" because the total gain is limited by the forager's ability to assimilate energy: Even if there were more time, the forager would be unable to increase its gain. "Time limitation" may occur in one of two ways. If K is not attained, the forager is able to forage until time T. Alternatively, there may be a time limit that limits foraging to a period t_L. The time limit may be imposed naturally as with day length in the northern winter (e.g. Systad 1996) or the tide (e.g. Robles et al. 1989), or the time limit may be experimentally imposed (e.g. Tooze and Gass 1985, Swenhen et al. 1989). In both cases, increases in the foraging time would allow the forager to increase its gain.

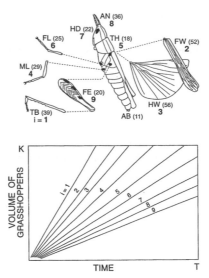

Figure 9.2. Prey-preparation tactics of grasshopper sparrows, based on Kaspari (1991). Diagram (*top*) shows an "exploded" view of a grasshopper and, specifically, the body parts that sparrows may remove. Each body part is labeled with the percentage of indigestible chitin (in parentheses) and the order in which sparrows tend to remove the body parts ($i = 1$ is the first body part removed in prey preparation.) Graph (*bottom*) shows how the removal tactics relate to the rate of intake and the K (maximum intake) and T (time available) boundaries.

9.2.6.1 The Relation between Time and Energy Constraints

The foraging of grasshopper sparrows (*Ammodramus savannarum*) illustrates how time and energy constraints interact (Kaspari 1990, 1991). These birds capture large insect prey such as grasshoppers and often expend considerable time handling them, discarding some parts before consuming the remainder. Why do grasshopper sparrows discard food they could eat? How much time should they invest in prey preparation? The options for preparing a grasshopper for consumption range from no preparation ($i = 1$) to removal of all body parts except the abdomen and thorax, which are always eaten ($i = 9$) (fig. 9.2). Each successive option requires more handling time ($h_{i+1} > h_i$) and produces a smaller item for ingestion (volume $v_{i+1} < v_i$) with less energy in it ($e_{i+1} < e_i$), although the energy density (e/v) of the item is higher.

Suppose that a sparrow can ingest and process a total volume K of grasshoppers over a period of time T, and that the encounter rate with grasshoppers during search is λ. (The encounter rate is the rate at which prey items are encountered during search. Search time does not include the handling time. The expected search time to find a prey item is $1/\lambda$.) With little or no prey preparation (low values of i on fig. 9.2), gut capacity is quickly reached and the sparrow will be forced to spend some of the day resting. As defined, I call this phenomenon energy limitation. The gross daily energy gain is $B_i = e_i n_i$, where n_i is the number of prey items consumed ($n_i = K/v_i$). The level of prey preparation that maximizes the gross daily gain can be found by locating the value of i such that at the next higher level of preparation ($i + 1$) the sparrow will reach T before K is fully processed. (In fig. 9.2 these are $i = 5$ and $i + 1 = 6$.) The total daily energy intake is the rate of intake multiplied by the time available, $B_{i+1} = T (\lambda \cdot e_{i+1})/(1 + \lambda h_{i+1})$. As defined, this is time limitation.

The intake-maximizing level of prey preparation occurs when no extra gain can be had by increasing preparation and obviously depends on λ. (Because the levels of preparation occur in quantum jumps, the best level will often lie in between i and $i + 1$ but will be unattainable. The sparrow must choose which of i and $i + 1$ yields greater gain). As prey are increasingly easy to find, the sparrow is able to spend more time in preparation of each item so that the material it processes is of higher quality. The encounter rate at which the sparrow should switch to the next higher level of prey preparation can be found by setting $B_i = B_{i+1}$ and solving for λ, which yields $\lambda = (e_i n_i)/(e_{i+1} T - n_i e_i h_{i+1})$.

Kaspari (1990, 1991) found that sparrows removed parts from captured grasshoppers in reverse order of nutrient density; this behavior supports his "nutrient concentration" hypothesis. He also found that the amount of preparation varied as predicted with the availability of large prey. Taghon et al. (1978) advanced a similar idea to explain the size selection of particles for ingestion by deposit feeders, which use another feeding mode that likely utilizes the full gut capacity.

One way to think about the sparrow's tactics (levels of preparation i) is that they differentially use up the time and processing capacity the sparrow has available. Little preparation quickly takes up processing capacity because the prey are bulky, but extensive preparation uses up time in the form of handling. The intake-maximizing tactic will occur at the level at which time and processing capacity are used as fully as possible.

9.2.7 A Simple Feeding Model

The following simple model of feeding behavior illustrates how rate and efficiency are related. It is presented in full by Ydenberg and Hurd (1998). When feeding, a forager chooses among a number of foraging options $i = 1, 2, 3, \ldots n$. (In any particular model, the types of options may be the prey sizes to capture, flight speeds, or patch times; however, for now option type is not specified). Each option has an associated rate of energy gain during foraging, b_i, which depends on the rate of energy expenditure, c_i, as $b_i = f(c_i)$. An

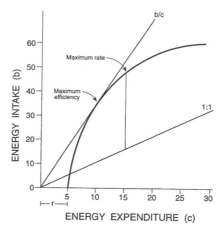

Figure 9.3. Example of the relation between the rate of energy expenditure on feeding (c) and the rate of energy intake [$b = f(c)$]. The specific function shown here [$b = 60 \, (1 - e^{-0.25(c-r)})$] is used in the example given in Ydenberg and Hurd (1998). The basic assumption is that harder work (higher c) shows diminishing returns (b). The efficiency (b/c) and rate ($b - c$) maxima are indicated.

example is shown in figure 9.3. The central and likely very general assumption about the form of $f(c_i)$ is that there are diminishing returns on higher rates of energy expenditure. The resting metabolic rate r is indicated on the x-axis and can be thought of as the rate of energy expenditure at which there is no energy gain ($b = 0$). The different rates of energy expenditure possible are the foraging options or tactics, and they are deployed by the forager in support of a strategy. The maximum net rate ($b_i - c_i$) and maximum efficiency (b_i/c_i) tactics are indicated on figure 9.3. Note that rate maximizing involves harder work (higher c) than efficiency maximizing.

To illustrate how rate and efficiency can be observed, we calculate the total gain during foraging (G_f) and the total net daily gain (G_d) under time and energy limitations. The equations are developed in appendix A. Their form is designed to illustrate the relation of the behavioral choices depicted in figure 9.3 to the two strategies. The basic results are straightforward. The most proficient tactic depends on the strategy (maximize net gain during foraging or total net daily gain) and the constraints (time or energy limitation). Efficiency, or the closely related "modified efficiency" (in which the energy expenditure rate is discounted by the resting metabolic rate r), is important when the energy constraint K operates. Simply put, the forager should choose the option that maximizes rate when there is a shortage of time and the option that maximizes efficiency when there is a shortage of energy. Indeed, this is what Stephens and Krebs (1986, 9) wrote in the single paragraph that is devoted to the idea of efficiency in their book.

This approach shows that recognizing a diversity of feeding contexts within a simple framework can accommodate a variety of outcomes. This approach also shows that distinguishing between what ecologists observe (rate or efficiency) and what the forager attempts (with its choice of behavior) to maximize is important. The framework also makes clear how rate-maximizing and efficient tactics are related and under which circumstances each is to be preferred. With the framework, it is also indicated what information the forager needs to be able to make a choice of behavior. The forager needs to evaluate the foraging context (feeding or provisioning), the strategic goal, the constraints (time or processing limitation), and the rate of energy intake as it relates to the rate of expenditure for each behavioral option. One way to think about the behavioral options is as a gradient of successively harder working tactics, and the feeder "gears up" its choice of tactic until the desired outcome can be achieved. Factors other than purely energetic considerations (notably the risk posed by predators) may also cause foragers to adjust their choice of behavior if the different tactics are differentially vulnerable. These factors will be considered further in the following.

9.2.8 A Provisioning Model

In provisioning, unlike in feeding, foragers deliver the prey they capture to offspring, nest mates, or a hoard. The costs and benefits differ from those in feeding situations because the delivered resource is not for the provisioner's immediate consumption; therefore, foraging proficiency must be measured in currencies different from those in feeding situations. In fact, the currency "net rate" seems reasonable for feeders, but in a provisioning context, the meaning of this currency is not so clear.

In a feeding context, net rate is calculated by subtracting the forager's energy expenditure from its energy intake. A provisioner, however, does not consume the delivered prey; the food items are delivered to and consumed by nestlings (for example). Hence, it usually does not make sense to subtract from the energy value of delivered prey the energy expenditure of the provisioner who captured and delivered the prey. Nevertheless, in many published models, this is exactly the calculation that is performed. It seems more sensible that the provisioner maximizes delivery over some period, such as a day, and maintains energy balance by spending some time and energy self feeding. Accordingly, I use the currency "total daily delivery" in provisioning situations.

In the basic provisioning model, the provisioner may choose among a number of foraging options $i = 1, 2, 3, \ldots, n$. Option i yields a rate of prey delivery d_i and has a rate of energy expenditure c_i. As noted, the rate (d_i) and efficiency maximizing (d_i/c_i) tactics can be identified. (Note that in a provisioning context these currencies consider delivery d_i instead of intake

b_i). During the day, the forager has time T to forage, and its total energy expenditure can not exceed K, which is set by the maximum amount of energy the forager can ingest and assimilate in a day. The provisioner is assumed to choose the provisioning tactic i that maximizes the energy the provisioner delivers each day; however, the provisioner is subject to the restriction that its energy budget is balanced. To remain in energy balance, the provisioner must recover by self feeding the energy it expends on delivery.

The equations are developed in appendix B. When the provisioner's assimilation capacity (K) is reached in self feeding (energy limitation), the behavioral tactic that maximizes delivery is the tactic with the highest "modified efficiency" ($d_i/[c_i - r]$); with this tactic, the energy-expenditure rate is discounted by the resting-metabolic rate (Houston 1995). Efficiency is also important when the provisioner runs out of time (equation B2). Ydenberg et al. (1994) showed that, in this situation, the tactic that maximizes total daily delivery lies between the efficiency-maximizing and the rate-maximizing tactics; the exact choice of behavior depends on the net self-feeding rate ($b_s - c_s$) (see Houston 1995 for further discussion.) The provisioner must feed itself to power the provisioning behavior, and equation B2 indicates that the provisioner's choice of behavior changes from use of the tactic with maximum efficiency when the self-feeding rate is low to use of the tactic with the maximum rate when the self-feeding rate is high.

The predicted behavior depends critically on the self-feeding rate, and the results would be sensible if the self-feeding rates were low in the studies that demonstrated efficiency and high in the studies that demonstrated rate maximizing. However, the only study that attempted to account for the self-feeding rate was that of Waite and Ydenberg (1994a, 1994b) on grey jays (*Perisoreus canadensis*) during late summer in central Alaska. The jays hoarded experimentally offered raisins, but fed themselves on blueberries, which were abundantly available on the forest floor. A jay takes only a few seconds to interrupt a flight and quickly gulp a few berries, and so the self-feeding rate was very high (although it was difficult to quantify exactly). Owing to the high self-feeding rate, we reasoned the behavior of hoarding grey jays should approach rate maximizing instead of efficiency maximizing. In our experiments, jays hoarded raisins proffered at a feeding board. On each visit to the board, the birds took the single raisin available on arrival and could have departed immediately or waited for a short (experimentally controlled) interval for two more simultaneously available raisins. When the intervals were short, it was worthwhile to wait and hoard three raisins; however, as the interval was experimentally lengthened, the single-raisin option eventually became more profitable. The critical time interval was very different under rate- and efficiency-maximizing, and we found that the time at which grey jays switched to single-raisin loads was predicted better with rate maximizing.

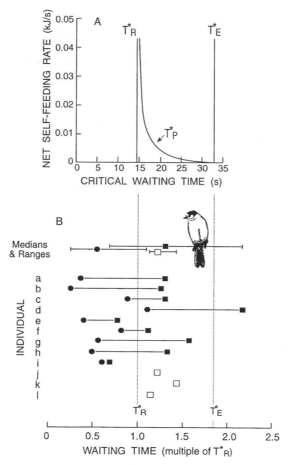

Figure 9.4. (*a*) The critical waiting time in the grey jay experiment is very different with rate (T_R^*) and efficiency (T_E^*) currencies. The critical waiting time with the provisioning model (T_P^*) depends on the self-feeding rate (see equation B2). The self-feeding rate is low near the efficiency level, and the self-feeding rate is high near the rate maximum. (*b*) The critical waiting time of individual grey jays differs between summer (*circles* = high self-feeding rate) and winter (*filled squares* = low self-feeding rate) and shifts toward the efficiency maximum in winter. Open squares show the waiting times measured only in winter of three individuals, so the result is not an order effect (from Waite and Ydenberg 1994b).

The critical waiting time, however, depends on the self-feeding rate, as shown in figure 9.4*a*. A test of the provisioning model would repeat the described experiment with a manipulated change in the self-feeding rate and the critical waiting time would shift as described in figure 9.4*a*. Such a test was designed by Waite (Waite and Ydenberg 1994b). He was unable to manipulate the self-feeding rate directly, so he instead repeated the experiment during winter when the self-feeding rate was reasonably assumed to be much lower than that during the peak of blueberry availability in late summer. The same individuals that behaved as rate maximizers during the summer changed their behavior significantly to that of efficiency maximizing during the winter (fig. 9.4*b*).

How are grey jays able to evaluate and perform such subtle shifts of behavior? To make a choice of behavior, each individual needs to evaluate the context, the constraints, the rate of delivery relative to the rate of expenditure of each behavioral option, and the self-feeding rate. Like feeding behavior, provisioning behavior can be represented as a range of successively higher levels of work; the level that can be sustained is dependent on the self-feeding rate.

9.3 FORAGING AND INFORMATION

9.3.1 Information Use in Patch Foraging

The simple models developed were useful for illustrating the relationships between currencies, strategies, tactics, and decision mechanisms; these models rely on highly simplified and general assumptions. Most relevant to the following discussion is the assumption that foragers know all of the relevant characteristics of their foraging habitat; foragers can recognize instantly the prey or patch types they have encountered. However, like the bumblebee flying to an inflorescence whose contents she cannot know, foragers are often likely to be uncertain about many aspects of a foraging situation. Their decision mechanisms must incorporate the certainties and uncertainties of any particular situation; the exact nature of this information is likely specific to the ecology of the species.

Consider a forager that feeds by traveling from patch to patch, and each patch consists of two holes (called "bits" by Green 1980). A forager cannot know if any particular hole contains food without examining it. There is a variety of tactics the forager may use: The forager may always open both holes (exhaustive search); it may open the second hole only if the first hole contains a prey (stay after find); it may open the second hole only if the first hole is empty (leave after find); or it may only visit the first hole (partial search) regardless of circumstances.

The most proficient foraging tactic depends on the way the prey are distributed among patches. Compare three cases (fig. 9.5). The overall average density of prey per patch is one in each case, but the prey are distributed differently among the patches. In the clumped habitat, patches contain no or two prey (both holes full or both holes empty), and in the regular habitat, all patches contain one prey (the first hole is full and the second is empty or vice versa).

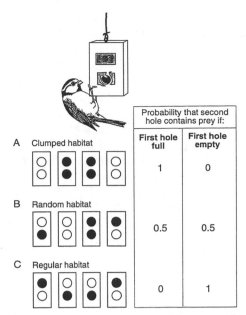

		Probability that second hole contains prey if:	
		First hole full	First hole empty
A	Clumped habitat	1	0
B	Random habitat	0.5	0.5
C	Regular habitat	0	1

Figure 9.5. Simple conceptual habitats illustrate the importance of information use in patch foraging. Circles indicate the two holes in each patch; a filled circle indicates the hole contains prey, although the prey is not visible to the forager before it opens the hole. The average density of prey in all habitats is one per patch, but the distributions are very different. In habitat *a*, prey are clumped; in habitat *b*, prey are randomly dispersed; and in habitat *c*, prey are spaced regularly. As indicated, the discovery of prey in the first hole provides information about the presence of prey in the second hole. Foragers can use this information to elevate the rate of prey capture.

In the random habitat, prey are distributed in each hole with p equal to .5, so that patches contain no, one, or two prey.

To understand the function of information in foraging problems, assume the forager knows the distribution (clumped, regular, or random) of prey, but that the forager cannot recognize patches externally. Examination of the first hole in a patch provides the forager with information about the presence of prey in the other hole. In the clumped habitat, finding a prey item in the first hole means that the second hole also has prey, and the absence of prey in the first hole means the second hole is also empty; therefore, examining the second hole would be a waste of time. The forager should stay in the patch after prey has been found and leave when prey has not been found.

In the regular habitat, however, a prey item in the first hole means the second hole is empty, and an empty first hole means prey is in the second hole. The forager should leave after finding prey and stay when no prey is found in the first hole. In the random habitat, the presence or absence of prey in a hole provides no information about the likelihood that the next hole contains food.

The best foraging tactic depends on the distribution of prey. Generally, the tactic "stay after find" is best when prey are clumped, and the tactic "leave after find" is best when prey are evenly dispersed (compare Iwasa et al. 1981). In the random habitat, exhaustive search is best and partial search (always leave after examining the first hole) is never favored because there is no diminishing return on patch residence. The exact solution depends on the proportions of patches with no, one, and two prey and the handling and travel times involved. By calculating the expected gain from further patch residence at each stage in the process (e.g. at patch arrival and after examination of the first hole), the conditional probability that the patch contains more prey is found. The expected gain is compared with the gain that is available by moving to the next patch. Behavior can be adjusted to give the maximum expected rate of gain. This was first derived by Oaten (1977). Green (1980) and McNamara and Houston (1987a) give clear quantitative examples (see also Stephens and Krebs 1986, chap. 4).

9.3.2 The Value of Information

How important is it for foragers to use information? One way to answer this question is to compare how well a forager performs with use of its best tactics when information is and is not used (Stephens 1989). In the example developed, use of information is advantageous when prey are clumped and regularly dispersed. The gain achieved with the best "no information" tactic (exhaustive search) is compared with that attained with use of the best information-use strategy ("leave after find" in the regular habitat and "stay after find" in the clumped habitat). Unfortunately, information use in foraging theory has received limited theoretical and experimental attention (but see Stephens 1987; Kamil et al. 1993; and Beachly et al. 1995 for examples of foraging). This is an essential area for further work because information use is perhaps the most relevant application of foraging theory in cognitive ecology.

A cognitive ecologist who is interested in knowing if foragers evaluate information as suggested by this simple framework could perform the following experiment: A forager is given successive choices between arms of an apparatus. Only one arm offers a food reward, and the forager does not know which arm. The choice is presented at (experimentally) controlled intervals, and at each trial, the forager must select one of the arms. With reward size e and interval duration t, the rate of gain, assuming that the forager's choices are made randomly (therefore proportion p of the choices are correct), is pe/t. The experiment, however, gives the forager another option: The forager may choose a key that reveals, after an additional wait, w, which arm will deliver the reward at that trial. The setup is similar to a paradigm in experimental psychology called "observing" (see Dinsmoor 1983). When information is used, the forager's gain is $e/(w + t)$, and the value of information use is $v = e/(w + t) - pe/t$, which is positive so long as $p < t/(w + t)$. The quantities p, t, and w can be manipulated, and the situations in which the forager ignores or uses information can be specified. The experiment may reveal how flexible foragers are and may yield insight into the types of mechanisms used by showing when performance matches expectation.

In most existing models of information use in patch foraging, it is assumed that foragers know the characteristics of the distribution of patch types. This assumption may be useful when as examining some types of foragers, but this assumption is likely to be too restrictive. For example, "area-intensive search" has been observed in thrushes (Smith 1974) and parasitoid wasps (Waage 1979). Based on the above, area-intensive search should occur when prey are clumped. Zach and Falls (1977), working with the ground-foraging ovenbird (*Seiurus aurocapillus*), showed that intensive search of the immediate locale followed prey discovery. However, when prey dispersion was altered experimentally to be regular, ovenbirds learned to delay area-intensive search after a prey was found and moved on a bit after a prey capture before beginning to search intensively. Evidently some foragers can learn the distribution of prey and update their knowledge as they go.

9.3.3 Information and Cognitive Ecology

How do foraging animals handle these complexities? Certain assumptions about the distribution or other attributes of prey are undoubtedly "hard-wired" into behavior, and individuals may be unable to adjust facultatively these aspects of their behavior. Bumblebees, for example, usually begin searching inflorescences at the bottom and work upward. This behavior seems sensible because nectar is typically concentrated in lower florets. Each species probably has certain properties hard-wired into behavior; at the same time, each species possesses mechanisms that enable adjustment of behavior to variable aspects of the foraging world. Which aspects are learned probably depends intimately on the ecology of the animal (Dukas, this vol. section 4.5). For example, honeybees are predisposed to use the leave-after-find tactic (often called win-shift tactic) (Demas and Brown 1995). Use of this tactic is sensible because florets are empty after visits. Experimental manipulations, however, revealed that honeybees can learn to use the stay-after-find tactic. By contrast, great tits (*Parus major*) seem predisposed to use the stay-after-find tactic; however, they can learn eventually to use the leave-after-find tactic (Ydenberg 1984). For great tits, use of the leave-after-find tactic is perhaps sensible because much of their insect prey is in a clumped distribution. General statements, however, about aspects of foraging behavior that are hard-wired and those that can be adjusted by the animal are not yet possible. An important goal of cognitive ecology is to develop a much more complete understanding of how hard-wired and adjustable mechanisms work together in the mechanics of foraging behavior.

Lima (1983) studied information use in the foraging of downy woodpeckers (*Picoides pubescens*) by creating a habitat of artificial patches (short lengths of branches with 24 holes each). He hung the patches from trees in a woodlot during winter. Each hole in each patch could contain a small seed, but each hole was covered by tape. The woodpeckers had to drill through the tape to obtain the seed. Woodpeckers flew from patch to patch and moved systematically from hole to hole within a patch. Lima stocked his experimental habitat with two types of patches, which could not be distinguished externally because of the tape. Patches were either empty or contained some seeds. The experiments tested the hypothesis that woodpeckers use their experience with the first few holes to discriminate between the patches.

This experiment is directly analogous to the simple experiment portrayed in figure 9.5; however, instead of two holes there were 24 in each patch. In the most directly comparable treatment, Lima's empty patches contained either no or 24 seeds; thus, opening the first hole revealed the patch type. A seed in the hole showed that the patch was full and should be exploited, and an empty hole showed that the patch was empty and should be abandoned. Lima (1983) further challenged the woodpeckers by presenting two other clumped distributions; one contained patches with no or 12 seeds, and another contained patches with no or six seeds. Now more holes had to be examined before patches could be classified as empty or full. Of course as soon as a seed was found, the patch was classified as full. But, what if no seeds were found in the first few holes? Lima calculated that a patch should be abandoned if the first three (half-full treatment) or six (quarter-full treatment) holes that were opened were empty.

Early in the morning, the patches that were carefully prepared the previous evening were set out. The exploited patches could be examined later in the day, and the pattern of opened and unopened holes could be used to measure the woodpeckers' foraging. The first prediction was that woodpeckers would open all of the holes or only a few holes in each patch. The second prediction was that the classification should be largely correct; all of the holes should be opened in the full patches and only a few holes should be opened in the empty patches. Finally, the number of holes opened in patches that were classified by woodpeckers as empty (i.e. abandoned after only a few holes were opened) should be one (0/24 distribution), three (0/12 distribution) or six (0/6 distribution).

The woodpeckers' performance exceeded all expectations. They showed that they quickly learned to classify patches correctly as empty or full because they opened only a few holes on some patches and all the holes on other patches. Their classification of the patches was highly accurate. Moreover, the number of holes sampled before accepting or abandoning a patch agreed quantitatively with the predictions. About the only thing the woodpeckers did not do was count the number of seeds and abandon patches if all the seeds were found before all the holes were opened. Once a patch was classified as full, it was completely exploited.

How is it that woodpeckers performed so well on an apparently difficult foraging task? One hypothesis is that the birds have a generalized foraging strategy that is complex enough to handle any food distribution. Another hypothesis is that woodpeckers are predisposed to categorize patches as empty or full. The task was not difficult because it resembled the type of prey distribution naturally faced. Not enough is known about woodpeckers and their natural history to determine which explanation is correct. It would be instructive to repeat Lima's experiment with clumped or regular prey distributions and see if woodpeckers perform as proficiently.

Sometimes patches have externally recognizable features that give information about their contents. Large and small inflorescences can be distinguished at a distance, for example. Why, as Pyke (1982) observed, should a bumblebee that arrives at a larger inflorescence be more persistent and investigate more florets before abandoning the inflorescence? The gain from a larger inflorescence would be relatively greater if its florets contained nectar. The potential reward in a larger inflorescence makes it worthwhile to spend a little extra time determining if the patch is empty or full. McNamara and Houston (1987a) gave a quantitative example of the calculations required.

In summary, problems regarding information use represent perhaps the most important application of foraging theory in cognitive ecology, but these problems have received as yet comparatively little examination. The decision mechanisms used likely combine hard-wired and learned aspects; the exact combination depends intimately on the ecology of the species under consideration. Foraging theory can be used to study decision mechanisms. First, the nature of the problems that have to be solved can be analyzed. Second, experimental situations can be created in which foragers reveal details of the function of decision mechanisms.

9.4 FORAGING UNDER THE RISK OF PREDATION

9.4.1 The Importance of Predation Risk

Ecologists have studied predation for almost a century, but the vast majority of this work has considered only how actual deaths influence demography. Until recently, the effects of the behavioral changes that individual animals may make when conditions are dangerous have been ignored. Now ecologists have begun to realize that the *risk* of predation is often more important than actual predation events (Lima and Dill 1990). The following anecdote illustrates. While I discussed with a student her research on molting geese in northern Québec, I wondered if foxes pose any danger when the geese are flightless for several weeks. The student replied that she had never seen a fox catch a goose; whenever a fox was anywhere in the vicinity, the geese stopped grazing and fled onto one of the many small lakes, where they were safe, and remained on the lake until the fox departed. This seemed a very sensible answer, and ecologists have often discounted the influence of predators in exactly this way. However, for a molting goose, the risk that a fox may approach presumably restricts grazing to areas near lakes; being flightless, any goose that grazed far from the safety of a lake would become very quickly a meal for a fox. Thus, foxes are most important in the ecology of molting geese—even if foxes never actually catch a goose. In the same way, many animals are always at some predation risk, and their behavior presumably reflects this basic fact.

The first formulation of a behavioral hypothesis that explicitly recognized the role of predation risk was made by Rosenzweig (1974). He suggested that a seasonal pattern in desert rodents of avoiding moonlight may be explained with individual animals weighing the risks of activity under moonlight (greater risk of capture by owls) against the benefits (access to mating opportunities), the balance of which changed and benefits changed as the breeding season approached. Steve Lima expanded this theme in a series of papers that explored phenomena such as body mass (1986) and clutch size (1987a). He (1986) also explicitly stated that the effect of behavioral alterations may have a much larger impact on demography than actual deaths. In more recent work (e.g. Peckarsky et al. 1993), these sublethal consequences have been addressed; the presence of predators reduces the activity, growth, and fecundity of animals.

To learn to interpret animal behavior that relates to predation risk, behavioral ecologists have tested hypotheses that are analogous to that of Rosenzweig (1974) described. Many experiments (reviewed by Lima and Dill 1990) have been created to observe how an individual animal behaves when placed in a situation in which the costs (predation risk) or benefits (access to food or matings) were manipulated. Several examples are given in figure 9.6. The results of these simple experiments are generally those expected. When risk is higher, animals adopt safer tactics; when the benefit to be gained is greater, animals adopt riskier tactics. The interpretation of the behavioral adjustments observed is often straightforward; however, because there are many subtle effects that cannot be so easily observed in simple, manipulative experiments, the role of behavioral adjustments to predation risk in large-scale ecological phenomena such as vertical migration (Clark and Levy 1988) and population

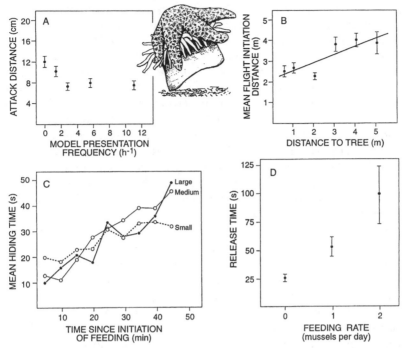

Figure 9.6. Experiments in which foraging responses of animals to risk manipulations were examined. (*a*) The distance from which juvenile coho salmon attack a drifting prey item decreases as the model of a predatory trout is presented more often (Dill and Fraser 1984). (*b*) Gray squirrels flee from an approaching predator at a greater distance as the feeding site is placed further from the safety of a tree (Dill and Houtman 1989). (*c*) After an experimental disturbance, barnacles take longer to reemerge if they have been able to feed for a time before the disturbance (Dill and Gillett 1991). (*d*) Anemones take longer to release their pedal disc from the substrate to escape the attack of a predatory starfish if they have experienced good feeding at the site (Houtman et al. 1997).

fluctuations (Dehn 1994; Norrdahl and Korpimäki 1996) is becoming clear only slowly. The role of predation risk in life-history factors such as in the costs of reproduction (Harfenist and Ydenberg 1995) and traits such as the timing of metamorphosis (Rowe and Ludwig 1991) are also receiving increased attention.

9.4.2 Predation Risk in Foraging Models

In the foraging models outlined, foragers choose among tactics that vary in the rate of gain and energy expenditure. If the tactics also vary in the extent of exposure to predators that is required, then predation risk should influence the decision. The interesting situation is one in which the gain and the risk trade off so that the largest gains come at the expense of the largest risks. (If the largest gain comes with the smallest risk, the choice is obvious.) An illustrative situation considers vigilance during feeding (Lima 1987b, Roberts 1996). Safety increases with the time spent in vigilance, but the rate of feeding slows.

Predation risk and foraging often seem to trade off, and many such situations have been described. For example, animals may have to leave the relative safety of cover or a refuge to feed in the open (Lima 1987b; Sih 1992). The most dangerous habitats may contain the best food (e.g. Prins 1989). By moving to forage, cryptic animals may lose their camouflage (Houtman 1995). Defenses that protect against predators, such as thick shells, reduce growth rate (Palmer 1985). Infaunal organisms may increase their safety by burying more deeply; however, by doing so, these organisms reduce their feeding rate (Zaklan and Ydenberg 1997). Vertical migrants are better able to feed near the surface because there is more light but they are more exposed to predators (Clark and Levy 1988). In each of these examples, the forager can increase the feeding rate only at the expense of increased exposure to predators.

One misconception about predation risk is that it is often so small that it can, for all practical purposes, be ignored. For example, Piersma et al. (1993) asserted that the risk posed to each knot (*Calidris canuta,* a species of sandpiper) in a flock of 30,000 by the occasional appearance of a distant falcon is minuscule. However, the point is not that the risk is small; it is that the occasional presence of falcons confines sandpipers to big flocks in open areas, because individual birds taking up other options would be at greater risk. The relevant quantity is how much the risk is reduced by the flocking habit.

Even a minuscule reduction in risk can be important if the risk reduction pertains to an action that is repeated many times during life. For the parental songbird, the predation danger per provisioning visit to its nest is tiny—perhaps on the order of 10^{-6} (Ydenberg 1994). However, because a parent songbird may make thousands of provisioning visits to its brood, the compounded risk is substantial and may have a strong influence on the clutch size (Lima 1987a).

The idea that tiny risks can be ignored has its complement; namely, tiny benefits should be ignored in the face of large risks. For example, it has been asserted that prey should flee as soon as predators are detected (e.g. Myers 1983). However, as Ydenberg and Dill (1986) showed, this should not and does not occur. Instead, the decision to flee is adjusted according to the costs and benefits of each situation; fleeing is delayed as the opportunity foregone by fleeing increases.

9.4.3 Information about Predation Risk

Foragers can never be certain that there is not a predator lurking just around the next corner, and presumably foragers always behave as though a predator may be present—even in laboratory experiments. With use of the same logic that predicts proficiency in feeding, we expect natural selection has led to the evolution of keen sensory abilities that enable detection of predators and clues about the presence of predators; this prediction is hardly powerful or surprising. More intriguing is the idea that nervous systems allow adjustment of behavior on the basis of information that is received proficiently. We expect proficiency because the obvious benefits of behavior that reduce predation risk must be balanced against the costs—the foregone opportunity to feed, seek mates, or care for offspring. Individuals that behave too cautiously on the basis of information received about predators forego too much opportunity, and animals that are too cavalier live less long. Bouskila and Blumstein (1992) considered how errors in the assessment of hazard influence foraging decisions; they showed that (over a certain range) errors have little impact on survival, but it is better to be too cautious than too bold. Sih (1992) developed a model on the basis of Bayesian statistics to show how foragers may update their estimate that a predator is still present. He also concluded that erring on the side of caution is selected for and provided an experimental test.

Mangel (1990) discussed three desirable attributes of an information-processing theory: (1) a decay of memory so that recent events have a bigger impact than remote events on the selection of current tactics; (2) a succinct estimate of the parameter of interest; and (3) flexibility of the estimate through a consistent treatment of uncertainty. There has been very little work in which any of these aspects as they relate to proficiency in reacting to information about predators has been examined. As discussed, relevant questions about foraging decisions include the nature of the decision mechanisms, the currencies, the possible tactics, and the strategies involved.

The sculpin, *Oligocottus maculosus,* inhabits tidepools along the Pacific coast of North America. Sculpins are cryptically colored and forage by lying motionless on the bottom of tidepools and occasionally lunging forward to capture small items. Their natural history creates a dilemma; camouflage works best when sculpins are still, but foraging requires movement. Houtman's (1995) study of their behavior is an instructive example of how cues about the presence of predators may be integrated into foraging decisions.

The model conceptualized foraging in a series of intervals, during each of which a decision was made. The decision was to reject an encountered prey item (i.e. remain still and cryptic) or attempt prey capture. Captures could be made at any of four distances; greater distances increased the rate of prey capture (because a larger area was searched) and predation risk (because more movement was required). At the end of each interval, the forager updated its estimate of the probability that a predator was present (called φ) and used that estimate in its decisions in the next interval.

The updating of φ took place in three steps, and three sources of information were incorporated. If a predator was detected during the previous interval (indexed as $t - 1$), the value of φ of course became 1.0. However, information was available even if there was no detection because one of three events must have happened: (1) There was no predator present. (2) A predator was present but did not detect the sculpin. (3) The predator was present, detected the sculpin, but did not attack. The forager had an estimate that a predator was present $[\varphi(t - 1)]$, and it was assumed that the forager knew the probabilities of detection and attack. On the basis of likelihoods of these events, $\varphi(t - 1)$ was updated to an intermediate value, $\varphi(t')$. The updating procedure used Bayes's theorem, and the formulae are given in the dissertation by Houtman (1995).

The second updating step accounted for predator movements. During any interval, a predator that was present may have departed or a predator may have arrived. In the model, predator movements were described by a first order Markov chain, and the transition probabilities were assumed to be known to the forager; these probabilities allowed the forager to update $\varphi(t')$ to $\varphi(t'')$. The final updating step incorporated information during vigilance, which was acquired during a scan of the surroundings. There were four possible outcomes: (1) No predator was detected and no predator was present. (2) A predator that was present was not detected. (3) A predator that was present was detected. (4) A predator was falsely detected. Either outcome (detecting or not detecting a predator) enabled the updating of $\varphi(t'')$ to $\varphi(t''')$, which became the estimate of φ used during period t.

This model contains many simplifying and possibly unreal assumptions. For example, the forager was assumed to know features of the habitat such as the arrival and departure rates of predators and how predators behaved (probabilities of attack and detection). In reality, the values of these features are more likely uncertain and must be estimated by the forager. Nonetheless, the model meets all three of Mangel's (1990) criteria for a model of information processing, especially with the consistent treatment of uncertainty, and the model, therefore, is valuable. Figure 9.7a shows an example that is calculated with the model of how φ changes after a forager has seen a predator.

Houtman (1995) integrated this information-processing procedure into a model of sculpin foraging. He assumed that foraging decisions were made to maximize survival—which required avoiding both predators and starvation—and he used dynamic programming (Mangel and Clark 1988) to calculate the strategy used to make attack decisions, relative to the current estimate of φ, and the amount of food reserves (fig. 9.7b). Sensibly, the model forager was more cautious when φ was high, and the forager took greater chances when its energy reserves were low.

9.5 CONCLUSIONS

In this chapter, I examined models of foraging and antipredator behavior to provide a basic framework that can be integrated into building the new disci-

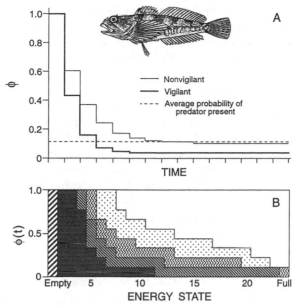

Figure 9.7. (*a*) Change in φ after a tide pool sculpin has detected a predator at time 0. Heavy solid line labeled "vigilant" represents the calculation in the text. Light solid line labeled "nonvigilant" represents the same calculation, except the forager did not use the third updating step (φ″ to φ‴). Horizontal dotted line represents the habitat average probability that a predator was present. (*b*) The survival-maximizing foraging strategy that governed prey-attack decisions depended on the energy state and the estimate that a predator was present [φ(*t*)]. In the hatched region, the forager died of starvation; in the clear region, the forager remained motionless without attacking prey. In the darkest shaded region, prey were attacked at the longest distance; in the lightest shaded region, prey were attacked at the shortest distance (from Houtman 1995).

pline of cognitive ecology. In the terminology used here, the alternative behaviors or options that a forager has in a particular situation are tactics. The choice among the tactics is made with a decision mechanism. I assume natural selection acts on the structure of the decision mechanism and favors variants that perform most proficiently. "Proficient" foraging best attains an objective, often termed a strategy, such as "maximize the net daily gain" or "minimize the foraging time to attain the requirement." Scientists measure the performance of foragers with a currency and often employ the procedure of optimization to predict the most efficient tactic. More careful use of these terms by foraging theorists will help resolve some of the difficulties that many ecologists have with optimal-foraging theory.

In this chapter, I distinguished two basic foraging processes, feeding and provisioning, and explored how time and energy limits affect strategies. In most theoretical work, rate (energy gain per unit time) currencies have been investigated; however, I concluded that there are many situations in which tactics that maximize the currency called efficiency (energy gain per unit energy expenditure) should be observed in feeding and provisioning situations. A review of the available empirical tests supported this conclusion, but much more careful quantitative work is required. Even in the well-studied area of foraging, this theory remains rudimentary.

A few foraging theorists have worked on the "problem" of information since Oaten (1977) published his article shortly after the publication of Charnov's (1976) marginal value paper. Despite its importance, Oaten's central point is not widely appreciated because perhaps of its apparent difficulty. In trying to give the essence of the problem, I have portrayed the basic idea in its most simple and broadest terms. Any theory of cognitive ecology will have to recognize the importance of information use in foraging behavior to a much greater extent than has been the case during the past twenty years.

In contrast to the reluctant adoption by researchers in foraging of information ideas, ecologists during the past decade have embraced quickly the idea that predation risk is important in foraging behavior. In this chapter, I stressed the points that even very small risks can be important and risks often trade off with foraging gains so that tactics with the highest gains are usually those with the greatest risk. This fact has general, fundamental implications in behavioral ecology and deserves more consideration. Because risk is pervasive, foragers presumably always behave as though a predator is present. Nevertheless, foragers can adjust their behavior to the perceived risk on the basis of information acquired. I closed the chapter by outlining one of the few models that considers the process of information use.

9.6 SUMMARY

In this chapter, I provided an introduction to models of foraging and antipredator behavior. I reviewed some issues that are currently relevant to optimality models of behavior and focused on general properties of these models (and especially the way "currency" is used). I analyzed feeding and provisioning situations and showed how the currency of efficiency (energy gain per unit energy expenditure) is related to the currency of rate (energy gain per unit time) and when each currency should occur. A review of the literature showed that rate and efficiency have been observed. I concluded that much more work must be done before the foundation of a good foraging theory is complete.

Foraging theory has concerned ultimate instead of proximate features of foraging behavior, and I discussed how foraging theory may contribute to the development of cognitive ecology, which concerns decision mechanisms. The analysis of information problems is likely to be central. Unfortunately, few ecologists have given much attention to this important area, and our understanding is such that even simple experiments can still be very instructive.

By contrast, over the last decade much attention has been focused on investigating ways that foragers deal with hazards posed by predators. I provided a brief overview of some current issues and closed by considering a model of how a forager may combine several sources of information to form an estimate of the likelihood that a predator is present and how this estimate may be incorporated into foraging decisions.

ACKNOWLEDGMENTS

Reuven Dukas, Peter Bednekoff, and Mark Abrahams provided valuable comments that improved greatly an earlier version of this chapter. Yoh Iwasa discussed the information experiments with me on a visit to Vancouver in 1984. Students in my graduate seminar, "Survival Strategies" (Spring 1996), enthusiastically discussed information aspects of foraging theory, and Liz Hui in that class developed the presentation of the grasshopper sparrow articles I used here. I thank all of these people for their help.

APPENDIX A: FEEDING MODELS

The time budget is composed of feeding and resting so total time, T, equals $t_f + t_r$. The forager has behavioral options $i = 1, 2, 3, \ldots n$. Energy expenditure during feeding is c_i and gives a rate of intake of b_i (fig. 9.3), and energy expenditure at rest is r. The maximum daily intake is K. Which behavior (choice of i) is best?

Energy Limitation

In energy limitation, the time spent feeding, t_f, is the time required to reach the intake capacity K at rate of intake b_i. If K is reached before T expires, $T - t_f$ must be spent resting. This is called energy limitation. The *total net gain while foraging*, G_f, is then $(b_i - c_i) t_f$. By substituting t_f with K/b_i,

$$G_f = K(1 - c_i/b_i). \tag{A1}$$

This equation corresponds to equation 1 in Ydenberg et al. (1994) and, as explained, is maximized by choosing the most efficient (i.e. highest b_i/c_i or, equivalently, lowest c_i/b_i) feeding option. The *total net daily gain*, G_d, includes the time spent resting. Therefore, we must subtract from G_f the energy spent resting $(t_r \cdot r)$. Then, $G_d = t_f(b_i - c_r) - (t_r \cdot r)$, which yields

$$G_d = K(1 - c_i/b_i + r/b_i) - Tr. \tag{A2}$$

An identical result is derived with equations 1 and 2a in Houston (1995), who showed further that G_d is maximized with the feeding option that provides the highest "modified efficiency," $b_i/(c_i - r)$.

Table 9A.1
Best Feeding Tactic Under Different Strategies and Constraints

	Feeding Strategy	
Constraints	Maximize Net Gain While Foraging (G_f)	Maximize Total Net Daily Gain (G_d)
No limits	Rate	Rate
Energy limitation	Efficiency	Modified efficiency
Time limitation	Rate	Rate

Time Limitation

As explained in the text, feeding is considered time limited when the available feeding time, t_L is less than t_f, so that K is not reached before t_L expires. With a time limit, the *total net gain while foraging* is

$$G_f = t_L(b_i - c_i). \tag{A3}$$

Taking resting time ($t_r = T - T_L$) into account, the *total net daily gain* is

$$G_d = t_L(b_i - c_i + r) - Tr. \tag{A4}$$

In both these conditions, the greatest total net daily gain is given with behavioral choice i that corresponds to the rate maximum.

The basic results are summarized in table A1.

APPENDIX B: PROVISIONING MODELS

A provisioner's time budget is composed of time spent in delivery, self-feeding, and resting. Therefore, total time, T, equals $t_d + t_s + t_r$. The time allocated to self-feeding, t_s, must be great enough to maintain positive energy balance. Energy intake and expenditure during self-feeding are b_s and c_s (so the net self-feeding rate is $b_s - c_s$), and energy expenditure at rest is r. The provisioner has behavioral options for delivery $i = 1, 2, 3, \ldots n$. Energy expenditure during delivery is c_i and gives rate of delivery d_i. The energy balance equation is $t_d \cdot c_i + t_s \cdot c_s + t_r \cdot r = b_s \cdot t_s$. The maximum daily expenditure that permits energy balance is limited by the maximum daily intake, K. I assume that the provisioning strategy maximizes the total daily delivery, D_i, while maintaining energy balance. Which behavior (choice of i) is best?

Time Limitation

If the K limit is not reached, no time needs to be spent resting (T = $t_d + t_s$) because the forager is able to "finance" delivery by self-feeding. The day's energy expenditure is recovered by self-feeding at rate b_s for time t_s, so that $b_s \cdot t_s = t_d \cdot c_i + t_s \cdot c_s$. Solving for t_d yields

$$t_d = \frac{t_s(b_s - c_s)}{c_i}. \tag{B1}$$

Equation B1 shows that the time available for delivery increases with the net self-feeding rate $(b_s - c_s)$ and declines as delivery becomes more expensive (higher c_i). Total daily delivery $D_i = d_i t_d$. Substituting from equation B1,

$$D_i = \frac{d_i t_s(b_s - c_s)}{c_i}. \tag{B2}$$

The term d_i / c_i indicates the importance of efficiency. Behavioral choices with a higher rate of energy expenditure increase the delivery rate, but also increase the amount of time required for self feeding. Hence, the attainable self-feeding rate determines the behavior that maximizes total daily delivery, which lies between the highest-efficiency and the highest-rate options (Ydenberg et al. 1994). Behavior changes from the most efficient option when the self-feeding rate is very low to the highest rate option when the self-feeding rate is high.

Energy Limitation

When K is reached, some time must be spent resting. Therefore, the energy balance equation is $t_d \cdot c_i + t_s \cdot c_s + t_r \cdot r = b_s \cdot t_s$. Solving for t_d and substituting into $D_i = d_i t_d$, we obtain

$$D_i = \frac{d_i[t_s(b_s - c_s) - t_r r]}{c_i}. \tag{B3}$$

Houston (1995) showed that under these conditions the tactic with highest "modified efficiency," $d_i / (c_i - r)$, gives the maximum delivery while maintaining energy balance.

Literature Cited

Alexander, R. M. 1982. *Optima for Animals*. London: Edward Arnold.

Beachly, W. M., D. W. Stephens, and K. B. Toyer. 1995. On the economics of sit-and-wait foraging: Site selection and assessment. *Behavioral Ecology* 6:258–268.

Bouskila, A., and D. T. Blumstein. 1992. Rules of thumb for predation hazard assessment: Predictions from a dynamic model. *American Naturalist* 139:161–176.

Charnov, E. L. 1976. Optimal foraging: The marginal value theorem. *Theoretical Population Biology* 9:129–136.

Cheverton, J. 1983. Bumblebees may use a suboptimal arbitrary handedness to solve difficult foraging decisions. *Animal Behaviour* 30:934–935.

Clark, C. W., and D. A. Levy. 1988. Diel vertical migrations by juvenile sockeye salmon and the antipredation window. *American Naturalist* 131:271–290.

Dehn, M. M. 1994. Optimal reproduction under predation risk: Consequences for life history and population dynamics in microtine rodents. Ph.D. diss., Simon Fraser University, Burnaby, British Columbia.

Demas, G. E., and M. F. Brown. 1995. Honeybees are predisposed to win-shift but can learn to win-stay. *Animal Behaviour* 50:1041–1045.

Dill, L. M., and A. H. G. Fraser. 1984. Risk of predation and the feeding behavior of juvenile coho salmon (*Oncorhynchus kisutch*). *Behavioral Ecology and Sociobiology* 16:65–71.

Dill, L. M., and R. Houtman. 1989. The influence of distance to refuge on flight initiation distance in the grey squirrel (*Sciurus carolinensis*). *Canadian Journal of Zoology* 67:233–235.

Dill, L. M., and J. F. Gillett. 1991. The economic logic of barnacle *Balanus glandula* (Darwin) hiding behavior. *Journal of Experimental Marine Biology and Ecology* 153:115–127.

Dinsmoor, J. A. 1983. Observing and conditioned reinforcement. *The Brain and Behavioral Sciences* 6:693–728.

Ens, B. J. 1992. The social prisoner: Causes of natural variation in reproductive success of the Oystercatcher. Ph.D. diss., Rijksuniversiteit Groningen, the Netherlands.

Green, R. F. 1980. Bayesian birds: A simple example of Oaten's stochastic model of optimal foraging. *Theoretical Population Biology* 18:244–256.

Gould, S. J., and R. C. Lewontin. 1979. The spandrels of San Marco and the Panglossian paradigm: A critique of the adaptationist programme. *Proceedings of the Royal Society of London,* ser. B 205:581–598.

Harfenist, A., and R. C. Ydenberg. 1995. Parental provisioning and predation risk in the rhinoceros auklet (*Cerorhinca monocerata*): Effects on nestling growth and fledging. *Behavioral Ecology* 6:82–86.

Hedenström, A., and T. Alerstam. 1995. Optimal flight speeds of birds. *Philosophical Transactions of the Royal Society of London,* ser. B 348:471–487.

Heinrich, B. 1978. *Bumblebee Economics.* Boston, Mass.: Belknap.

Houston, A. I. 1987. Optimal foraging by parent birds feeding their young. *Journal of Theoretical Biology* 124:251–274.

———. 1995. Energetic constraints and foraging efficiency. *Behavioral Ecology* 6:393–396.

Houtman, R. 1995. The influence of predation risk on within-patch foraging decisions of cryptic animals. Ph.D. diss., Simon Fraser University, Burnaby, British Columbia.

Houtman, R., L. R. Paul, R. V. Ungemach, and R. C. Ydenberg. 1997. Feeding and predator avoidance in the rose anemone *Tealia piscivora. Marine Biology* 128:225–229.

Iwasa, Y., M. Higashsi, and N. Yamamura. Prey distribution as a factor determining the choice of optimal foraging strategy. *American Naturalist* 117:710–723.

Juanes, F. 1992. Why do decapod crustaceans prefer small-sized prey? *Marine Ecology Progress Series* 87:239–249.

Juanes, F., and E. B. Hartwick. 1990. Prey size selection in the Dungeness crab: The effect of claw damage. *Ecology* 71:744–758.

Kacelnik, A. 1984. Central place foraging in starlings (*Sturnus vulgaris*). I. Patch residence time. *Journal of Animal Ecology* 53:283–299.

Kamil, A. C., J. R. Krebs, and H. R. Pulliam, eds. 1987. *Foraging Behavior.* New York: Plenum.

Kamil, A. C., R. L. Misthal, and D. W. Stephens. 1993. Failure of optimal foraging models to predict residence time when patch quality is uncertain. *Behavioral Ecology* 4:350–363.

Kaspari, M. 1990. Prey preparation and the determinants of handling time. *Animal Behaviour* 40:118–126.

———. 1991. Prey preparation as a way that grasshopper sparrows (*Ammodramus savannarum*) increase the nutrient concentration of their prey. *Behavioral Ecology* 2:234–241.

Kasuya, E. 1982. Central place water collection in a Japanese paper wasp (*Polistes chinensis antennalis*). *Animal Behaviour* 30:1010–1014.

Krebs, J. R., and A. Kacelinik. 1991. Decision making. In J. R. Krebs and N. B. Davies, eds., *Behavioral Ecology: An Evolutionary Approach,* 105–136. Oxford: Blackwell Scientific.

Kühnholz, S. 1994. Regulation der wassersammelaktiviteit bei honigbienen (*Apis mellifera*). Diplomarbeit, Julius-Maximilians-Universität, Wurzbürg, Germany.

Lima, S. L. 1983. Downy woodpecker foraging behavior: Efficient sampling in simple stochastic environments. *Ecology* 65:166–174.

———. 1986. Predation risk and unpredictable feeding conditions: Determinants of body mass in birds. *Ecology* 67:377–385.

———. 1987a. Clutch size in birds: A predation perspective. *Ecology* 68:1062–1070.

———. 1987b. Vigilance while feeding and its relation to the risk of predation. *Journal of Theoretical Biology* 124:303–316.

Lima, S. L., and L. M. Dill. 1990. Behavioral decisions made under the risk of predation: A review and prospectus. *Canadian Journal of Zoology* 68:619–640.

Mangel, M. 1990. Dynamic information in uncertain and unchanging worlds. *Journal of Theoretical Biology* 146:181–189.

Mangel, M., and C. W. Clark. 1988. *Dynamic Programming in Behavioral Ecology.* Princeton, N.J.: Princeton University Press.

Marrow, P., J. M. McNamara, A. I. Houston, I. R. Stevenson, and T. H. Clutton-Brock. 1996. State-dependent life history evolution in Soay sheep: Dynamic modeling of reproductive scheduling. *Philosophical Transactions of the Royal Society of London,* ser. B 351:17–32.

McNamara, J. M., and A. I. Houston. 1987a. Foraging in patches: There's more to life than the marginal value theorem. In M. L. Commons, A. Kacelnik, and S. J. Shettleworth, eds., *Quantitative Analyses of Behavior,* Vol. 6, 23–39. Hillsdale, N.J.: Erlbaum.

———. 1987b. A general framework for understanding the effects of variability and interruptions on foraging behaviour. *Acta Biotheoretica* 36:3–22.

McLaughlin, R. L., and R. D. Montgomerie. 1990. Flight speeds of parental birds feeding dependent nestlings: Maximizing foraging efficiency or food delivery rate? *Canadian Journal of Zoology* 68:2269–2274.

Montgomerie, R. D., J. M. Eadie, and L. D. Harder. 1984. What do foraging humming-birds maximize? *Oecologia* 63:357–363.

Myers, J. P. 1983. Commentary. In A. H. Brush and G. A. Clark, Jr., eds., *Perspectives in Ornithology,* 216–221. New York: Cambridge University Press.

Mylius, S. D., and O. Diekmann. 1995. On evolutionarily stable life histories and the need to be specific about density dependence. *Oikos* 74:218–224.

Norrdahl, K., and E. Korpimäki. 1996. Do nomadic avian predators synchronize popu-lation fluctuations of small mammals?: A field experiment. *Oecologia* 107(4):478–483.

Oaten, A. 1997. Optimal foraging in patches: A case for stochasticity. *Theoretical Population Biology* 12:263–285.

Orians, G. H., and N. E. Pearson. 1979. On the theory of central place foraging. In D. J. Horn, R. D. Mitchell, and R. D. Stairs, eds., *Analysis of Ecological Systems,* 154–177. Columbus: Ohio State University Press.

Palmer, A. R. 1985. Adaptive value of shell variation in *Thais lamellosa:* Effect of thick shells on vulnerability to and preference by crabs. *Veliger* 27:349–356.

Peckarsky, B. L., C. A. Cowan, M. A. Penton, and C. Anderson. 1993. Sublethal conse-quences of stream-dwelling predatory stoneflies on mayfly growth and fecundity. *Ecology* 74:1836–1846.

Piersma, T., R. Hoekstra, A. Dekinga, A. Koolhaas, P. Wolf, P. Battley, and P. Wiersma. 1993. Scale and intensity of intertidal habitat use by knots *Calidris canuta* in the western Wadden Sea in relation to food, friends, and foes. *Netherlands Journal of Sea Research* 31:331–357.

Price, K., and R. C. Ydenberg. 1995. Begging and provisioning in broods of asynchro-nously-hatched yellow-headed blackbird nestlings. *Behavioral Ecology and Socio-biology* 37:201–208.

Prins, H. H. T. 1989. Condition change and choice of social environment in African buffalo bulls. *Behavior* 108:297–324.

Pyke, G. H. 1981. Hummingbird foraging on artificial inflorescences. *Behaviour Analy-sis Letters* 1:11–15.

———. 1982. Optimal foraging in bumblebees: Rule of departure from an inflores-cence. *Canadian Journal of Zoology* 60:417–428.

Rapport, D. J. 1991. Myths in the foundations of economics and ecology. *Biological Journal of the Linnean Society* 44:185–202.

Real, L. A. 1993. Toward a cognitive ecology. *Trends in Ecology and Evolution* 8:413–417.

Roberts, G. 1996. Why individual vigilance declines as group size increases. *Animal Behaviour* 51:1077–1086.

Robles, C., D. A. Sweetnam, and D. Dittman. 1989. Diel variation of intertidal foraging by *Cancer productus* L. in British Columbia. *Journal of Natural History* 23:1041–1049.

Rosenzweig, M. L. 1974. On the optimal aboveground activity of bannertail kangaroo rats. *Journal of Mammology* 55:193–199.

Rowe, L., and D. Ludwig. 1991. Size and timing of metamorphosis in complex life cycles: Time constraints and variation. *Ecology* 72:413–427.

Salant, S. W., K. L. Kalat, and A. Wheatcroft. 1996. Deducing fitness implications of fitness maximization when a trade-off exists among alternate currencies. *Behavioral Ecology* 6:424–434.

Schmid-Hempel, P. 1987. Efficient nectar collection by honeybees. I. Economic models. *Journal of Animal Ecology* 56:209–218.

Schmid-Hempel, P., A. Kacelnik, and A. I. Houston. 1985. Honeybees maximize efficiency by not filling their crop. *Behavioral Ecology and Sociobiology* 17:61–66.

Schaffner, F. E. 1990. Food provisioning by white-tailed tropicbirds: Effects on the developmental pattern of chicks. *Ecology* 71:375–390.

Sih, A. 1992. Prey uncertainty and the balancing of antipredator and feeding needs. *American Naturalist* 139:1052–1069.

Smith, J. M. N. 1974. The food searching behaviour of two European thrushes. II. The adaptiveness of the search patterns. *Behaviour* 49:1–61.

Soler, M., J. J. Soler, A. P. Møller, J. Moreno, and M. Lindén. 1996. The functional significance of sexual display: Stone carrying in the black wheatear. *Animal Behaviour* 51:247–254.

Stephens, D. W. 1987. On economically tracking a variable environment. *Theoretical Population Biology* 32:15–25.

Stephens, D. W. 1989. Variance and the value of information. *American Naturalist* 134:128–140.

Stephens, D. W., and J. R. Krebs. 1986. *Foraging Theory.* Princeton, N.J.: Princeton University Press.

Sutherland, G. D., and C. L. Gass. 1995. Learning and remembering of spatial patterns by hummingbirds. *Animal Behaviour* 50:1273–1286.

Swennen, C., M. F. Leopold, L. L. M. de Bruijn. 1989. Time-stressed oystercatchers, *Haematopus ostralegus,* can increase their intake rates. *Animal Behaviour* 38:8–22.

Systad, G. H. 1996. Effects of reduced daylength on the activity patterns of wintering seaducks. Candidate scientist thesis, University of Tromsø, Norway.

Taghon, G. L., R. F. L. Self, and P. A. Jumars. 1978. Predicting particle selection by deposit feeders: A model and its implications. *Limnology and Oceanography* 23:752–759.

Thorpe, W. H. 1979. *The Origins and Rise of Ethology.* New York: Praeger.

Tooze, Z. J., and C. L. Gass. 1985. Responses of rufous hummingbirds to midday fasts. *Canadian Journal of Zoology* 63:2249–2253.

Waage, J. K. 1979. Foraging for patchily distributed hosts by the parasitoid *Nemeritis canescens. Journal of Animal Ecology* 48:353–371.

Waite, T. A., and R. C. Ydenberg. 1994a. What currency do scatter-hoarding grey jays maximize? *Behavioral Ecology and Sociobiology* 34:43–49.

———. 1994b. Shift towards efficiency maximizing by grey jays hoarding in winter. *Animal Behaviour* 48:1466–1468.

———. 1996. Foraging currencies and the load size decision of scatter-hoarding grey jays. *Animal Behaviour* 51:903–916.

Weiner, J. 1992. Physiological limits to sustainable energy budgets in birds and mammals: Ecological implications. *Trends in Ecology and Evolution* 7:384–388.

Welham, C. V. J., and R. C. Ydenberg. 1988. Net energy versus efficiency maximizing by foraging ring-billed gulls. *Behavioural Ecology and Sociobiology* 23:75–82.

———. 1993. Efficiency maximizing flight speeds in parent black terns. *Ecology* 74: 1893–1901.

Wolf, T., and P. Schmid-Hempel. 1990. On the integration of individual foraging strategies with colony ergonomics in social insects: Nectar collection in honeybees. *Behavioural Ecology and Sociobiology* 27:103–111.

Ydenberg, R. C. 1984. Great tits and giving-up times: Decision rules for leaving patches. *Behaviour* 90:1–24.

———. 1994. The behavioral ecology of provisioning in birds. *Ecoscience* 1:1–14.

Ydenberg, R. C., and L. M. Dill. 1986. The economics of fleeing from predators. *Advances in the Study of Behavior* 16:229–249.

Ydenberg, R. C., C. V. J. Welham, R. Schmid-Hempel, P. Schmid-Hempel, and G. Beauchamp. 1994. Time and energy constraints and the relationships between currencies in foraging theory. *Behavioral Ecology* 5:28–34.

Ydenberg, R. C., and P. Hurd. 1998. Simple feeding models with time and energy constraints. *Behavioral Ecology* (in press).

Zach, R., and J. B. Falls. 1977. Influence of capturing a prey on subsequent search in the ovenbird (Aves:Parulidae). *Canadian Journal of Zoology* 55:1958–1969.

Zaklas, S., and R. C. Ydenberg. 1997. The body size–burial depth relationship in the infaunal clam *Mya arenaria. Journal of Experimental Marine Biology and Ecology* 215:1–17.

CHAPTER TEN

Evolutionary Ecology of Partner Choice

LEE A. DUGATKIN AND ANDREW SIH

10.1 INTRODUCTION

If we were to plot out virtually any animal's time budget, one thing would become strikingly obvious. Animals are constantly faced with choices—what to eat, where to live, with what individuals to live, with what individuals to associate, and what individual to choose as a mate. The list is almost endless. Many of these decisions concern *choice of partners* (i.e. other individuals with which animals interact) for various activities. How such choices are made can have an obvious effect on an individual's fitness. To the extent that choice behavior is cognitive, investigation of partner choice falls under the rubric of cognitive ecology, and we expect that natural selection has shaped partner choice behavior.

Even a cursory scan of the behavioral ecology literature demonstrates that mate choice, particularly female mate choice, is an area of great interest. Hardly an issue of any major journal in the field is published without at least one article on female mate choice. Andersson's (1994) monograph on sexual selection cites 243 studies on female mate choice (versus 30 references on male mate choice) (see Table 6.2.2; Andersson 1994, 128). Why the fascination with female mate choice? There are no doubt many answers to this question. On one level, female mate choice is important because it clearly influences the fitness of preferred and unpreferred males and could affect the fitness of the female. In particular, from an evolutionary perspective, the dynamics of female mate choice may explain the evolution of bizarre male sexual ornaments and behavior; thus, there is considerable interest in the coevolution of female preference and male traits.

Choice of mates, however, is only one of a number of decisions regarding association patterns that animals make on a regular basis. With regard to time budgets, mate choice may only account for a small portion of the total time that an animal spends on partner choice. Animals associate with other conspecifics and choose partners for a wide array of activities besides mating, such as foraging, antipredator activities, sharing territories, and dividing various labors. Depending on the associated costs and benefits, partner choice in these contexts may be more or less important than mate choice.

Despite the likely importance of partner choice, very little attention has been paid to how individuals choose partners other than mating partners. Why partner choice has received so little attention is difficult to explain. One possibility is that researchers simply have not started with the premise that animals are capable of complex behaviors such as partner choice (except in the limited sense of female choice). Once data began to accrue that animals are indeed capable of such actions, numerous follow-up studies were inevitable. Another possibility, before the emergence of game-theoretical thinking (which is based on the notion that an individual's fitness is contingent on its actions and those of others) is that behavioral researchers did not focus on how others affect one's fitness. Regardless of the reason that partner choice received little attention until lately, the subject is interesting because of its behavioral, evolutionary, and cognitive aspects, and the area of partner choice is replete with conceptually intriguing issues of general interest. In this chapter, we will examine some of these issues: (1) the definition of partner choice; (2) examples of partner choice; (3) different types of partner choice; (4) atomistic approaches to the study of partner choice; (5) the cost or benefit perspective of partner choice; and (6) cognitive ecology and partner choice. A fair share of the work on partner choice has been undertaken in the lower vertebrates; because this work is particularly interesting from a cognitive perspective, we will concentrate on this group of animals. This is not to say that partner choice has not been documented many times in mammals and birds (e.g., see the volume of Harcourt and de Waal 1992 on coalitions in animals); however, to tackle the questions we wish to address in a single chapter, we have opted to focus on lower vertebrates (particularly fish) and include discussion of the occasional invertebrate study.

10.2 WHAT IS PARTNER CHOICE?

The term "choice" is somewhat amorphous, and so we begin with a brief discussion of definitions (see Bateson 1990 for further discussion). In its most stringent sense, choice implies a decision made among alternative options (Hutto 1985). For example, in the context of habitat choice, Rosenzweig (1990) suggested that choice means nonrandom behavior that is guided by the reception and processing of information by the nervous system. Although we agree that information processing is part of the process of choice, we define choice independently of the neural mechanisms that govern it (see Byers and Bekoff 1986 for more discussion on recognition, nonrandom behavior, and the nervous system). We define partner choice as a nonrandom tendency for an individual—the chooser—to interact with some individuals (or types of individuals) instead of other encountered individuals (or types of individuals).

Continuing with our discussion of definitions, we think it is worthwhile to distinguish between partner choice and apparent partner choice. Apparent partner choice occurs when an individual interacts with a particular partner at a higher-than-expected frequency simply because it encounters that potential partner with greater frequency than that of random occurrence. Apparent partner choice does not mean interaction at a higher than random probability with that partner given an encounter. For example, size-assortative mate choice— the tendency for like-sized individuals to mate—can be due to true size-assortative mate choice by both males and females or size-assortative microhabitat use (Crespi 1989). The former possibility is mate choice; the latter possibility may be apparent mate choice. The existence of size-biased microhabitat use, however, does not rule out true partner choice. Microhabitat choice itself may reflect mate choice. For example, during the mating season, large individuals may share the same microhabitat because they prefer to encounter and mate with each other (see Hutto 1985 for a discussion of apparent habitat selection).

We distinguish a hierarchy of levels of decision making and partner choice. On one level, organisms choose activity patterns in space and time (i.e. they choose when and where to be active and a set of potential activities, such as foraging, antipredator, and mating activities). While active, organisms encounter and choose among potential partners. After selecting a partner, individuals choose particular behaviors that are often in response to a partner's past or current actions. We will focus on the middle step—partner choice.

This hierarchical view is analogous to that used in studies of foraging and antipredator behavior (Chesson 1983; Lima and Dill 1990; Ydenberg, this vol. section 9.4). For example, suppose that a forager displays a preference to feed on prey type 1 over prey type 2. A hierarchical perspective suggests three reasons why such a preference may be found: (1) Preference is determined by the forager's search mode. Does the forager ambush or actively search for prey? If the forager ambushes its prey and prey type 1 is more active and thus encountered more frequently by the ambush predator, then these conditions may result in an apparent preference for prey type 1. (2) The forager may exhibit a preference for a particular patch choice or a particular diurnal foraging habit, which may include prey 1 but not prey 2. (3) The forager experiences both prey types but displays an active choice for prey type 1. Active diet choice for type 1 occurs when a forager has a higher probability of attacking 1 than 2, given equal probability of an encounter with each.

If we now consider partner choice in the context of foraging, a similar scenario unfolds. Suppose that while foraging, individual X tends to be in the vicinity of individual Y but not in that of individual Z. This situation may occur because X and Y share a common preference for some type of prey and Z does not. Alternatively, all three share the same food preference, but X and Y are active at the same time and place and Z is active at a different time or place. Finally, X may be equally likely to encounter Y or Z; however, given an encounter, X prefers to forage with Y instead of Z. This last situation clearly involves partner choice. The other two situations result in a tendency for X to forage near Y (but not Z) because there are more encounters between X and Y than between X and Z.

Most laboratory experiments on partner choice are designed to force X to encounter Y and Z; thus, some of the difficulties inherent in studying partner choice in the field are avoided. Nonetheless, laboratory experiments can be constructed to examine the possible complexities associated with the hierarchical view outlined. To study partner choice within the hierarchical framework, we suggest the following methodology. First, quantify nonrandom associations (e.g. X interacts with Y more frequently than X interacts with other individuals) with use of statistics developed for foraging experiments (Chesson 1983; Van der Meer 1992). Then examine activity patterns: Does X simply encounter Y more often than X encounters other individuals? If X and Y interact with greater frequency than that expected at random, remove Y to see if X's activity pattern is caused primarily by partner choice. Finally, directly observe encounters to see if X preferentially chooses to interact with Y (over others) when X and Y are near each other.

10.3 EXAMPLES OF PARTNER CHOICE

To be a truly interesting subject to cognitive and behavioral ecologists, partner choice needs to be a general phenomenon (i.e. one that takes place in a wide array of contexts). The evidence on most types of partner choice in lower vertebrates is still relatively scanty; however, we argue in this section that partner choice may be found in many different social milieus.

10.3.1 Imitation and Female Mate Choice

While there have been literally hundreds of studies on how females choose their mates (Andersson 1994), most of these studies, although interesting in their own right, say little about cognitive ecology. Recent work on imitation and female mate choice, however, directly addresses cognitive and evolutionary aspects of partner choice (Dugatkin 1996a).

Using the guppy (*Poecilia reticulata*), Dugatkin (1992) performed the first controlled experiments that examined female mate copying. A "focal" female observed another "model" female near one of two males, and the focal female was subsequently given the choice to spend time near either male. A significant proportion of the focal females preferred to associate with the male that was near the model female. In five control experiments, alternative hypotheses were examined regarding group size, position effects, male behavior, and partner choice outside the context of mating. These control experiments ruled out alternative explanations and strongly supported the hypothesis that the focal female remembered the identity of the male near the model and preferred that

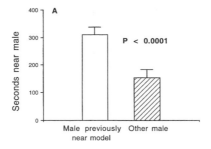

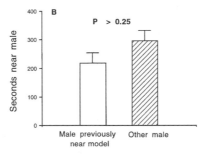

Figure 10.1. (*a*) Mean number of seconds (±standard error [SE]) younger focal females spent near each male when one male was observed near an older model female. (*b*) Mean number of seconds (±SE) older focal females spent near each male when one male was observed near a younger model female (adapted from Dugatkin and Godin 1993).

male in mate-choice tests. Subsequent work established that a female's choice of mates could be reversed with social cues (Dugatkin and Godin 1992a) and that young females were more likely to imitate older females than vice versa (Dugatkin and Godin 1993) (fig. 10.1). Furthermore, female guppies with a heritable preference for orange body color in males preferred less orange males if (1) the focal female observed the less colorful male near a model female; and (2) the difference in male coloration was small or moderate. If, however, the focal female observed the less orange male near a model but the difference in male color was very large, the focal female preferred the more orange male (Dugatkin 1996a).

10.3.2 Antipredator Behavior and Partner Choice

Because of the interest predator-inspection behavior has drawn as a possible case of Tit-for-Tat cooperation (see Dugatkin 1997 for a review), some of the most direct evidence for partner choice comes from studies on Tit-for-Tat cooperation in fish. During predator inspections, fish break away from their school and approach a putative predator in a slow, directed, saltatory manner;

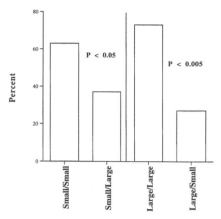

Figure 10.2. Two bars on the left show the percentage of time small sticklebacks spent near schools composed of small or large conspecifics. Two bars on the right represent the results when large fish were tested in a similar manner (adapted from Ranta and Lindstrom 1990).

fish appear to attempt to gain information about this potential danger (Pitcher 1993; Dugatkin and Godin 1992b). Dugatkin (1992) found that predator inspectors suffer relatively high rates of predation; therefore we may expect natural selection to select for "trustworthy" partners during this dangerous task (Milinski et al. 1990; Dugatkin and Alfieri 1991). As such, we might predict that inspectors tend to associate with others that have shown a tendency to inspect in the recent past. The effect of this type of partner choice would presumably decrease the risk of death because of the dilution effect. Recent work in guppies (*Poecilia reticulata*) and sticklebacks (*Gasterosteus aculeatus*) supports this notion; inspectors preferred co-inspectors that had, during prior inspection sorties, stayed by the inspector's side and approached a predator (Milinski et al. 1990; Dugatkin and Alfieri 1991). Experience, however, is not the only factor that influences how one chooses partners for inspecting a predator. Kulling and Milinski (1992) found that inspecting sticklebacks also seem to prefer large individuals as coinspectors.

One emergent effect of partner choice is the formation of groups that are segregated with the criteria used in such a choice. A widely discussed hypothesis for the production of such size-assorted groups is the oddity effect (Ohguchi 1981), which occurs when individuals sort into groups according to size because odd-sized individuals are most apparent to predators (Landeau and Terborgh 1986; Theodorakis 1989). For example, Ranta and Lindstrom (1990) and Ranta et al. (1992) found that large sticklebacks preferred to associate with schools composed of large instead of small fish, although large fish foraged more effectively when associating with smaller conspecifics (fig. 10.2). Apparently for larger sticklebacks, the decreased predation pressure experienced in same-sized groups outweighs any foraging benefits accrued in the presence of smaller conspecifics.

10.3.3 Partner Choice and Foraging

When food is scarce, individuals may prefer foraging partners that enhance their feeding rates. For example, bluegill sunfish (*Lepomis macrochirus*) appeared to choose partners on the basis of long-term expected foraging returns associated with those partners instead of on aggressive history or relative size (Dugatkin and Wilson 1992). In addition, Metcalfe and Thomson (1995) found that European minnows (*Phoxinus phoxinus*) preferred to associate with groups composed of poor food competitors over groups consisting of good competitors (fig. 10.3). The most intriguing aspect of this study was that minnows were capable of making this distinction even when they had no experience foraging with fish in the groups from which they chose! In other words, minnows were able to use other (nonforaging related) cues to determine the foraging abilities of conspecifics.

In the wasp *Cerceris arenaria*, relative size (and not experience) seems to guide the choices of foraging (and more generally nesting) partners (Willmer 1985). Large wasps prefer small individuals as nest mates and vice versa. This mutual preference is likely to benefit both parties because large and small wasps have complementary foraging specializations. Large individuals forage well when temperatures are not extremely hot (and overheating is not a prob-

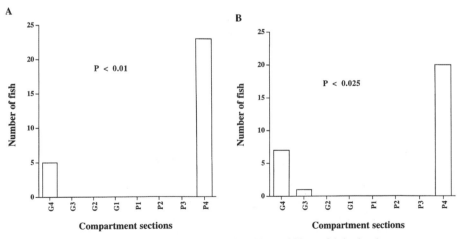

Figure 10.3. Frequency distribution of preferred positions of 28 test fish in the absence (*a*) and presence (*b*) of food. The preferred position was defined as the most frequently occupied zone in the compartment. Section G4 was nearest the shoal of fish with good competitors, and section P4 was closest to the shoal of poor competitors (adapted from Metcalfe and Thomson 1995).

386   Lee A. Dugatkin and Andrew Sih

lem), and small individuals forage more efficiently during very hot times of day. Large, territorial striped parrotfish (*Scarus iserti*) also display a preference for smaller nonterritorial conspecifics (Clifton 1991).

10.3.4 Alternative Reproductive Strategies and Partner Choice

Given that males typically compete directly or indirectly with one another for females, one might expect males to display partner-choice rules that increase their probability of mating with a female. The particular rule used, however, may depend critically on the ecology of the species being studied. For example, in natterjack toads (*Bufo calamita*), satellite males prefer to associate with other males that attract the most females to their territory (Arak 1988); this association increases the satellite's encounter rate with females. By contrast, male guppies prefer to associate with less attractive males (Dugatkin and Sargent 1994) (figs. 10.4 and 10.5). This rule may be favored because male

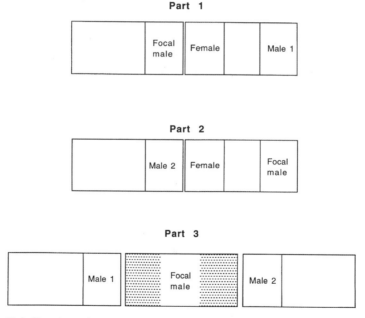

Figure 10.4. Experimental apparatus used by Dugatkin and Sargent (1994) in winner/loser experiment. In parts 1 and 2 of a trial, female proximity to a male was staged by placing the female next to one of two males (with use of clear Plexiglas divider). (*a*) Male 1 was a "loser" with respect to the focal male because the female was nearer to the focal male. (*b*) Male 2, however, was a "winner" with respect to the focal male because the female was closer to male 2 than the focal male. (*c*) In the third part of a trial, the focal male chose between the loser male in part 1 and the winner male in part 2. Results of control experiments ruled out any side preferences (adapted from Dugatkin and Sargent 1994).

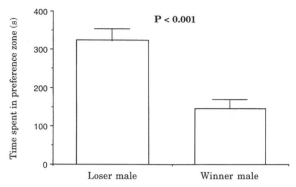

Figure 10.5. Mean number of seconds focal males spent near winner and loser males (adapted from Dugatkin and Sargent 1994).

guppies are not territorial, and females do not mate numerous times in a given male's territory. Thus for guppies, associating with high-quality males would probably decrease another male guppy's chance of copulating.

10.3.5 Familiarity and Partner Choice

Even when one can experimentally document partner choice, the precise rules used may remain a mystery (see section 10.7); (the mechanisms are somewhat better understood for partner choice among kin; Fletcher and Mitchner 1987; Waldman 1987; Hepper 1991). For example, a week after being captured, bluegill sunfish (*Lepomis macrochirus*), preferred to associate with individuals that were caught in the same trap as they were (Brown and Colgan 1986). Familiarity may be what guides partner choice in this case, but other alternatives must be ruled out. The familiarity hypothesis is, however, supported with work on adult bluegill sunfish. Dugatkin and Wilson (1992) found that when two groups of six bluegill sunfish were kept in separate communal tanks for three months and then given choice tests, the sunfish preferred fish from their own tank in 35 of 36 trials!

10.4 TYPES OF PARTNER CHOICE

Partner choice has been studied in numerous contexts (see section 10.3). To help organize and understand similarities and differences among these contexts, we distinguish several types of partner choice. From a functional view, we characterize choice by asking the following questions: (1) Is choice one-sided or reciprocal? (2) Do both sides or does only one side benefit from the interaction? (3) Will potential partners have one, a few, or many interactions? From a cognitive view, we ask the following: (4) Do individuals remember

Table 10.1
Scenarios of Partner Choice

Scenarios	Characteristics	Examples
I	+/−, one-sided	Producer/scrounger, satellite males
II	+/+, one-sided	Female mate choice
III	+/+, two-sided	Two-sided mate choice, predator, inspection, cooperative foraging

the actions of others and choose on the basis of previous experience or are partner-choice rules innate? (5) Do individuals choose partners on the basis of individual differences or do they divide potential partners into simple categories and choose on the basis of a category? We will discuss these types of partner choice in greater detail.

By combining questions 1 and 2, we can identify three main scenarios (table 10.1). In scenario I, choice is strictly one-sided and the chooser alone benefits from the interaction; the chosen partner suffers (i.e. scenario I is a positive/negative [+/−] interaction). Scenario I depicts the well-known producer/scrounger interaction; examples include dominants parasitizing the foraging efforts of subordinates (e.g. Barnard 1984; Hockey and Steele 1990), subordinate satellite males sneaking copulations in the territories of dominant males (Arak 1988; Gross and Charnov 1980), and brood parasitism of parental care (Rothstein 1990; Davies 1992). The +/− nature of this type of interaction creates a conflict that is absent from interactions in which both individuals benefit.

In scenario II, partner choice is still one-sided, but both individuals benefit (positive/positive [+/+]). The most obvious example of this scenario is female mate choice. In many systems, males are generally not choosy because they benefit from mating with any female (i.e. males are almost always willing to be chosen), but females are choosy about their mating partners. A large literature exists on the evolution and adaptive significance of single-sided mate choice (Andersson 1994; Møller 1994).

Finally, in scenario III, both individuals potentially choose their partners. Examples of +/+ two-sided choices include simultaneous male and female mate choice and partner choice for cooperative foraging (Dugatkin and Wilson, 1992), cooperative predator inspection (Milinski 1987; Milinski et al. 1990; Dugatkin and Alfieri 1991, 1992), and the formation of various coalitions (Harcourt and de Waal 1992). The dynamics of the three scenarios depend on the extent of individuals' interactions; repeated interactions tend to favor stable pair bonds and cooperation among partners (Trivers 1971; Axelrod and Hamilton 1981; Mesterton-Gibbons and Dugatkin 1992; Dugatkin 1997). Two-sided +/− partner choice should be relatively rare because the

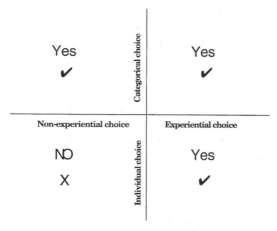

Figure 10.6. Cognitive axes for partner choice. A nonexperiential, individual choice is not possible under the schema devised.

individual that suffers should try to avoid the interaction entirely. Of course, if one defines avoiding a partner as a choice (and it seems reasonable to do so at times), then two-sided $+/-$ partner choices may be more common than stated. Such avoidance may even cause the partner that suffers to move to less preferred areas (e.g. more dangerous spots) to avoid the partner that benefits. In addition, if this partner is making the "best of a bad job," then the frequency of $+/-$ interactions may increase.

From a cognitive view, we distinguish two axes of partner choice—experiential versus nonexperiential and individual versus categorical (fig. 10.6). Experiential choice can be individual or categorical, but nonexperiential choice must be categorical (see the following). When invoking nonexperiential choice, individuals do not base their decision on experience with potential partners; instead, a decision is made on the basis of the categorization of potential partners. For example, a category may be size, and sizes may be large and small. Individuals that prefer to forage in the company of large conspecifics or females that prefer males on the basis of tail length may be displaying categorical partner choice. A correlate of nonexperiential categorical partner choice (or for that matter, categorical choice in general) is that if choice is based on n categories, then all individuals that have the same characteristic in each of these n categories are identical in the eyes of the chooser (i.e. when choosing between two of these individuals, the chooser should be indifferent). On the surface, nonexperiential choice appears equivalent to genetically encoded choice; however, this need not be the case. Partner-choice rules can, in theory, be "inherited" by means of cultural transmission (Boyd and Richerson 1985).

In experiential choice, the chooser uses its own experience to guide partner choice. For example, an individual may prefer a partner that has been proved to be adept at finding profitable food patches in the past (e.g. Dugatkin and Wilson 1992). Such a choice may be indicative of individual recognition, but need not be. Experiential partner choice can also be categorical (e.g. if an individual has consistently foraged successfully when paired with large partners, its experiential rule may be "choose larger individuals as foraging partners").

Experiential choice may be represented best with points on a continuum instead of a true dichotomy because precise, narrowly defined categories can result in few or even only one individual per category. If individuals have a partner-choice rule such as "choose big blue male individuals that have found profitable food patches in prior foraging bouts," then categorical choice occurs (the categories are "size," "color," "sex," and "foraging history"); however, the composite of these categories may be so narrow that categorization essentially yields individual recognition (Barnard and Burk 1979; also see Barlow et al. 1975 for more on this issue in the context of kin recognition).

Partner choice may also entail a combination of experiential and nonexperiential criteria. For instance, Kulling and Milinski (1992) showed that sticklebacks prefer larger conspecifics as partners during predator inspection. This preference suggests that inspectors classify potential partners on the basis of categories (e.g. large versus small). On the other hand, individual recognition can also occur in the choice of coinspectors; sticklebacks prefer to inspect with partners that have stayed by their side during prior inspection visits (Milinski et al. 1990). More generally, the experiential component of partner choice may dictate "choose a partner that did x and not y the last time you interacted," and the nonexperiential component may be "choose individuals that fall in bin 1, not 2." If the chooser has no prior experience with potential partners, it should simply choose the largest partner. If two potential partners are of equal size, the chooser should select a partner on the basis of prior experience. If, however, one potential partner falls in bin 2, but did x and the other falls in bin 1, but did y, partner choice necessitates use of an algorithm that balances the relative importances of experiential and nonexperiential criteria. Experiments that address these sorts of questions should be easy enough to construct and would provide insight into the decision rules that underlie partner-choice behavior.

10.5 THE ATOMISTIC APPROACH TO STUDYING PARTNER CHOICE

Partner choice has been studied in only a very limited context; focal individuals have almost always been given a choice between particular partners who differed in only a single trait (e.g. foraging abilities or antipredator qualities) (see section 10.3 for specific examples). Furthermore, tests of partner choice typically have been done in a single context—foraging only or antipredator qualities only (see section 10.5 for specific examples). In studies in other areas (e.g. patch choice or activity patterns), animals often appear to use multiple cues to guide behavior. For example, Sih and Krupa (1992) found that predation and hunger level affected mating patterns in *Aquarius remigis* in complex ways that would have remained hidden if an atomistic approach had been taken. We need to move away from this atomistic approach and examine multiple traits and multiple fitness demands in the context of partner choice in general (Sih and Krupa 1992; Sih 1994).

Use of multiple cues appears to allow animals to balance multiple demands (e.g. foraging and mating) that often conflict (Sih 1987; Abrahams and Dill 1989; Lindstrom and Ranta 1993). For example, consider two partner-choice scenarios: (1) Females prefer to forage with males that have been successful at finding profitable food patches despite unusually aggressive behavior in males; or (2) females choose to forage and then mate with males that share food discovered during foraging bouts. In the first case, partner choice requires a balancing of the benefits of foraging against the costs of encountering aggression. In the second case, the same partner may be beneficial in two contexts. Thus, it is difficult (if not impossible), in the second case to delineate the proximate reason females choose particular males (e.g. female choice may depend on their energy intake or the food requirements of future offspring). Single-trait, atomistic partner-choice experiments may be a necessary starting point, but these experiments clearly only initially address the scenarios described.

How may experiments be constructed so that the effect of multiple cues on partner choice are elucidated? One possibility is a titration experiment in which the strength of a given cue is raised or lowered while the other cue remains constant. Such tests may provide evidence for the relative importance of different cues in relation to one another. Different cues may even provide contrasting information. For example, female guppies from the Paria River prefer to mate with more orange males; however, Dugatkin (1996b) found that if an observer female sees another female prefer the less orange of two males, the observer female will, under certain conditions, switch her preference to less orange males. Furthermore, the difference in male orange content can be titrated to examine the relative importance of color and social cues. Similar experiments could easily be constructed so that multiple cues in various partner-choice contexts can be examined.

10.6 THE COSTS AND BENEFITS OF PARTNER CHOICE

10.6.1 Variation in Partner Choice within a Given Context

Three interrelated questions about partner choice that can be analyzed by a cost/benefit approach are the following: (1) Should an individual, in a given situation have a partner or not? (2) If an individual interacts with a partner, should the individual invoke some sort of partner choice? (i.e. Should the individual prefer to interact with some potential partners over others?) (3) If an individual shows partner choice, what criteria should the individual use to guide its choice? For each question, we can also ask how environmental factors or individual states ought to affect partner choice. Our primary focus is on question 2—Should an individual show partner choice?

In some contexts, an individual must have a partner; in other words, the individual cannot perform a particular task without a partner. In most contexts, however (e.g. foraging or antipredator behavior), one option is no partner. There has been some theoretical work on conditions that favor solitary action over performance of the same action with a partner, particularly in producer/scrounger scenarios (Barnard 1984; Vickery et al. 1991; Caraco and Giraldeau 1991) and during predator inspection (Lima 1989); however, more theoretical and empirical work on this issue is needed in most contexts.

To organize our thinking on adaptive partner choice, we return to the three main types of partner choice (see section 10.5). The best studied scenario is the one-sided $+/+$ interaction (i.e. single-sided mate choice in which both individuals benefit). By following Parker (1978) and Janetos (1980), Real (1990) used a theory that is analogous to optimal diet-choice theory to predict when mate choice (as opposed to random mating) is more likely. Real (1990) argued that mate choice should be relatively common when (1) the benefit of potential partners varies more to the chooser (i.e. there is a greater benefit to partner choice); (2) there is a higher density of high-value partners (i.e. there is a lower time cost of rejecting low-value partners to search for better options); and (3) there are decreased costs per unit time of searching for or choosing partners. These basic predictions should be true in other scenarios (e.g. $+/-$ scenarios or two-sided choices); however, there may be some added complexities (see Real 1991 and Weigmann et al. 1996 for more).

For producer/scrounger scenarios, the $+/-$ nature of the interaction introduces an inherent conflict. Scroungers choose the best producers, and producers presumably attempt to avoid or resist being chosen. Nonetheless, the basic predictions listed for $+/+$ interactions should guide partner choice by scroungers. The value of each potential partner (producer) should depend not only on the partner's inherent quality as a producer but on the probability that the scrounger is able to parasitize that partner (Giraldeau and Lefebvre 1987) and bear the costs of overcoming the partner's resistance/avoidance. Furthermore, if producers can modulate the amount of effort used to resist scroungers, then the interaction can be analyzed as a game (Maynard Smith 1982). Existing theory, however, on adaptive behavior by scroungers focuses not on partner choice among potential partners but on the first question in partner choice: Does an individual scrounge or act alone? Explicit theories (including game theory) and tests of adaptive partner choice and behavioral responses by potential partners in $+/-$ interactions should prove rewarding.

In other scenarios, individuals show reciprocal (two-sided) choice. Game theory on male/female, simultaneous mate choice (Parker 1983; Crowley et al. 1991) predicts that partner choice should depend on an individual's quality relative to other members of its sex. Higher-quality individuals should display the basic predictions listed for single-sided, +/+ mate choice. Lower-quality individuals, however, must account for the possible rejection by their favorite partners. For example, mate choice by lower-quality males should depend on the availability of high-quality females, which depends on the abundance of high-quality females and the likelihood that high-quality females will accept lower-quality males as mates. In turn, mate choice by high-quality females should depend on the availability of high-quality males, which depends on mate choice by high-quality males. Obviously, adaptive partner choice in reciprocal games can be very complex.

Both +/− and +/+ games become even more interesting if individuals can recognize one another and remember the actions that were taken by particular individuals during prior interactions. In general, theory suggests that, with repeated interactions, individuals should be more likely to cooperate (Trivers 1971; Axelrod and Hamilton 1981; Mesterton-Gibbons and Dugatkin 1992; Dugatkin 1997). In producer/scrounger scenarios, with repeated interactions, it can be beneficial for scroungers to give some benefits to producers to reduce resistance by producers. For example, dominant individuals may allow subordinates to enjoy some feeding or mating opportunities in return for greater loyalty in the producer/scrounger interaction (Stacey 1982; Reeve 1989; Keller and Reeve 1994).

To date, theory on adaptive behavior in repeated reciprocal games has focused on the cooperative or uncooperative behavior of individuals (i.e. on the balance between cooperation and cheating/deception) and not on the partner chosen. Clearly, individuals should prefer partners that cooperate over ones that cheat; however, explicit theory must address more subtle partner-choice decisions that involve partners with variable inherent qualities (e.g. foraging abilities) and traits such as "honesty," "gullibility," or "forgiveness."

10.6.2 Comparisons of Partner Choice across Contexts

We have considered aspects of adaptive partner choice within a given context (e.g. foraging or mating). A more global view would compare partner choice across contexts. Does an adaptationist view help us understand which contexts (e.g., mating, antipredator, feeding) are more likely to feature partner choice?

In the absence of obvious constraints, we expect partner choice to be more likely to occur in situations in which the identity of one's partner has a large effect on fitness. For example, if all else is the same, partner choice for antipredator activities should be highly likely to occur (i.e. one mistake can result in death); choice of mating partners should also be common, but choice of foraging partners should be less common—especially when food is abundant. This ranking of contexts is based on the supposition that predation and mating are most often multiplicative in terms of selection analyses (Caswell 1989; Lande and Arnold 1983); foraging rate is less tightly related to fitness and more likely additive in nature (we recognize that our ranking system is debatable). Indeed, when food is not limiting, foraging rate (and thus choice of foraging partner) may have little or no measurable effect on fitness. One should not always expect partner choice to be more common in anti-predator activities than in foraging. For example, suppose a physically strong partner is needed to bring down potential prey items, but most potential partners in the context of antipredator activities can warn of oncoming dangers. In such a case, partner choice would be more likely in the context of foraging than antipredator activities. Our argument regarding the occurrence of partner choice can be extended to address how much is invested in partner choice (a more continuous version of the "be choosy" versus "don't be choosy" argument outlined).

We may also expect partner choice to be more likely when it is associated with a larger future commitment in time and resources. In theory, individuals should continuously reevaluate partners and contrast their values against apparent values of potential partners (Peck 1993); however, the costs of switching partners may often force individuals into relatively large, long-term commitments (Lima 1989; Peck 1993).

Finally, $+/+$ interactions should be more likely to include partner choice than $+/-$ interactions. In antagonistic producer/scrounger interactions ($+/-$), a scrounger's encounter rate with available producers may be low because the producer should avoid or resist being chosen by a scrounger. As a result, scroungers may not be choosy about producers; scroungers may "parasitize" any producer they find. By contrast, $+/+$ interactions should result in a greater availability of potential partners for a chooser because both sides benefit; thus, choosers can be choosier. Also, partner choice should be more likely in situations with a one-sided choice versus a two-sided choice. If both sides choose, then most individuals face the risk of being rejected by some fraction of their potential, desirable partners. This risk of rejection, in effect, reduces the availability of potential partners for all but the most valuable individuals and thus reduces the likelihood of nonrandom partner choice.

10.6.3 New Horizons for the Study of Adaptive Partner Choice

Two relatively recent approaches within the optimality framework hold promise for expanding our understanding of partner choice: dynamic optimization (Mangel and Clark 1988; Houston et al. 1988) and the function of uncertainty and information processing in governing behavior (Stephens and Krebs 1986;

Sih and Krupa 1992; Abrams 1994; Bouskila et al. 1995). These approaches have been admirably reviewed elsewhere; we will discuss the sorts of major insights that could emerge from the integration of these approaches into analyses of partner choice.

Dynamic optimization incorporates several major elements of reality that are missing from static models including (1) the role of variation in individual state (e.g. hunger level, developmental stage, reproductive value); (2) the existence of time horizons (i.e. organisms face a finite end for each bout, day, season, or lifetime); and (3) explicit assumptions about the relationship between behavior, state, and fitness. These assumptions are required to express the effects of conflicting demands (e.g. foraging, antipredator, or mating needs) in the common currency, fitness. Some attempts have been made to use dynamic optimization techniques to understand mate choice (Crowley et al. 1991); to our knowledge, however, these techniques have not yet been applied to other aspects of partner choice. Given that variation in individual state, time horizons, and conflicting demands are probably important factors in many partner choice situations, the application of dynamic optimization to partner choice should prove insightful.

Uncertainty is another ubiquitous aspect of reality that is ignored in most analyses of partner choice. Adaptive partner choice requires estimates of the value of each potential partner, of the costs and benefits associated with not having a partner or searching for a partner (e.g. energy costs or predation risk), and assessments of relevant aspects of individual state. In addition, when games are involved, subjects must know not only their own conditions but the options and conditions experienced by other individuals to predict the behavior of perspective partners. Finally, in a variable environment, organisms must assess changes in all of these conditions in time and space. (See Bateson and Kacelnik, this vol. chap. 8; Ydenberg, this vol. chap. 9.)

Analyses of behavior in other contexts suggest that uncertainty serves an important function in shaping foraging and antipredator behavior (Stephens and Krebs 1986; Lima 1989; Sih and Krupa 1992). At a minimum, animals must sample their environment (both potential partners and surrounding conditions) to generate estimates of initial and changing conditions. Sampling usually implies suboptimal behavior in the short-term in exchange for information that allows a closer approach to an optima in the long term. To use information, organisms must obviously remember their past experiences. How should organisms weigh the relative importance of various past experiences? Depending on the pattern of change in conditions in space and time, some memories can be useless or even counterproductive. In addition, memory itself may have neural costs (Dukas, this vol. section 3.5.3). These trade-offs can be balanced to yield an optimal pattern of memory (Healey 1992). In some situations, the cost of sampling can be prohibitively high (e.g. uncertainty about predation risk) (Sih and Krupa 1992). If organisms cannot afford to gather the information necessary to respond to changing conditions, then organisms may be forced to show inflexible use of an option that is, on average, satisfactory—even if, at times, the option is suboptimal. Some studies on the cognitive aspects of partner choice have shown that prior experience can influence choice (see section 10.5). To our knowledge, however, partner choice studies have not addressed explicitly optimal-sampling regimes or other adaptive responses to uncertainty (see Bekoff et al. 1989 for a discussion in the context of habitat choice). To understand how information processing affects partner choice, future studies should integrate cognitive and adaptationist approaches.

10.7 COGNITIVE ECOLOGY AND PARTNER CHOICE

Some of the most interesting unexplored issues in the field of partner choice concern the new discipline of cognitive ecology, in which the adaptive value of memory and cognition, from proximate and ultimate perspectives is studied (see Sherry this vol. chap. 7).

One issue addressed by cognitive psychologists concerns "domains of learning"—i.e. Do animals have a single, all-purpose learning "algorithm" or do various, separate (but not necessarily unconnected) domains exist in the brain? In terms of partner choice, we may ask if an animal uses categorical-choice criteria in some contexts but individual recognition in others. For example, Dugatkin and Wilson (1992) found that bluegill sunfish use experiential choice on the basis of individual recognition when choosing foraging partners. The question then arises whether sunfish use experiential choice on the basis of individual recognition when choosing partners in other contexts such as antipredator activities or mating. With the domain-general perspective it is suggested that bluegills should use the same type of partner choice across contexts. The domain-specific view allows for the possibility of employing different types of choices in different contexts. Domain specificity, however, does not necessarily predict the use of different types of choices. If natural selection favors different types of partner choices in different contexts and the species has the cognitive prerequisites to perform different types of choices, the domain-specific perspective predicts variation in types of choices across domains. If selection favors the same type of choice across domains, however, the domain-specific view predicts no variation across domains.

With the cognitive view of partner choice, the possibility of "crossover" effects between domains is also suggested. By a crossover effect, we mean the possibility that choices made in one domain (e.g. the foraging domain) affect choices made in a different domain (e.g. the antipredator domain). Imagine an experiment in which individuals use an experiential, individually-based partner choice in one domain (domain 1). With another group of subjects, however, the experiment uncovers that no partner choice is used in a second domain (individuals choose randomly between potential partners in domain-

2 related trials). To test for the possibility of crossover effects, the experiment can be repeated with a single set of subjects, tested first in the domain-1 task and then in the domain-2 task. A given individual chooses between two potential partners in a domain-1 test; then the individual chooses between the same two potential partners in a domain-2 test. Crossover effects are indicated if individuals prefer the same partner in the domain-2 task as in the domain-1 task. Of course, appropriate controls should be run to test for various confounding factors.

A particularly exciting aspect of cognitive ecology is its potential for identifying cognitive constraints on adaptive partner choice (see Dukas, this vol. section 3.2). The ability to make adaptive decisions in a complex, uncertain world can obviously be limited by memory constraints. For example, even if optimality theory shows that an omniscient organism should base its partner choices on the outcome of its last 10 interactions with each of 50 potential partners, the organism cannot if it can only remember the last two interactions with five players. Experiments in other contexts have indeed indicated that some animals base their decisions primarily on their most recent experiences (Logue 1988; Real 1994; Kacelnik and Bateson, this vol. section 9.4.3). If this represents a memory constraint (as opposed to an adaptive decision to ignore information from the more distant past), then this constraint could certainly limit adaptive partner choice.

Another possible constraint arises if individuals are domain general, even when it is adaptive to be domain specific. For example, in some situations it may be beneficial to use categorical, nonexperiential criteria for mate choice (e.g. prefer larger males) but use experience to choose partners for predator inspection (i.e. prefer individuals that have demonstrated that they are cooperative coinspectors). A domain-general individual that always used categorical, nonexperiential criteria to choose partners may then be incapable of adaptive choice of coinspectors. As discussed, the opposite situation does not pose a constraint; a domain-specific individual can, in theory, use the same partner-choice rules across domains if use of these rules is optimal.

Finally, crossover effects (which presumably reflect the organism's cognitive architecture) (compare Real 1994) can also result in nonadaptive partner choice. Any correlation between decisions in different contexts implies a potential constraint on independent adaptive choice in different contexts (Sih 1994). If a correlation (a crossover effect) does not match the optimal pattern of choice across contexts, then the correlation causes suboptimal behavior. For example, if different partners are best in different contexts (mating versus foraging versus predator inspection), then it may not be beneficial to show a tendency to prefer to forage or inspect predators with a favorite mating partner.

Conventional wisdom (no doubt reflecting our natural, unavoidable anthropocentric biases) suggests that organisms with more highly developed brains should be less limited by cognitive constraints (e.g. Harcourt, 1992; Seyfarth and Cheny, 1994). Most of the extant studies on cognition, however, have concerned several model systems (e.g. rats, mice, pigeons, jays, bees, and primates). A broader comparative survey, with use of modern phylogenetic methods (e.g. Harvey and Pagel 1991), should help identify phylogenetic patterns of cognitive architecture. (See Dukas, this vol. section 4.5.1. Sherry, this vol. section 7.4.4.)

At present, few studies have melded methods in behavioral ecology, evolutionary biology, and cognitive ecology to address the possibility of cognitive constraints on adaptive behavior in any context (but see Real 1994; Kamil 1994; chap. in this vol.). Further work blending ideas in these fields in the study of partner choice should prove rewarding.

10.8 SUMMARY

Choice of mates is only one of a number of decisions regarding partner choice that animals make on a regular basis. Despite its likely importance, until recently, very little attention has been paid to how individuals choose partners outside the context of mating partners. Regardless of the reason that partner choice has received little attention until lately, the subject is interesting in its own regard from behavioral, evolutionary, and cognitive perspectives, and the study of partner choice is replete with conceptually intriguing issues that are of very general interest. In this chapter, we put forth the following ideas, definitions, and challenges:

1. Partner choice is a nonrandom tendency for an individual to associate with some individuals or categories of individuals over other *encountered* potential partners.

2. To date, most investigators of partner choice have focused on female mate choice; however, several studies have documented partner choice in other contexts (e.g. choice of foraging partner). More studies are required to determine the generality of partner choice in nonmating contexts.

3. To examine the adaptive importance of partner choice, it is useful to distinguish four categories of interactions on the basis of whether choice is one-sided or reciprocal and whether both sides $(+/+)$ or only one side $(+/-)$ benefits from the interaction. Female mate choice is an example of a one-sided, $+/+$ interaction. Reciprocal $+/+$ games include reciprocal mate choice and various cooperative interactions. Some models exist on mate choice; however, for other types of partner choice, existing models address primarily whether animals should engage in partner choice at all and not the adaptive consequences of such a choice. To further test the adaptive importance of partner choice, we need more empirical studies that document the fitness consequences of interacting with different potential partners. In the future, optimality-based studies of partner choice should incorporate dynamic opti-

mization, game theory, and the function of uncertainty in shaping partner choice.

4. From a cognitive view, it is useful to distinguish between partner choice on the basis of individual recognition versus categories of individuals (e.g. large versus small) and investigate the role of experience and memory in governing partner choice. More detailed cognitive issues that should prove informative include studies on the domain specificity of partner choice (e.g. Does a given organism use similar rules in different partner choice contexts?) and the possibility of crossover effects. Limited memory, domain-general rules, and crossover effects may serve to generate cognitive constraints that limit the potential for optimal partner choice.

5. Both cognitive and optimality views of partner choice should become less atomistic; they should account more for the possibility that partner choice may be based on multiple traits that balance multiple fitness demands. With both views, comparative methods should also eventually be used to identify phylogenetic patterns of partner choice.

ACKNOWLEDGMENTS

We wish to thank D. Blumstein, A. Dugatkin, D. Dugatkin, R. Dukas, L. Sih, and M. Sih.

LITERATURE CITED

Abrahams, M., and L. M. Dill. 1989. A determination of the energetic equivalence of the risk of predation. *Ecology* 70:999–1007.

Abrams, P. A. 1994. Should prey overestimate the risk of predation? *American Naturalist* 144:317–328.

Andersson, M. 1994. *Sexual Selection*. Princeton, N.J.: Princeton University Press.

Arak, A. 1988. Callers and satellites in the natterjack toad: Evolutionarily stable decision rules. *Animal Behaviour* 36:416–432.

Axelrod, R., and W. D. Hamilton. 1981. The evolution of cooperation. *Science* 211: 1390–1396.

Bateson, P. 1990. Choice, preference, and selection. In M. Bekoff and D. Jamieson, eds., *Interpretation and Explanation in the Study of Animal Behaviour*, 149–156. Boulder, Colo: Westview.

Barnard, C. J., ed. 1984. *Producers and Scroungers*. London: Croom Helm/Chapman Hall.

Barnard, C. J., and T. Burk. 1979. Dominance hierarchies and the evolution of individual recognition. *Journal Theoretical Biology* 81:65–73.

Bekoff, M., A. Scott, and D. Conner. 1989. Ecological analyses of nesting success in evening grosbeaks. *Oecologia* 81:647–674.

Bouskila, A., D. T. Blumstein, and M. Mangel. 1995. Prey under stochastic conditions should probably overestimate predation risk: A reply to Abrams. *American Naturalist* 145:1015–1019.

Boyd, R., and P. J. Richerson. 1985. *Culture and the Evolutionary Process.* Chicago: University of Chicago Press.

Brown, J., and P. Colgan. 1986. Individual and species recognition in centrachid fishes: Evidence and hypotheses. *Behav. Ecol. Sociobiol.* 19:373–379.

Byers, J. A., and M. Bekoff. 1986. What does "kin recognition" mean? *Ethology* 72: 342–345.

Caraco, T., and L.-A. Giraldeau. 1991. Social foraging: Producing and scrounging in a stochastic environment. *Journal Theoretical Biology* 153:559–583.

Caswell, H. 1989. *Matrix Population Models.* Sunderland, Mass.: Sinauer.

Chesson, J. 1983. The estimation and analysis of preference and its relationship to foraging models. *Ecology* 64:1297–1304.

Clifton, K. 1991. Subordinate group members act as food-finders within striped parrot-fish territories. *Journal of Experimental Marine Biology and Ecology* 145:141–148.

Crespi, B. J. 1989. Causes of assortative mating in arthropods. *Animal Behaviour* 38: 980–1000.

Crowley, P. H., S. Travers, M. Linton, A. Sih, and R. C. Sargent. 1991. Mate density, predation risk, and seasonal sequence of mate choices: A dynamic game. *American Naturalist* 137:567–596.

Davies, N. B. 1992. *Dunnock Behavior and Social Evolution.* Oxford: Oxford University Press.

Dugatkin, L. A. 1992. Sexual selection and imitation: Females copy the mate choice of others. *American Naturalist* 139:1384–1389.

———. 1996a. Imitation and female mate choice. In G. Galef and C. Heyes, eds., *Social Learning in Animals: The Roots of Culture,* 85–106. New York: Academic.

———. 1996b. The interface between culturally-based preferences and genetic prefer-ences: female mate choice in *Poecilia reticulata. Proceedings of the National Acad-emy of Sciences,* U.S.A. 93:2770–2773.

———. 1997. *Cooperation among Animals: An Evolutionary Perspective.* Oxford: Oxford University Press.

Dugatkin, L. A., and M. Alfieri. 1991. Guppies and the TIT FOR TAT strategy: Prefer-ence based on past interaction. *Behavioral Ecology and Sociobiology* 28:243–246.

———. 1992. Interpopulational differences in the use of the Tit for Tat strategy during predator inspection in the guppy. *Evolutionary Ecology* 6:519–526.

Dugatkin, L. A., and J.-G. Godin. 1992a. Reversal of female mate choice by copying. *Proceedings of the Royal Society of London,* ser. B 249:179–184.

———. 1992b. Prey approaching predators: A cost-benefit perspective. *Annales Zoo-logica Fennici* 29:233–252.

———. 1993. Female mate copying in the guppy, *Poecilia reticulata:* Age dependent effects. *Behavioral Ecology* 4:289–292.

Dugatkin, L. A., and R. C. Sargent. 1994. Male-male association patterns and fe-male proximity in the guppy, *Poecilia reticulata. Behavioral Ecology and Sociobiol-ogy* 35:141–145.

Dugatkin, L. A., and D. S. Wilson. 1992. The prerequisites of strategic behavior in the bluegill sunfish. *Animal Behaviour* 44:223–230.

Fletcher, D. J., and C. D. Mitchner, eds. 1987. *Kin Recognition in Animals.* New York: Wiley.

Giraldeau, L. A., and L. Lefebrve. 1987. Scrounging prevents cultural transmission of food-finding in pigeons. *Animal Behaviour* 35:387–394.

Gross, M., and R. Charnov. 1980. Alternative male life histories in bluegill sunfish. *Proceedings of the National Academy of Sciences, U.S.A.* 77:6937–6940.

Harcourt, A. H. 1992. Coalitions and alliances: are primates more complex than non-primates? In A. H. Harcourt, and F. B. M. de Waal, eds., *Coalitions and Alliances in Humans and Other Animals,* 445–472. Oxford: Oxford University.

Harcourt, A. H., and F. B. M. de Waal, eds. 1992. *Coalitions and Alliances in Humans and Other Animals.* Oxford: Oxford University.

Harvey, P. H., and M. D. Pagel. 1991. *The Comparative Method in Evolutionary Biology.* Oxford: Oxford University Press.

Healey, S. 1992. Optimal memory: Toward an evolutionary ecology of animal cognition. *Trends in Ecology and Evolution* 8:399–400.

Hepper, P. G., ed. 1991. *Kin Recognition.* Cambridge: Cambridge University Press.

Hockey, P. A. R., and W. K. Steele. 1990. Intraspecific kleptoparasitism and foraging efficiency as constraints on food selection by kelp gulls *Larus dominicanus.* In R. N. Hughes, ed., *Behavioural Mechanisms of Food Selection,* 679–706. Berlin: Springer-Verlag.

Houston, A. I., C. Clark, J. M. McNamara, and M. Mangel. 1988. Dynamic models in behavioral ecology. *Nature* 332:29–34.

Hutto, R. L. 1985. Habitat selection by nonbreeding: Migratory land birds. In M. L. Cody, ed., *Habitat Selection in Birds,* 455–476. Orlando, Florida: Academic.

Janetos, A. C. 1980. Strategies of female choice: A theoretical analysis. *Behavioral Ecology and Sociobiology* 7:107–112.

Kamil, A. 1994. A synthetic approach to the study of animal intelligence. In L. A. Real, ed., *Behavioral Mechanisms in Evolutionary Ecology,* 11–45. Chicago: University of Chicago Press.

Keller, L., and H. K. Reeve. 1994. Partitioning of reproduction in animal societies. *Trends in Ecology and Evolution* 9:98–102.

Kulling, D., and M. Milinski. 1992. Size-dependent predation risk and partner quality for predator inspection in sticklebacks. *Animal Behaviour* 44:949–955.

Lande, R., and S. J. Arnold. 1983. The measurement of selection on correlated characters. *Evolution* 37:1210–1226.

Landeau, L., and J. Terborgh. 1986. Oddity and the "confusion effect"—in predation. *Animal Behaviour* 34:1372–1380.

Lima, S. 1989. Iterated Prisoner's Dilemma: An approach to evolutionarily stable cooperation. *American Naturalist* 134:828–834.

Lima, S., and L. Dill. 1990. Behavioral decisions made under the risk of predation: a review and prospectus. *Canadian Journal of Zoology* 68:619–640.

Lindstrom, K., and E. Ranta. 1993. Social preferences by male guppies, *Poecilia reticulata* based on shoal size and sex. *Animal Behaviour* 46:1029–1031.

Logue, A. W. 1988. Research on self-control: An integrating framework. *Behavioral and Brain Science* 11:665–709.

Mangel, M., and C. Clark. 1988. *Dynamic Modeling in Behavioral Ecology,* Princeton, N.J.: Princeton University Press.

Maynard Smith, J. 1982. *Evolution and the Theory of Games.* Cambridge: Cambridge University Press.

Mesterton-Gibbons, M., and L. A. Dugatkin. 1992. Cooperation among unrelated individuals: Evolutionary factors. *Quarterly Review of Biology* 67:267–281.

Metcalfe, N. B., and B. C. Thomson. 1995. Fish recognize and prefer to shoal with poor competitors. *Proceedings of the Royal Society of London,* ser. B 259:207–210.

Milinski, M. 1987. TIT FOR TAT and the evolution of cooperation in sticklebacks. *Nature* 325:433–435.

Milinski, M., D. Kulling, and R. Kettler. 1990. Tit for Tat: Sticklebacks "trusting" a cooperating partner. *Behavioral Ecology* 1:7–12.

Møller, A. P. 1994. *Sexual Selection and the Barn Swallow.* Oxford: Oxford University Press.

Ohguchi, O. 1981. Prey density and selection against oddity by three-spined sticklebacks. *Zeitschrift für Tierpsychologie* 23 (suppl.).

Parker, G. A. 1978. Searching for mates. In J. R. Krebs and N. B. Davies, eds., *Behavioral Ecology,* 214–244. Sunderland, Mass.: Sinauer.

———. 1983. Mate quality and mating decisions. In P. Bateson, ed., *Mate Choice,* 141–166. Cambridge: Cambridge University Press.

Peck, J. R. 1993. Friendship and the evolution of cooperation. *Journal of Theoretical Biology* 162:195–228.

Pitcher, T. 1993. Who dares wins: The function and evolution of predator inspection behaviour in shoaling fish. *Netherlands Journal of Zoology* 42:371–391.

Ranta, E., and K. Lindstrom. 1990. Assortative schooling in three-spined sticklebacks? *Annales Zoologica Fennici* 27:67–75.

Ranta, E., K. Lindstrom, and N. Peuhkuri. 1992. Size matters when three-spined stickleback go to school. *Animal Behaviour* 43:160–162.

Real, L. 1990. Search theory and mate choice. I. Models of single-sex discrimination. *American Naturalist* 136:376–404.

———. 1991. Search theory and mate choice. II. Mutual interaction, assortative mating, and equilibrium variation in male and female fitness. *American Naturalist* 139: 901–917.

———. 1994. Information processing and the evolutionary ecology of cognitive architecture. In L. A. Real, ed., *Behavioral Mechanisms in Evolutionary Ecology,* 99–132. Chicago: University of Chicago Press.

Reeve, H. K. 1989. The evolution of conspecific acceptance thresholds. *American Naturalist* 133:407–435.

Rosenzweig, M. 1990. Do animals choose habitats? In M. Bekoff and D. Jamieson, eds., *Interpretation and Explanation in the Study of Animal Behaviour,* 157–179. Boulder, Colo.: Westview.

Rothstein, S. I. 1990. A model system for coevolution: Avian brood parasitism. *Annual Review of Ecology and Systematics* 21:481–508.

Seyfarth, R. M., and D. L. Cheny. 1994. The evolution of social cognition in primates. In L. Real, ed., *Behavioral Mechanisms in Evolutionary Ecology,* 371–389. Chicago: University of Chicago Press.

Sih, A. 1987. Predators and prey lifestyles: An evolutionary and ecological overview. In W. C. Kerfoot and A. Sih, eds., *Predation: Direct and Indirect Impacts on Aquatic Communities,* 203–224. Hanover, N.H.: University Press of New England.

———. 1994. Predation risk and the evolutionary ecology of reproductive behavior. *Journal Fish Biology* 45:111–130.

Sih, A., and J. J. Krupa. 1992. Predation risk, food deprivation, and non-random mating by size in the stream water strider, *Aquaris remigis. Behavioral Ecology and Sociobiology* 31:51–56.

Stacey, P. B. 1982. Female promiscuity and male reproductive success in social birds and mammals. *American Naturalist* 120:51–64.

Stephens, D., and J. Krebs. 1986. *Foraging Theory.* Princeton, N.J.: Princeton University Press.

Theodorakis, C. 1989. Size segregation and the effects of oddity on predation risk in minnows. *Animal Behaviour* 38:496–502.

Trivers, R. L. 1971. The evolution of reciprocal altruism. *Quarterly Review of Biology* 46:189–226.

Van der Meer, J. 1992. Statistical analysis of the dichotomous preference test. *Animal Behaviour* 44:1101–1106.

Vickery, W. L., L. A. Giraldeau, J. J. Templeton, D. L. Kramer, and C. C. Chapman. 1991. Producers, scroungers, and group foraging. *American Naturalist* 137:847–863.

Waldman, B. 1987. Mechanisms of kin recognition. *Journal Theoretical Biology* 128:159–185.

Weigmann, D. D., L. Real, T. A. Capone, and S. Ellner. 1996. Some distinguishing features of models of search behavior and mate choice. *American Naturalist* 147:188–204.

Willmer, C. 1985. Thermal biology, size effects, and the origin of communal behaviour in Cerceris wasps. *Behavioral Ecology and Sociobiology* 17:151–160.

CHAPTER ELEVEN

Cognitive Ecology: Prospects

REUVEN DUKAS

All contributors to this volume express the central philosophy of cognitive ecology namely, that cognition must be studied with regard to an animal's ecology and evolutionary history, and that knowledge of cognitive mechanisms can help us explain behavioral, ecological, and evolutionary phenomena. In various chapters, however, the depth of analysis of cognitive and evolutionary-ecological mechanisms differs. This variation reflects the need for further cognitive-ecological research in ways I will discuss.

11.1 NEURONS AND BEHAVIOR

The building blocks of all cognitive systems are neurons. Hence, a comprehensive theory of cognitive ecology should explain how natural selection has acted on individual neurons and neural networks to produce various cognitive capacities. Research on the way networks of neurons mediate information processing and behavior is still in its infancy. The major theoretical tool presently is simulation of hypothetical or real neural networks (e.g. Rumelhart et al. 1986; Schwartz 1990; Gardner 1993). It is sometimes tempting to criticize a first generation of models for being unrealistic. For example, in artificial neural networks such as the ones described in chapter 2, there are very few neuronal components, and various known properties of biological neurons are not considered. All models, however, are somewhat simplistic because they are supposed to capture essential properties and omit numerous details. Thus, a simple theoretical neural network can be instructive *because* it is simplistic. As Enquist and Arak (chap. 2) showed, one can use theoretical models of neural networks to illuminate important behavioral and ecological phenomena such as the design of signals and the choice of mates.

One measure of the success of a theoretical tool, such as neural-network simulation, is its capacity to generate novel testable predictions. Neural-network models have indeed inspired exciting neurobiological research (e.g. Zipser and Anderson 1988; Gardner 1993); however, it is too early to evaluate the models' contributions to behavioral ecology and related disciplines. Nevertheless, it is likely that neural-network simulations can guide behavioral research by identifying (1) what networks of neurons can and cannot do; and

405

(2) what neuronal characteristics must be included in order to generate realistic models of certain cognitive functions and behaviors. Perhaps, however, neural-network simulations will inspire the generation of new analytic or simulation tools for network modeling. Such models should capture more of the fundamental properties of individual neurons, synapses, and neurotransmitters than the first generations of neural-network simulations (see Gerstein and Turner 1990; Gardner 1993).

11.2 CONSTRAINTS

In many chapters of this volume, various cognitive constraints are referred to implicitly or explicitly. In chapter 2, Enquist and Arak suggest that the basic design of neural networks creates some perceptual biases that affect signal design. In chapter 3, I discuss limits on the amount of information that the brain can process simultaneously (attention) and over short periods of time (working memory), constraints on long-term memory, and an inability to sustain vigilance indefinitely. In chapter 4, I raise the issue of learning rate limits, which may prevent animals from learning tasks that are crucial for survival quickly enough. In chapter 6, Dyer mentions computational limitations that may have shaped navigational capacities of animals with little brains. Finally, perceptual and computational limitations are also invoked in chapter 8 to explain animal sensitivities to variable rewards.

One can readily accept the notion of constraints on design if their only effects are some perceptual biases, which may have little or no effect on fitness. However, when constraints seem to influence survival and reproduction, a thorough analysis is required to determine the fitness costs and benefits and the fundamental system design that determine a certain cognitive capacity (sections 3.2, 3.6). For example, we all know that it takes time for animals (including humans) to learn certain tasks. Does this learning rate represent some optimal value, or is there a basic neuronal limit to learning rate? To answer these and similar questions, we need to know more about the way neurons and networks generate various cognitive capacities. The pursuit of such knowledge may be enhanced by integrating neurobiological knowledge with evolutionary-ecological considerations and with use of theoretical tools such as optimality analyses and refined models of neural networks.

11.3 COMPUTATION

Many animals must acquire knowledge about their location in space and time and the relative qualities of alternative resources such as food patches, shelters, and social partners. To these ends, mere acquisition of raw information is insufficient because this information must be somehow filtered and analyzed to evaluate environmental variables and compare available alternatives. Although some of the analyses may require little or no computation, it seems that information processing typically involves, or at least can be described with, various algorithms. For example, to evaluate the relative qualities of various food sources, one can consider a few factors such as net energy content, handling time, and associated risk of injury or mortality caused by predation or other hazards. What algorithm should an animal use when considering these factors? In chapters 8 and 9, it is suggested that the answer is not obvious even if risks of injury and mortality are ignored. Currently, behavioral ecologists conduct behavioral experiments to test what algorithms animals may use to make foraging decisions. This somewhat indirect method can be augmented by direct measurement of neuronal activity, which can tell us how computation is indeed carried out in the brain.

Hatsopoulos et al. (1995) nicely demonstrated the feasibility of such an approach (section 1.4). First, predictions of the feasible algorithms that the locust *could* use to recognize objects on a collision course were tested and these predictions were rejected on the basis of observed neuronal activity. Second, they constructed a new algorithm that agreed with their neurobiological data. Thus, by considering feasible algorithms and testing them directly against neuronal activity, Hatsopoulos et al. (1995) gained more insight than that achieved with behavioral experimentation alone. Of course, to measure neuronal computation, one must first identify the neurons and networks involved. Although much of the relevant neuronal information is yet unknown, the rapid pace of neurobiological research provides an increasing number of model systems for integrative analyses of neuronal computations and the behavioral decisions they control (e.g. Hammer 1993; Ferrus and Canal 1994; Hall 1994; Montague et al. 1995).

11.4 THE NATURAL ENVIRONMENT

A principal message in this volume is that knowledge of proximate cognitive mechanisms can help us understand behavior of individuals, populations, and ecological communities. However, research on cognitive mechanisms, as it is practiced by many neurobiologists and psychologists, cannot be an end to itself. Not only it is important to consider how these mechanisms determine an individual's fitness in its natural settings; studies on cognitive mechanisms must also be relevant to an animal's environment in the wild. Otherwise, conclusions about mechanisms that are reached in highly artificial laboratory settings can simply be wrong. For example, patterns of song learning by song sparrows in the field are different from the patterns observed in laboratory experiments with taped tutors (chapter 5). For another example, in most cognitive psychology texts, it is asserted that working memory has a very limited capacity. When tested with random information, humans indeed show a limited memory capacity; however, when experts are tested in their area of expertise, they demonstrate a much greater working-memory capacity (section 3.5). Hence, these examples demonstrate that the empirical search for mechanisms must remain within boundaries determined with the animals' relevant tasks as they are experienced in the natural environment.

11.5 How Do Brains Work?

Despite the outpouring of new information about brain functioning, we still do not understand how brains work. The central message in this volume is that insights from evolutionary biologists and behavioral ecologists can significantly contribute to the solution of this biological puzzle. After all, "nothing in biology makes sense except in the light of evolution" (Dobzhansky 1973). The central reasoning of ecological and evolutionary analyses is that a biological trait usually has some costs and benefits, which can be expressed in units of fitness (section 1.3). This rational is mostly absent in many neurobiological and psychological studies. As discussed in the chapters of this volume, such a functional approach for studying cognition provides fresh understanding and novel testable predictions. This volume cannot yet answer how the brain works; however, in this volume is indicated the way for a broader interdisciplinary field of cognitive science, which considers cognition an evolved characteristic that is shaped by numerous ecological factors that determine survival and reproduction of animals in their natural settings. The quest for understanding the brain's workings should proceed along two interconnected paths of investigation; in one, the focus on evolutionary-ecological questions must include fundamental neurobiological knowledge, and in the other, neurobiological mechanisms must be sought with use of established evolutionary-ecological principles.

Literature Cited

Dobzhansky, T. 1973. Nothing in biology makes sense except in the light of evolution. *American Biology Teacher* 35:125–129.

Ferrus, A., and I. Canal. 1994. The behaving brain of a fly. *Trends in Neuroscience* 17:479–485.

Gardner, A., ed. 1993. *The Neurobiology of Neural Networks,* Cambridge: MIT Press.

Gerstein, G. L., and M. R. Turner. 1990. Neural assemblies as building blocks of cortical computation. In E. L. Schwartz, ed. *Computational Neuroscience,* 179–191. Cambridge: MIT Press.

Hall, J. C. 1994. The mating of a fly. *Science* 264:1702–1714.

Hammer, M. 1993. An identified neuron mediates the unconditioned stimulus in associative olfactory learning in honeybees. *Nature* (London) 366:59–63.

Hatsopoulos, N., F. Gabbiani, and G. Laurent. 1995. Elementary computation of object approach by a wide-field visual neuron. *Science* 270:1000–1003.

Montague, P. R., P. Dayan, C. Person, and T. J. Sejnowski. 1995. Bee foraging in uncertain environments using predictive hebbian learning. *Nature* (London) 377: 725–728.

Rumelhart, D. E., G. E. Hinton, and R. J. Williams. 1986. Learning internal representations by back-propagating errors. *Nature* (London) 323:533–536.

Schwartz, E. L., ed. 1990. *Computational Neuroscience.* Cambridge: MIT Press.

Zipser, D., and R. A. Anderson. 1988. A back-propagation programmed network that simulates response properties of a subset of posterior parietal neurons. *Nature* (London) 331:679–684.

CONTRIBUTORS

Anthony Arak
Archway Engineering (UK) Ltd.
Ainleys Industrial Estate
Elland, HX5 9JP
U.K.

Melissa Bateson
Department of Zoology
University of Oxford
South Parks Road
Oxford OX1 3PS
U.K.
and
Department of Psychology:
Experimental
Duke University
Durham, NC 27708
U.S.A.

Michael D. Beecher
Box 351525
Animal Behavior Program
Departments of Psychology and
Zoology
University of Washington
Seattle, Washington 98195
U.S.A.

S. Elizabeth Campbell
Box 351525
Animal Behavior Program
Departments of Psychology and
Zoology
University of Washington
Seattle, Washington 98195
U.S.A.

Lee Alan Dugatkin
Department of Biology
University of Louisville
Louisville, KY 40292
U.S.A.

Reuven Dukas
Nebraska Behavioral Biology Group
School of Biological Sciences
University of Nebraska
Lincoln, NE 68588
U.S.A.

Fred C. Dyer
Department of Zoology
Michigan State University
East Lansing, MI 48824
U.S.A.

Magnus Enquist
Department of Zoology
University of Stockholm
S-106 91
Stockholm
Sweden

Alex Kacelnik
Department of Zoology
University of Oxford
South Parks Road
Oxford OX1 3PS
U.K.

J. Cully Nordby
Box 351525
Animal Behavior Program
Departments of Psychology and
Zoology
University of Washington
Seattle, Washington 98195
U.S.A.

David F. Sherry
Department of Psychology
University of Western Ontario
London, Ontario N6A 5C2
Canada

Andrew Sih
Department of Biology
University of Kentucky
Lexington, KY 40506
U.S.A.

R. C. Ydenberg
Behavioral Ecology Research Group
Department of Biological Sciences
Simon Fraser University
Burnaby BC
Canada

INDEX